MATHEMATICAL
RECREATIONS
IN THE MIDDLE AGES

Cover: *Uno che trae **3** dadi in su una tavola*, from an arithmetical treatise by Filippo Calandri, *c.* 1480 (Biblioteca Riccardiana di Firenze, MS. 2669, fol. 105^v). See below, pp. 236 & 264.

MATHEMATICAL
RECREATIONS
IN THE MIDDLE AGES

Jacques Sesiano

EPFL PRESS

EPFL PRESS is an imprint owned by the Presses polytechniques et universitaires romandes, a Swiss academic publishing company whose main purpose is to publish the teaching and research works of the Ecole polytechnique fédérale de Lausanne (EPFL). PPUR, EPFL – Rolex Learning Center, CM Station 10, CH-1015 Lausanne, info@epflpress.org, tel. : +41 21 693 21 30, fax : +41 21 693 40 27.

www.epflpress.org

© 2023, First edition in English, EPFL Press
ISBN 978-2-88915-512-5
Printed in France

Preface

This book is an adaptation of the one published in French a few years ago under the title *Récréations mathématiques au Moyen Âge*. There have been major changes. The earlier edition, being for French-speaking readers, included many extracts from late mediaeval French manuscripts, together with a glossary and some grammatical indications. Such sources have now been reduced, the problems kept being translated or summarized, and the same has been done for texts in other languages. Of all that there now remain only a few cases where the original formulation needed to be stressed. As to additions, they include notably a chapter on linear systems of equations, of great importance for the appearance of negative numbers, and another on the most common geometrical recreations, generally in connection with the Pythagorean theorem.

Although there are allusions to problems of the same kind found in antiquity, our focus is on mediaeval times. There are nevertheless frequent examples from the 16th century; indeed, the tradition just continues, whether allusion is made to earlier examples or not. The time limit could be set at the collection printed (1612, 1624) by Claude-Gaspar Bachet de Méziriac, who alludes in the title to predecessors and may be seen as the point of transmission to modern times.

As to a general reference book, we used for the French version the last edition (1980) of Tropfke's *Geschichte der Elementarmathematik*, which has a rich collection of references. The English-speaking reader may in addition consult the books or studies by W. Rouse Ball (general), D. Singmaster and A. Heeffer (particular subjects).

Finally, we have included as *addenda* two studies, originally presented at the St. Petersburg seminars on the history of mathematics in 2017 and 2018. The first, here translated into German, describes Euler's empirical way to solve the problem of the knight's move on the chessboard. The second, the original Russian version, is a short history of continued fractions, both finite and infinite, with many quotations from the original sources. Footnotes in English will enable the general reader to follow these studies.

The numerous illustrations will, hopefully, provide relief from what might otherwise be a rather dry work. They have been taken from late mediaeval manuscripts or early printed works. In the first case, the following libraries have granted me permission to reproduce illustrations from manuscripts or devices belonging to them, either graciously or for a moderate fee (BNC Firenze, BCI Siena). I should thus like to express my gratitude to the following institutions:

In France (Paris), the *Bibliothèque Nationale de France* (MS. fr. 1346, for Fig. 38, 71, 81, 115, 120, 226, 254, 255; MS. fr. 2050, for Fig. 156; MS. fr. AF 7481, for Fig. 233); the *Bibliothèque de l'Arsenal* (MS. fr. 5107 rés., for Fig. 69, 106, 107, 173); for all, Source gallica.bnf.fr.

In Germany (Heidelberg), the *Universitätsbibliothek* (MS. Pal. germ. 848, for Fig. 105).

In Great Britain (London), the *British Museum* (MS. Reg. 13 A.XVIII, for Fig. 180, 182; © British Library Board).

In Italy, in Bologna, the *Biblioteca Universitaria* (MS. it. 250, for Fig. 148; © Alma Mater Studiorum Università di Bologna - Biblioteca Universitaria di Bologna); then, in Florence, the *Biblioteca Medicea Laurenziana* (MS. Acquisiti e doni 154, for Fig. 53; ©, su concessione del Ministero della Cultura), the *Biblioteca Nazionale Centrale* (MS. Magl. XI 86, for Fig. 54, 84, 90, 151, 223; MS. Conv. soppr. G 7 1137, for Fig. 114; ©, su concessione del Ministero della Cultura), the *Biblioteca Riccardiana* (MS. Ricc. 2669, for Fig. 62, 88, 104, 119, 153, 154, 218, 241, 242, 243, 244, 245, 251, 252; ©, su concessione del Ministero della Cultura); in Rome, the *Accademia dei Lincei* (MS. 1875, for Fig. 42, 43, 85, 92, 160, 219, 224, 237); in Siena, the *Biblioteca Comunale degli Intronati* (MS. L.IV.21, for Fig. 103; © Archivio storico, Istituzione del Comune di Siena).

In Russia (St. Petersburg), the *Archive of the Russian Academy of Sciences* (Fig. 176, 177; *Addendum* 1, Abb. 2, 32, 36, 44; © APAH).

In Switzerland (Lausanne), the *Musée monétaire cantonal* (Fig. 36; © MCAH).

In the United States, in New York, the *University of Columbia* (MS. Plimpton 167, for Fig. 44, and MS. X511.AL3, for Fig. 137, 139); then, in Ann Arbor, the *University of Pennsylvania* (MS. LJS 27, for Fig. 51, 52, 57, 61, 66, 67, 70, 80, 89, 91, 93, 159, 161, 232, 246).

Table of contents

Chapter I. Sharing liquid by decanting

Chapter II. Sharing liquid and vessels

Chapter III. Minimal set of weights

Chapter IV. Successive distributions

Chapter V. Mutual borrowing

Chapter VI. Filling and emptying cisterns

Chapter VII. Messengers

Chapter VIII. Large numbers

Chapter IX. Arrangements

Chapter X. Crossing a river

Chapter XI. Miscellanea

Chapter XII. Family relationships

Chapter XIII. The knight's tour

Chapter XIV. Magic squares

Chapter XV. Infinite sets

Chapter XVI. Geometrical recreations

Chapter XVII. Hidden numbers

Appendices

A. Brief outline of the history of mathematics

B. Sources used

Addenda

Euler and the knight's move
(in German; English summary)

Short history of continued fractions
(in Russian; English summary)

I. Finite continued fractions

II. Infinite continued fractions

Chapter I: Sharing liquid by decanting

§ 1. First appearance

The first known occurrence of this problem is found in the *Annales Stadenses* (Annals of Stade, a town near Hamburg), just after notification of the saint Hildegard von Bingen's death (1179).[1] The monk Albert, who wrote these chronicles in the middle of the 13th century, relates a conversation which took place on Christmas Eve between two well-read young people, named Tirri and Firri. Tirri is quite clever, whereas Firri is rather slow-witted. This is reflected not only in the problems they propose, subtle in Tirri's case and simpler or even banal in Firri's, but also in their formulation —as seen from (*i*) below.

[1] (*i*) *Firri said: There were in Cologne three brothers, having nine jars of wine. The first contained one bucketful, the second two, the third three, the fourth four, the fifth five, the sixth six, the seventh seven, the eighth eight, the ninth nine. Share this wine into equal parts among them without opening any jar.*

Tirri said: I shall do it. I give to the eldest the first, the fifth and the ninth; this will make fifteen bucketfuls. I give to the middle one the third, the fourth and the eighth; he too will have fifteen. I give to the youngest the second, the sixth and the seventh. Since he also has fifteen, the wine has been shared, without opening any jar.

(*ii*) *Now, Firri, it is my turn to ask you something else, and indeed more subtle. My master, having to receive a guest, sent me to the neighbouring town fetch wine. I took with me one jar, containing eight measures. I filled it, there was no more in the shop. Whilst I was returning, you meet me on the way, fetching wine as well. You ask me where I come from. I say: 'I come from the market, having fetched wine for my master'. You ask me how much. 'Eight measures', I say. You tell me: 'I go there for wine, too'. I answer: 'You will not find any!'. You ask me to share with you. I ask you whether you have jars. You tell me that you have two, one of five measures, the other of three. 'I shall give you half, thus four measures, if you manage to share using these jars; either you share it, or you will have to do without wine'. Firri remained silent, being incapable of performing the sharing. Tirri told him: 'I shall myself perform the sharing, since you do not know it. Put down your jars. Here is mine, of eight measures, and your larger one, five measures, and smaller one, three'. Tirri pours into the middle jar five measures, thus filling it, and from it three into the smallest, filling it in turn. Then he pours from the smallest jar its three measures into the largest, of eight measures. There are then six measures in the largest and two in the middle. Then he pours those two measures from the*

---

[1] *Monum. Germ. Hist.* XVI (ed. Pertz), p. 333.

middle into the smallest, thus leaving the middle one empty. Then he again pours five measures into the middle vessel, which was empty, thus filling it, then pours right away one measure from the latter into the smallest. 'This is it: you have four measures in your larger vessel'. Pouring the three measures from the smallest vessel into his largest one, that of eight measures, he obtains four. 'This is how you were to share the wine, Firri'.

(iii) With revenge in mind, Firri said: 'You are to take across a river a wolf, a goat and cabbages using a small boat, able to contain only one of them, in such a way that neither the wolf eats the goat nor the goat the cabbages'.

Tirri answered: 'Pooh! That is a child's poetry:

    A child barely five years old first took the goat,
    Came back, took the wolf and returned with the goat
    Then, having taken across the cabbages, returned for the goat'.[2]

Let us examine these three problems, first Firri's two simple ones then Tirri's which is the subject of this chapter.

(i) Firri's first question, on sharing among three nine jars of wine with contents from 1 to 9, thus 15 for each, has in fact two solutions, which are easy to remember with the arrangement of the 3×3 magic square in mind (Fig. 1): Tirri's answer corresponds to the vertical rows, the other to the horizontal ones. We may also observe that one condition is omitted in the question as asked: each of the three brothers is also to receive the same number of jars, thus three; otherwise other solutions (such as $6, 9$; $3, 5, 7$; $1, 2, 4, 8$) would also be acceptable.

4	9	2
3	5	7
8	1	6

Fig. 1

Perhaps because this problem is ancient in origin, or perhaps because it is simple, it is found here and there around the Mediterranean. Thus, in a late Arabic manuscript (copied 1571), but evidently reporting traditional material, we find a similar question.[3] Instead of jars of wine we have nine purses bequeathed by a father to his three sons, containing sums from 1 to 9. As in the above problem, there is a single solution

---

[2] *Quidam puerulus, vix quinque annorum, primo hircum transvexit,*
   et rediens lupum attulit, et hircum reduxit,
   tunc, caulibus transvectis, iterum pro hirco perrexit.
[3] MS. Paris BNF ar. 4441, fol. 43$^\text{v}$ – 44$^\text{r}$.

given (the same) and no mention of the equal repartition of the purses. Surely by carelessness: two 14th- and 15th-century texts reporting the same problem, but on sharing barrels, explicitly state that each must receive an equal quantity of liquid and an equal number of barrels.[4]

(*iii*) Firri's second question is the well-known problem of crossing a river, the solution of which was equally well known on account of the three Latin verses mentioned by Tirri. See also below, Chapter X.

(*ii*) As said by Tirri himself, his question is indeed more subtle. It is the problem of sharing by decanting. Following the Latin text translated above there is a table to illustrate the steps; it corresponds, in a slightly different form, to our Fig. 2. Summarizing, we first fill the middle jar, from it the third, which is then poured into the first; after pouring the liquid left in the second into the third, we repeat the previous steps: from the first into the second, from the second filling the third. Pouring then the contents of the third into the first, we obtain the required sharing. This same problem is again found in 15th-century French manuscripts.[5]

8	5	3
8	0	0
3	5	0
3	2	3
6	2	0
6	0	2
1	5	2
1	4	3
4	4	0

Fig. 2

§ 2. Decanting operations

Considering this procedure, two questions come immediately to mind. First, whether we might have chosen the smallest of the three jars as intermediate; second, what would happen if we were to continue this procedure after dividing the liquid into halves. Our figures 3 and 4 give the answer, at least for this particular numerical example: Fig. 3 takes the smallest jar as intermediate (as indicated by the asterisk) and Fig. 4 sets out the previous example to the end, which brings us back to the initial situation, thus with all the liquid in the largest jar. Note that this last table comprises the two previous ones, the steps of the first, ending

---

[4] *Subtilitates*, No. 24; *Algorismus Ratisbonensis*, No. 351.

[5] MSS. Tours 399, fol. 136^v – 137^r; Paris BNF fr. 1339, fol. 76^v – 77^r; Nantes 456, fol. 80^r – 80^v. For their text, see the French edition (*Récréations*), pp. 10–11.

with the two equal parts, being followed by the steps of the second, but taken in reverse order.

8	5	3*
8	0	0
5	0	3
5	3	0
2	3	3
2	5	1
7	0	1
7	1	0
4	1	3
4	4	0

Fig. 3

8	5*	3
8	0	0
3	5	0
3	2	3
6	2	0
6	0	2
1	5	2
1	4	3
4	4	0
4	1	3
7	1	0
7	0	1
2	5	1
2	3	3
5	3	0
5	0	3
8	0	0

Fig. 4

Let us now consider the treatment of this problem in a more general form. We suppose that we have three vessels A, B, C, of whatever shape, with no intermediary graduation marks, and the different, but always integral, contents a, b, c at full capacity ($a > b > c$). At the outset, A is completely filled while B and C are empty. At the end of the procedure, a must be divided into two equal parts. To this end, the following steps must be performed.

(i) Filling the intermediate vessel: one of the two smaller vessels, chosen as the intermediate one, is filled *completely* from A.

(ii) Filling the third vessel: From the intermediate one we pour as much as we can into the third one.

(iii) Pouring again into the first: When the third vessel has been *completely* filled, we pour its contents into the first, and largest, vessel A.

(iv) Emptying the intermediate one: Before filling the intermediate vessel once again, we pour its remainder into the third.

After emptying thus the intermediate, repeat the same steps.

We may already note three points:

— The intermediate vessel is always filled *completely*, and at once, by A.

— Therefore, this intermediate vessel must always be *empty* before being filled by A.

— The other of the two smaller vessels is poured into A only when *filled completely*.

We have represented, with the same numerical values as before, the sequence of these steps in Fig. 5 and 6. Note in the second case that, since the smallest vessel C, now the intermediate, cannot fill B at once, there is repetition of the first two steps. Conversely, there is one step less at the end.

(*step*)	8	5*	3
	8	0	0
(*i*)	3	5	0
(*ii*)	3	2	3
(*iii*)	6	2	0
(*iv*)	6	0	2
(*i*)	1	5	2
(*ii*)	1	4	3
(*iii*)	4	4	0

Fig. 5

(*step*)	8	5	3*
	8	0	0
(*i*)	5	0	3
(*ii*)	5	3	0
(*i*)	2	3	3
(*ii*)	2	5	1
(*iii*)	7	0	1
(*iv*)	7	1	0
(*i*)	4	1	3
(*ii*)	4	4	0

Fig. 6

§3. Other examples

An ancient origin of this problem may explain why it appeared at very different times and places in mediaeval Europe. The first occurrence is, as said, in 13th-century Germany, and it is later found, as mentioned, in 15th-century France. But also, in the 14th and 15th centuries, in Spanish and Italian texts. There is not much change: the values chosen are the same, but the intermediate vessel is often the smallest.[6] Of interest are some remarks about the problem itself and its solving. According to the Spanish text, finding a solution is more a question of intuition than computation (*por pensiamento de sotileza de coraçón e non por arte del alguarismo*); as to the Italian manuscript, it considers the problem interesting, even though it does not involve any computation (*e chosì questa proposta è di grande chonsideratione, ma in verità non è d'abacho; ma buona è da sapere*); finally, Paolo dell'Abbaco is more direct: its solving is (mathematically) of little use for it is attained only by trial and error (*questa ragione è axxaj pocho d'utile, perochè si fae solamente per avixo*). Five centuries later, and independently, the same opinion is repeated in a work on mathematical recreations (*problems like this can be*

---

[6] *Rascioni d'algorismo* (14th c.), ed. Vogel, No. 123; *Subtilitates* (14th c.), Nos. 18 & 29; treatise attributed to Paolo dell'Abbaco (14th c.), No. 66; *Arte del alguarismo* (14th c.), p. 156; MS. Florence BNC Palat. 573 (15th c.), fol. 242^v–243^r. On the terms 'alguarismo' and 'abacho', see pp. 245–246.

worked out only by trial).[7] As we shall see, this is simply not true. But let us first consider a few other late-mediaeval or Renaissance occurrences which extend or restrict the problem.

[2] In his *Triparty* (written in French, in 1484) Nicolas Chuquet takes in turn the receptacles of 5 and 3 as intermediate while the larger vessel is of undefined contents.[8] Thus here the aim is not sharing, but merely obtaining the quantity 4 in the vessel of 5. In fact, this does not change much from the previous solution: we shall just omit in Fig. 2 and 3 the first column and the first row. Note only that the vessel of indeterminate contents must contain at least 7 (case 5*) or 6 (case 3*).

[3] In his *De viribus quantitatis* (written ca. 1500, but with text in Italian, unlike the title), Luca Pacioli considers four examples.[9] The first is the common case (8, 5, 3*). The second (12, 7, 5*) —just inferred from the first by a common addition of liquid in the two smaller vessels— increases the number of steps and seems, for this reason, to be highly considered by the author (*et sia facta più bella asai che la precedente et con più mutationi*); see Fig. 7. At first sight, the third example seems to be a more general case: there are now four vessels, of respective contents 18, 7, 6, 5, and we are to divide up the liquid of the first equally among three people, thus each with 6; but this extension is misleading: despite the author telling us that it is a *più bella proposta*, we are simply brought back to the previous problem since he fills the third vessel, that of 6, right away. The last example, or rather case since it includes two sets of data, is said to be *un altro sotil caso*: first, we are to halve 10 using vessels of 6 and 4, then 12 by means of vessels of 8 and 4. Pacioli's attitude is curious: on the one hand he urges the reader to solve them, since he is by now familiar with the procedure, on the other he warns him that only an ignorant person will wear himself out attempting to reach an impossible goal (*el modo et sua solutione, inteso già le precedente, lasceremo al lectore, che son certo la intenderà* (...); *et alo idiota proposto s'afatigarà, invano cercando lo impossibile*). We thus encounter with Pacioli a (specious) generalization and an (unexplained) impossible case. We have added the (unsuccessful) procedures in Fig. 8–11.

---

[7] Rouse Ball, *Mathematical recreations*, pp. 16 (ed. 1905), 18 (ed. 1914), 27–28 (ed. 1947 —numerous editions).

[8] *Appendice*, p. 460; MS. Paris BNF fr. 1346, fol. 209^{v}.

[9] Part I, Nos. LIII–LV, MS. Bologna BU it. 250, fol. 97^{r}–99^{r} (or the printed —but uncommented— transcription of the MS. by M. G. Peirani), or Agostini's summary, p. 186.

12	7	5*
12	0	0
7	0	5
7	5	0
2	5	5
2	7	3
9	0	3
9	3	0
4	3	5
4	7	1
11	0	1
11	1	0
6	1	5
6	6	0

Fig. 7

10	6*	4
10	0	0
4	6	0
4	2	4
8	2	0
8	0	2
2	6	2
2	4	4
6	4	0
6	0	4
10	0	0

Fig. 8

10	6	4*
10	0	0
6	0	4
6	4	0
2	4	4
2	6	2
8	0	2
8	2	0
4	2	4
4	6	0
10	0	0

Fig. 9

12	8*	4
12	0	0
4	8	0
4	4	4
8	4	0
8	0	4
12	0	0

Fig. 10

12	8	4*
12	0	0
8	0	4
8	4	0
4	4	4
4	8	0
12	0	0

Fig. 11

[4] In all previous examples the sum of the contents of the two smaller vessels was equal to that of the largest; furthermore, in the case of three vessels with respective contents a, b, c, we always had $b > \frac{a}{2}$, whereby the halves, when finally obtained, were in each of the two larger vessels. Not so in J. Widman's book (first printed in 1489), where 14 must be halved using vessels 5* and 3. The answer is given directly: the second vessel will hold 5 and the last, 2, so that the second half will be found in the two smaller ones.[10] In fact, this is related to the classical example 8, 5*, 3, as may be seen in Fig. 5: take at the outset 14 instead of 8 and stop at the third step before the last.

Further occurrences are found in 16th-century texts, sometimes with a tentative explanation for the choice of numbers.

[5] In his *Practica arithmetice* (published in 1539) Cardan considers the problem 8, 5*, 3, and mentions explicitly that the middle vessel (*maior ampula*) is always filled completely by the first (*urna*), and the third

---

[10] *Behende und hubsche Rechenung*, ed. 1489 (last problem; also ed. 1508, fol. 161ᵛ, & ed. 1526, fol. 189ᵛ – 190ʳ).

poured into the first again. In this way, he adds, we shall solve as many problems as we wish.[11] Though he implies that other quantities may be chosen for the contents, he considers a single possibility for the intermediate vessel.

24	13	11	5*
24	0	0	0
19	0	0	5
19	0	5	0
14	0	5	5
14	0	10	0
9	0	10	5
9	0	11	4
20	0	0	4
20	0	4	0
15	0	4	5
15	0	9	0
10	0	9	5
10	0	11	3
21	0	0	3
21	0	3	0
16	0	3	5
16	0	8	0
16	8	0	0

Fig. 12

24	11	5*
16	0	0
11	0	5
11	5	0
6	5	5
6	10	0
1	10	5
1	11	4
12	0	4
12	4	0
7	4	5
7	9	0
2	9	5
2	11	3
13	0	3
13	3	0
8	3	5
8	8	0

Fig. 13

15	9*	4	7
15	0	0	0
6	9	0	0
6	5	4	0
6	0	4	5
10	0	0	5
1	9	0	5
1	5	4	5
5	5	0	5

Fig. 14

[6] With Cardan the context was not sharing wine, but sharing balsam contained in an urn found by two diggers; with Tartaglia the object is the same, a flask of balsam, but the two diggers become two thieves, while the data are again 8, 5, 3*.[12] Tartaglia next solves the case of four vessels (24, 13, 11, 5*) and three thieves. The generalization is at least less banal than Pacioli's since a few steps are required to fill with 8, the required part, one of the smaller receptacles (that of 13), which leaves us, as before, with three vessels (Fig. 12 and 13; with the latter corresponding, in fact, to 16, 11, 5*).

[7] Like some of his mediaeval precursors, J. Buteo (J. Borrel) expresses his admiration for the problem 8, 5*, 3 even though some might consider it as having little to do with reckoning.[13] He also considers dividing up

---

[11] *Regula autem est ut procedas semper eodem ordine, & quot volueris solves questiones; nam semper a maiore urna impletur maior ampula, ab ea autem minor, a minore autem proiicitur in maiorem, nec ordo hic pervertitur* (Ch. LXVI, No. 33).

[12] *Sono duoi che hanno robbato una ampoletta di balsamo a uno signor* (*General trattato*, I, fol. 255ᵛ, No. 132).

[13] *Huiusmodi quæstio, si quibusdam fortasse parum videatur esse Logistica, nunquam*

the liquid among three people, namely 15, 9*, 4, 7, with the vessel of 7 disregarded when containing 5 (leaving 10, 9*, 4, see Fig. 14).

[8] Pacioli's example 12, 7, 5 is solved by P. Forcadel in its two ways.[14] Cardan had mentioned, without specifying, other possible numerical choices. Forcadel is more explicit: we may take any two consecutive odd numbers for the contents of B and C and their sum for A. It is true that this will always be solvable but, as seen from Widman's example and the above 10, 9*, 4, this is by no means a necessary condition.

Fig. 15

[9] Forcadel's contemporary J. Trenchant considers 8, 5*, 3, the only difference being that bushels of wheat are involved. He asserts the same condition as Forcadel, but is less absolute (10, 7, 3, with a non-consecutive pair, is admissible); he adds (wrongly) that the number of decantings will be one less than the contents of the largest vessel, thus $a - 1$.[15] He provides a rhyming mnemonic, in French, to keep in mind the steps of the solution.[16]

[10] It was to be expected that such a common mediaeval problem would be included in Bachet de Méziriac's early 17th-century classical work on recreational problems (see p. 261). He solves 8, 5, 3 in its two ways. He shows Forcadel's restriction to be false with an example, namely 12, 8*, 5 (Fig. 16; he apparently did not know Trenchant's assertion). He too asserts straightaway that the only way to solve such a problem would seem to be by trial and error.[17] We have added Fig. 17, which solves the problem using the other intermediate vessel, but with more steps.

---

tamen non ingeniosa censeri posset (*Logistica*, Ch. IV, No. 73, p. 283).

[14] *Arithmetique*, II, fol. 102^v – 103^r, with his illustration (Fig. 15).

[15] *Aritmetique*, III, Ch. x ('diverses questions'), No. 22 (ed. 1561, pp. 260–261). Obviously, he has not considered the case 8, 5, 3* (above, Fig. 6).

[16] This was common practice in mediaeval recreational problems: it could serve to demonstrate one's ability in society (we have already seen such a trick used by Tirri, see p. 2, and we shall meet with other instances —see Index, 'mnemonics').

[17] *Or bien qu'il semble que ceste question ne se puisse soudre par regle certaine, & qu'il faille necessairement proceder à tastons* (...); see *Problemes plaisans*, p. 135 (ed. 1612), p. 207 (ed. 1624).

12	8	5*
12	0	0
7	0	5
7	5	0
2	5	5
2	8	2
10	0	2
10	2	0
5	2	5
5	7	0
0	7	5
0	8	4
8	0	4
8	4	0
3	4	5
3	8	1
11	0	1
11	1	0
6	1	5
6	6	0

12	8*	5
12	0	0
4	8	0
4	3	5
9	3	0
9	0	3
1	8	3
1	6	5
6	6	0

Fig. 16   Fig. 17

§4. Solving conditions

The importance of Bachet's work for spreading and studying mediaeval recreational problems is perfectly illustrated in this case. It is in the new edition of his work by Labosne in the 19th century that the general conditions for halving are first set out.[18] There are four, the first two of which are evident.

Since the initial quantity a contained in the larger vessel A must be equally divided between two persons, and the half is supposed to be an integer, then

CONDITION I: The initial quantity a must be even.

The second concerns the final situation. In almost all above examples, the second half is found in B, thus $\frac{a}{2} < b$; but, as we have seen in Widman's example, this is not obligatory. We are thus only to require

CONDITION II: $\frac{a}{2} < b + c$.[19]

The third condition is less evident. It has to do with the uninterrupted procedure of decanting. Supposing B to be the intermediate vessel, we

---

[18] *Problèmes plaisants*, pp. 144–147; completed by Schubert, *Mathematische Musse-stunden*, I, §6.

[19] Leaving out the banal case of equality.

imagine that its remainder has just been poured into the empty vessel C (our step iv). If it does not fill C completely, this means that the quantity just poured from B is $c - 1$ or less. Now if A contains $b - 1$ or less, it cannot be used to fill B completely as required; therefore the quantity in C cannot be completed, thus C cannot be poured into A again and the procedure stops. The same happens if, with C the intermediate vessel, A contains $c - 1$ or less while there remains in B the quantity $b - 1$ or less. In order to *ensure* that the procedure ends with two halves, we are to suppose that a be greater than $(b - 1) + (c - 1)$, that is,

CONDITION III: $a \geq b + c - 1$.

As suggested by our *ensure*, the halving may sometimes be attained without this condition being fulfilled. Thus, in the case of 20, 13, 9: with the smallest vessel as intermediate the halving will be reached, while the procedure will stop before halving in the other case (Fig. 18 and 19). Note that the procedure will stop in the first case as well, but later, after halving. The same would happen in Buteo's case (Fig. 15), either after halving (10, 9*, 4) or before (10, 9, 4*).

20	13	9*
20	0	0
11	0	9
11	9	0
2	9	9
2	13	5
15	0	5
15	5	0
6	5	9
6	13	1
19	0	1
19	1	0
10	1	9
10	10	0

Fig. 18

20	13*	9
20	0	0
7	13	0
7	4	9
16	4	0
16	0	4
3	13	4
3	8	9
12	8	0
12	0	8

Fig. 19

Let us now consider the quantity of liquid found in A during decanting. Initially, it is a. Then (case B^*) it is successively reduced by b and increased by c, or (case C^*) successively reduced by c and increased by b. Thus the quantity *outside* A, which is eventually to reach $\frac{a}{2}$, is always of the form $\pm(mb - nc)$ with m, n natural numbers indicating how many times, respectively, we pour into B^* and empty C, or pour into C^* and empty B. Now an equation of the form $\alpha x - \beta y = \gamma$ with α, β, γ given natural numbers is always solvable in natural numbers if α and β are relatively prime. Applying this to our case $\frac{a}{2} = \pm(mb - nc)$ gives the last condition:

CONDITION IV: b and c must be relatively prime.

Here again, the halving may sometimes be attained without this condition being fulfilled. As seen with the decanting procedure carried on to the initial situation (Fig. 4), the quantity found in or outside A will display a continuous set of integral values between 0 and a. However, if b and c have a common divisor, the values displayed will be multiples of this common divisor; see Fig. 8–9 and 10–11, displaying the multiples of 2 and 4, respectively. Thus, if $\frac{a}{2}$ is among these values, that is, if $\frac{a}{2}$ is a multiple of the common divisor of b and c, we shall reach the desired halving. This is why the last two examples proposed by Pacioli were impossible: $\frac{a}{2} = 5$ was not among the multiples of 2, nor 6 among those of 4.

Summarizing, the problem of halving a given integral quantity a by decanting using two vessels of known contents b and c will always be possible if a, $\frac{a}{2}$, b, c are natural numbers while b, c are relatively prime with $2(b + c) \geq a \geq b + c - 1$.

We have seen that pursuing the decanting procedure brings us back to the initial situation —assuming that the procedure is not interrupted— with the whole quantity of liquid being in A again. We may estimate the total number of steps by considering individually the two types of decanting. Since the quantity of wine outside A takes the form $y = \pm(mb - nc)$ while b and c are relatively prime, the smallest solution for returning to the initial situation, and thus $y = 0$, is $m = c$ and $n = b$. Consider this when first B, then C, is the intermediate vessel.

(1) B intermediate.

According to what has just been seen, we return to the initial situation after pouring liquid from A into B c times, and from C into A b times. Since liquid from C is never poured into B, this leaves us with considering the number of times B is poured into C. Now this occurs in two situations. First, when B completes the liquid in C; since this just precedes pouring C into A, their number must equal b. Second, when the remainder in B is poured into C; since this just precedes the filling of B by A, their number must equal $c - 1$ since we are to omit the first filling of B.

This adds up to $2b + 2c - 1$ steps altogether; including, as in our tables, the initial situation, a total of $2(b + c)$ triads of numbers will describe the whole procedure.

(2) C intermediate.

This should lead to the same result since, as we have seen (pp. 3–4), the number of steps altogether is the sum of those in the two ways of halving (less the common one). Indeed, the number of times liquid from A has been poured into C is b while the number of times liquid from B has been poured into A is c. Next, the number of times C is poured into B is, here again, twofold. First, when C, filled completely, is poured into B; since this follows the filling of C, there are b of them. Next, the remainder in C is poured into B as many times as B has been, just before, poured into A, thus $c - 1$ times since we are to omit the last occurrence.

This adds up once again to $2b + 2c - 1$ steps, or $2(b + c)$ triads of numbers if we include the initial situation. As regards their order, it is just the reverse of the previous one.

Remark. If b and c have a common divisor d, the number of steps before returning to the initial situation will be $\frac{2(b+c)}{d}$.

§ 5. Generalization

As first suggested by Schubert,[20] we may consider ending with any two given integral quantities a_1 and a_2, with $a > a_1 > a_2$. (Thus, by varying a_1, a_2, we shall be able to *graduate* our vessels by representing in them the various integral quantities they can contain.) Of our four conditions, the first is now irrelevant, the second becomes $b + c > a_2$, but the last two are obligatory. Completing the procedure we shall have found in B in turn all integral quantities from 0 to b, in C from 0 to c, and thus, in one or both, the required quantity a_2. As to A, it will contain all integral quantities of liquid from $a - b - c + 1$ to a, the first limit arising from the fact that B and C cannot be both full simultaneously.

Depending on the values of a, b, c, the least possible quantity remaining in A, $l = a - b - c + 1$, may vary. On this depend the various integral values of liquid we shall obtain in the vessels by decanting.

— If $l = 0$, thus $a = b + c - 1$, we shall have the integral values from 0 to a in A, and the integral values from 0 to b in B as well (Fig. 20).

— If $0 < l < b + 1$, thus $b + c - 1 < a < 2b + c$, A will contain the integral values from l to a, B those from 0 to b, with those from l to b common to A and B (Fig. 21, $l = 9$).

— If $l = b + 1$, thus $a = 2b + c$, A and B have no measures in common, and this is the optimal situation since between A and B we shall reach all integral measures from 0 to a, without repetition (Fig. 22).

---

[20] *Mathematische Mussestunden*, I, end of § 6.

— If $l > b + 1$, thus $a > 2b + c$, the quantities from $b + 1$ to $b + c - 1$, not found in B, will be missing in A as well. Still, if $b + 1 < l \leq b + c$, the measures missing in A may be found in B and C taken together (Fig. 23, case $l = b + c$, thus $a = 2(b + c) - 1$). But, if $l > b + c$, intermediate values will in any case be missing.

24	11	5*
24	0	0
19	0	5
19	5	0
14	5	5
14	10	0
9	10	5
9	11	4
20	0	4
20	4	0
15	4	5
15	9	0
10	9	5
10	11	3
21	0	3
21	3	0
16	3	5
16	8	0
11	8	5
11	11	2
22	0	2
22	2	0
17	2	5
17	7	0
12	7	5
12	11	1
23	0	1
23	1	0
18	1	5
18	6	0
13	6	5
13	11	0
24	0	0

Fig. 21

11	4	3*
11	0	0
8	0	3
8	3	0
5	3	3
5	4	2
9	0	2
9	2	0
6	2	3
6	4	1
10	0	1
10	1	0
7	1	3
7	4	0
11	0	0

Fig. 22

13	4	3*
13	0	0
10	0	3
10	3	0
7	3	3
7	4	2
11	0	2
11	2	0
8	2	3
8	4	1
12	0	1
12	1	0
9	1	3
9	4	0
13	0	0

Fig. 23

10	7*	4
10	0	0
3	7	0
3	3	4
7	3	0
7	0	3
0	7	3
0	6	4
4	6	0
4	2	4
8	2	0
8	0	2
1	7	2
1	5	4
5	5	0
5	1	4
9	1	0
9	0	1
2	7	1
2	4	4
6	4	0
6	0	4
10	0	0

Fig. 20

Chapter II. Sharing liquid and vessels

§ 1. First appearance and solution

This problem is first met with in Alcuin's 8th-century 'Propositions to sharpen the minds of young people'.[21]

[11] *A dying father bequeathed to his three sons thirty glass bottles of which ten were completely filled with oil, ten further to half, and the last ten left empty. Let whoever can do it share the oil and the bottles in such a way that each of the three sons shall have the same quantity, of both bottles and oil.*

There are thus three sons and thirty bottles, of which ten are full, ten half full and ten empty. Multiply three by ten, it gives thirty; each of the three sons receives ten bottles as share. Divide up then (the oil quantity) by three; that is, give to the first son ten bottles half filled, then give to the second five full and five empty, and give likewise to the third. The sharing out among the three brothers will be equal, both in oil and in bottles.

	full	half	empty
1st	0	10	0
2nd	5	0	5
3rd	5	0	5

Fig. 24

We thus find in Alcuin's text only the results (Fig. 24), without any indication about the way of solving. These being rather evident, they may well have been attained by trial and error. Let us see how we would reach the solution mathematically.

Suppose generally that n be the quantity of (identical) bottles and p the number of heirs. Since there is the same number of bottles in each of the three states (full, half full, empty), this gives us a first condition:

CONDITION I: The number of bottles n must be divisible by 3.

Next, we know that each of the p heirs must receive an equal number of bottles. This gives us another condition:

CONDITION II: The number of bottles n must be divisible by p.

We also know that each of the heirs must receive the same quantity of oil. Now, since, of the n bottles, $\frac{n}{3}$ are full, $\frac{n}{3}$ half full, $\frac{n}{3}$ empty, the whole quantity of oil must be three times $\frac{n}{3}$ half bottles, thus n half bottles. So each heir is to receive the oil quantity of $\frac{n}{p}$ half bottles, an integer by Condition II.

---

[21] *Propositiones*, No. 12.

Let us now operate the sharing. Suppose the ith beneficiary receives x_i full bottles. He then has x_i bottles and a quantity $2x_i$ of oil, measured in half bottles. In order to have his due of $\frac{n}{p}$ half bottles of oil, he is still to receive $\frac{n}{p} - 2x_i$ half bottles, while, in order to receive $\frac{n}{p}$ bottles altogether, he is then to receive $\frac{n}{p} - (x_i + \frac{n}{p} - 2x_i) = x_i$ bottles, thus the same number of empty bottles as full ones. In order for the quantity of half-filled bottles to be positive, we must set:

CONDITION III: $x_i < \frac{n}{2p}$.

If we admit the solution 0 (as does Alcuin, but mostly not in mediaeval times), this condition reduces to $x_i \leq \frac{n}{2p}$.

The same reasoning being valid for all other partners, we may now set out the complete solution:

	full	half	empty
1st	x_1	$\frac{n}{p} - 2x_1$	x_1
2nd	x_2	$\frac{n}{p} - 2x_2$	x_2
3rd	x_3	$\frac{n}{p} - 2x_3$	x_3
$\vdots$	$\vdots$	$\vdots$	$\vdots$
p^{th}	x_p	$\frac{n}{p} - 2x_p$	x_p

Fig. 25

We are left with finding the actual numerical solution. Let us consider, for a while, only the number of full bottles, thus $\frac{n}{3}$, and let us represent it as the sum of p natural numbers fulfilling condition III:

$$\frac{n}{3} = x_1 + x_2 + \cdots + x_p.$$

That will determine the number of full bottles for each beneficiary, thus also his other parts: the ith beneficiary having x_i full bottles ($x_i \neq 0$), he is also to receive, as seen above, x_i empty bottles, and therefore $\frac{n}{p} - 2x_i$ half-filled ones. Furthermore, to each admissible representation of $\frac{n}{3}$ as the sum of positive integers there will be one solution of our problem —and just one since the order of the beneficiaries is irrelevant.

As said, a zero solution is rarely seen in mediaeval problems —which is logical since each of the types mentioned in the question has to appear in the answer. Alcuin's solution is one exception, which is all the more surprising since there are solutions here with all parts different from zero. Indeed, since $10 = 0+5+5 = 5+4+1 = 5+3+2 = 4+4+2 = 4+3+3$, with each triad giving the number of full bottles respectively distributed, there are four solutions other than Alcuin's, two of which have all terms different from zero (Fig. 26–27; $x_i \le \frac{n}{2p} = 5$ and $x_i < \frac{n}{2p} = 5$, respectively).

	full	half	empty	full	half	empty
1st	5	0	5	5	0	5
2nd	4	2	4	3	4	3
3rd	1	8	1	2	6	2

Fig. 26

	full	half	empty	full	half	empty
1st	4	2	4	4	2	4
2nd	4	2	4	3	4	3
3rd	2	6	2	3	4	3

Fig. 27

§ 2. Further examples

Alcuin's problem is not an isolated occurrence. First, we find the same problem in a French manuscript of the 15th century, with just the answer given.[22] Next, in a (seemingly) very different form, it occurs in a 14th-century Byzantine text:[23]

[12] *Someone had three sons and, before dying, ordered that they share equally his good. It consisted in 300 ewes, of which 100 had three lambs each, a further 100 two lambs each, the last 100 one lamb each —so that there were altogether 900 (animals). They must be divided up in such a way that no mother be separated from her offspring and that each of the three receive the same part.*

Each of the three sons will then receive 300 animals, namely 100 ewes and 200 lambs. The solution is directly given, without any explanation. It is just Alcuin's, except that the parts are multiplied by 10, as are the data (Fig. 28, the indices indicating the numbers of lambs). The comparison of Fig. 28 with Fig. 24 also shows that the number of bottles in the same state is now the number of ewes with the same number of lambs. Leaving the bottles unopened corresponds here to maintaining the integrity of the families.

	S_3	S_2	S_1
1st	0	100	0
2nd	50	0	50
3rd	50	0	50

Fig. 28

---

[22] MS. Tours 399, fol. 136^r. See (full text) *Récréations*, p. 30.

[23] Vogel, *Byz. Rechenbuch*, No. 40.

A first reasoned attempt at solving is seen, some seven centuries after Alcuin, in two examples treated by Tartaglia on dividing up wine barrels.[24] 'Attempt' is indeed the right word, for Tartaglia proceeds by trial and error: he first operates a sharing giving each an equal quantity of wine, then a further distribution provides each with an equal quantity of barrels; this last sharing leaving as many barrels in the same state as there are beneficiaries, distributing them equally among the participants will ensure that each receives his due. This is applied by Tartaglia in his first problem, which he presents as follows:

[13] *A burgess, close to death, makes his last will. He leaves as universal heir one of his sons, but makes also numerous donations to churches, the poor, hospitals, and pawnbrokers. Among other things he possesses are 21 barrels of wine of the same size, of which 7 are full, 7 half full and 7 empty. He bequeaths 7 to the monastery of Saint Mary of the Graces, 7 others to the monastery of Saint Mary of the Angels, and 7 to the church of Saint Mary of the Miracles, requiring that each receive as many barrels and as much wine, but wishing that no wine be removed from the barrels. I ask you how it will be done.*

In Tartaglia's answer the first monastery receives one barrel full, five half full and one empty, and the other two 3 full, 1 half full and 3 empty each. With $n = 21$, $p = 3$ and since, by our third condition, $x_i \leq \frac{n}{2p} = \frac{21}{6}$, thus $x_i \leq 3$, there are two possible solutions, corresponding to the representations $7 = 1 + 3 + 3 = 2 + 2 + 3$, of which Tartaglia's is thus the first one (Fig. 29).

	full	half	empty	full	half	empty
1st	1	5	1	2	3	2
2nd	3	1	3	2	3	2
3rd	3	1	3	3	1	3

Fig. 29

As said above, Tartaglia explains his way of solving. He represents his five steps in a table similar to that of our Fig. 30.

1st	2nd	3rd	result
4 half	2 full	2 full	equality of wine
0 empty	2 empty	2 empty	twice equality
1 full	1 full	1 full	twice equality
1 half	1 half	1 half	twice equality
1 empty	1 empty	1 empty	twice equality

Fig. 30

---

[24] *General trattato*, I, fol. 255ᵛ, Nos. 130 & 131.

Tartaglia adds that we may likewise solve the case of $21k$ barrels subject to the same conditions (indeed, the solutions in Fig. 29 will just be multiplied by k). But, since in the previous case he gave only one solution, he does not mention that the number of possible solutions will increase together with k.

[13′] His subsequent problem is about sharing 27 barrels, subject to analogous conditions, between three monasteries. The respective parts will be, he says: 4 full, 1 half, 4 empty; 3 full, 3 half, 3 empty; 2 full, 5 half, 2 empty. Now since the third condition is $x_i \leq \frac{n}{2p} = \frac{27}{6}$, thus $x_i \leq 4$, there are, besides $9 = 4+3+2$ two further admissible partitions, namely $9 = 4+4+1 = 3+3+3$ (Fig. 31). It is surprising that Tartaglia does not mention the (indeed, banal) last possibility.

	full	half	empty	full	half	empty	full	half	empty
1st	4	1	4	4	1	4	3	3	3
2nd	3	3	3	4	1	4	3	3	3
3rd	2	5	2	1	7	1	3	3	3

Fig. 31

§3. Solving conditions according to Bachet de Méziriac

Whereas Tartaglia does not tell us how he learned about these problems, Bachet explicitly refers to Tartaglia.[25] He criticizes this otherwise 'so skilful' (*si habile*) mathematician for not giving any general rule for solving such problems. He will not, he says, make the same mistake. Considering Tartaglia's example of 21 barrels, he first observes that we could not share them among four beneficiaries, since each would have $5 + \frac{1}{4}$. Thus, for the question to be possible, $\frac{n}{p}$ must be an integer (*je dis qu'il convient diviser le nombre des tonneaux par celuy des personnes, & si le quotient ne vient nombre entier, la question est impossible*); this is thus our Condition II. Next, we are to divide this quantity $\frac{n}{p}$, here $\frac{n}{3}$, into as many parts as there are beneficiaries, with each part being less than half the quotient (*il convient prendre ledit quotient, & en faire autant de parties qu'il y a de personnes, observant toutesfois que chascune d'icelles parties soit moindre que la moitié du susdict quotient*); that is, $\frac{n}{3} = x_1 + x_2 + x_3$, with (this is our condition III) $x_i < \frac{1}{2}\frac{n}{3}$. Applying it to Tartaglia's case of 21 barrels with three beneficiaries, he finds 3, 3, 1 and 2, 2, 3, this being, as he says, the number of full (or empty) barrels, from which the number of half full ones is determined.

He then likewise solves Tartaglia's second problem of 27 barrels and

[25] See his 1612 ed., p. 161; ed. 1624, p. 233.

three beneficiaries, and gives the three solutions, unlike Tartaglia who, he says, 'gives only one since he did not know the general rule for solving all such problems'.

Full of self-confidence after establishing his rules and solving the above examples, Bachet turns to the case of more than three beneficiaries, here four with 24 barrels of wine. There are thus 8 full, 8 half full and 8 empty barrels. Dividing 24, the number of barrels, by 4, the number of beneficiaries, he obtains 6, which he then divides into four parts (our x_i) each less than $3 = \frac{n}{2p}$, and finds 2, 2, 1, 1; each recipient will then receive altogether 6 barrels and the quantity of wine of 6 half full barrels. This is the situation shown in our Fig. 32.

	full	half	empty
1st	**2**	2	2
2nd	**2**	2	2
3rd	**1**	4	1
4th	**1**	4	1

Fig. 32

However, the attentive reader will observe that there are altogether six full and six empty barrels, but twelve half full, which counters the hypotheses. Indeed, whereas the quotient of condition II is $\frac{n}{p}$, that of condition I is *always* $\frac{n}{3}$, whatever the number of beneficiaries. Such confusion is all the more surprising in that Bachet had explicitly mentioned, at the beginning of the problem, the existence of eight barrels in each state.

The correct answer would thus be to divide $\frac{n}{3} = 8$ into a sum of four integers each less than $\frac{n}{2p} = 3$. The only possibility is $8 = 2 + 2 + 2 + 2$, each beneficiary receiving two barrels in each state, altogether six barrels, and the quantity of wine of six half-filled barrels —as before, but this time in accordance with the initial conditions. Admitting 3 in the partition, thus the absence of some types of barrel in the parts, we have the three further possibilities $8 = 3 + 3 + 1 + 1 = 3 + 2 + 2 + 1 = 3 + 3 + 2 + 0$, represented in Fig. 33.

	full	half	empty	full	half	empty	full	half	empty
1st	**3**	0	3	**3**	0	3	**3**	0	3
2nd	**3**	0	3	**2**	2	2	**3**	0	3
3rd	**1**	4	1	**2**	2	2	**2**	2	2
4th	**1**	4	1	**1**	4	1	**0**	6	0

Fig. 33

§4. Generalization

Consider the more general problem involving barrels in the same three states, but which now may be in different quantities, and maintaining the requirement that each beneficiary should receive as much wine and as many barrels.[26] So let there be n_1 full barrels, n_2 half full and n_3 empty ones.

Since the number of barrels is altogether $n = n_1 + n_2 + n_3$ and each of the p beneficiaries is to receive the same number of them, we are first to impose

CONDITION I: $n_1 + n_2 + n_3$ must be divisible by p.

Since the whole quantity of wine is, in half-barrels, $2n_1 + n_2$ and each of the p beneficiaries is to receive the same quantity of it, we are then to impose

CONDITION II: $2n_1 + n_2$ must be divisible by p.

We are left with operating the equal distribution of barrels and wine. Consider, as before, x_i to be the number of full barrels received by the ith beneficiary. We shall therefore have

$$\sum_{i=1}^{p} x_i = n_1$$

Assuming we have found a set of p suitable integers x_i, the quantities of half full and empty barrels will be deduced from that. Indeed, since the ith has received the quantity of wine corresponding to $2x_i$ half full barrels, he is still to obtain

$$\frac{2n_1 + n_2}{p} - 2x_i$$

half-full barrels to get his due share of wine and

$$\frac{n}{p} - x_i - \left(\frac{2n_1 + n_2}{p} - 2x_i\right) = \frac{n_3 - n_1}{p} + x_i$$

empty barrels in order to have his due quantity of barrels. (Note that it appears from Conditions I and II that $n_3 - n_1$ is divisible by p.)

Since each of the x_i may exceed neither the number of barrels nor the quantity of wine to be received by each beneficiary, we shall further impose

CONDITION III: $x_i \le \frac{2n_1+n_2}{2p}$ (satisfies, *a fortiori*, $x_i \le \frac{n}{p}$).

Solving the problem thus amounts to representing n_1 as the sum of p non-negative integers x_i satisfying Condition III, and each acceptable

---

[26] This generalization is considered by Labosne in his reedition of Bachet's work; see *Problèmes plaisants*, p. 171, with the same numerical example as here below.

representation will give a solution (and a single one since the order of the p beneficiaries is irrelevant). We may admit the possibility $x_i = 0$ since a beneficiary may receive no barrel completely filled. The general solution will then be

	full	half	empty
1st	x_1	$\frac{2n_1+n_2}{p} - 2x_1$	$\frac{n_3-n_1}{p} + x_1$
2nd	x_2	$\frac{2n_1+n_2}{p} - 2x_2$	$\frac{n_3-n_1}{p} + x_2$
3rd	x_3	$\frac{2n_1+n_2}{p} - 2x_3$	$\frac{n_3-n_1}{p} + x_3$
$\vdots$	$\vdots$	$\vdots$	$\vdots$
p^{th}	x_p	$\frac{2n_1+n_2}{p} - 2x_p$	$\frac{n_3-n_1}{p} + x_p$

Fig. 34

Consider next, with Labosne, the case with $n_1 = 5$, $n_2 = 11$, $n_3 = 8$ $(n_3 > n_1)$, $p = 3$, thus with $n = n_1 + n_2 + n_3 = 24$ and $2n_1 + n_2 = 21$ both divisible by 3. Since, by condition III, $x_i \leq \frac{7}{2}$, we must satisfy the relation $x_1 + x_2 + x_3 = 5$ using the four numbers 0, 1, 2, 3. There are three possibilities, namely $3 + 2 + 0$, $3 + 1 + 1$, $2 + 2 + 1$, giving the solutions of Fig. 35.

	full	half	empty	full	half	empty	full	half	empty
1st	3	1	4	3	1	4	2	3	3
2^{nt}	2	3	3	1	5	2	2	3	3
3rd	0	7	1	1	5	2	1	5	2

Fig. 35

Chapter III. Minimal set of weights

§1. Introduction

This problem is often called 'Bachet's problem of weights' because, like the two above, it became widely known through his work.[27] It too is already found in mediaeval times and surely goes back to antiquity.

Fig. 36

One of the main tools of goldsmiths was a set of small weights for measuring their merchandise (see Fig. 36[28]). The question thus arising is to find the smallest possible set that will measure the continuous sequence of integral weights from 1 on to some given limit. (Although this might be of practical interest, the problem was to remain purely theoretical since no set fulfilling that condition is known to have existed.) The answer will depend on the type of scales used.

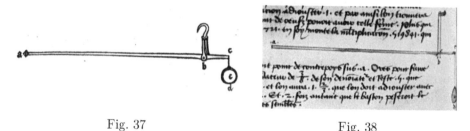

Fig. 37   Fig. 38

Now there were three types of scales in use since antiquity.

— First, the scales with one weight moving on one arm, the quantity to be measured being hung on the other (also called 'Roman scales'; see Fig. 37 & 38[29]): we have on one side the arm ab, on the other the merchandise to be weighed c, at a fixed distance d from the hook; it must

---

[27] *Problemes plaisans*, pp. 143–146 (ed. 1612), pp. 215–219 (ed. 1624).

[28] Dated 1759; in the Musée monétaire cantonal, Lausanne.

[29] Illustration from de la Roche *Arismethique*, fol. 107ʳ and original drawing in Chuquet's treatise, MS. Paris BNF fr. 1346, fol. 202ᵛ.

be counterbalanced by a movable constant weight p hanging on the arm ab. If equilibrium takes place when p is at the distance x from the hook, then $c : p = x : d$, and this determines the weight c (whence also the possibility of graduating the arm by varying c). But these Roman scales are irrelevant here since there is a single weight.

— The second type was the scales with arms of equal length at the ends of which were hooks, or pans, one for the merchandise and the other for the weights (Fig. 39 & 40).[30]

— The third type was similar, but weights could be put on both sides, thus also on the same side as the merchandise in order to counterbalance those in the other pan.

Fig. 39   Fig. 40

Remark. In the 'Mechanical problems' attributed to Aristotle, we are told that some sellers of purple put the cord supporting the scales off-centre and used for the smaller arm wood from a root or with a knot (which is heavier), or even filled this arm with lead.[31]

§ 2. Arabic references

Although the antique origin of this problem can be considered as certain, the earliest references occur in Arabic texts. First, we find it in the 11th-century 'Key to commercial transactions' by Muḥammad ibn Ayyūb Ṭabarī.[32] We are to find weights (سنگ, 'stone', for سنگ ترازو, thus stone used as a weight standard), all different and all being integral multiples of the same unit (pound), which will enable us to determine all weights continuously from 1 to (say) 10 000 pounds. The author gives us the answer: ten stones with respective weights 1, 3, 9, 27, 81, 243, 729, 2187, 6561, 19 683 pounds. He observes that each of these numbers arises from the previous one by multiplying it by 3 (they are indeed the

---

[30] Illustrations (woodcuts) from Tagliente's *Componimento*, ed. 1547 and 1554. (Spot the differences.)

[31] *Mechanica* I.1 (*in fine*), 849 b 34 – 850 a 2.

[32] *Miftāḥ al-mu'āmalāt*, IV, 43 (pp. 125–127 in the edition of the Persian text).

consecutive powers of 3 from 3^0 to 3^9). He then indicates how to combine the first of these stones to obtain the weights from 1 to 22, thus by balancing and counterbalancing.

His text gives no theoretical explanations, unlike al-Khāzinī in his 'Balance of wisdom' (ميزان الحكمة), completed 1121/22.[33] As he writes in the first section of Ch. VI, the 'usual weights' (الصنجات المعهودة), at least those in use at the time (المستعملة في زماننا), were nine altogether, namely three among each of the first three decimal orders of natural numbers, all occupying the same ranks: 1, 2, 5 among the units; 10, 20, 50 among the tens; 100, 200, 500 among the hundreds.[34] With them, one could determine each integral weight from 1 to 888, which is the sum of all these weights. But, as observed by the author himself, with this system there is no uniqueness in the combination of these weights; thus, 3 could equally well be weighed as $1 + 2$, $5 - 2$, $10 - (5 + 2)$ or $20 - (10 + 5 + 2)$.

We may therefore add a new condition to our problem, namely that to each given weight will correspond a single combination. A second condition will be for the set of weights at our disposal to be minimal. A third condition restricts the size of this set: its weights must just measure all the (integral) weights measured by the 'usual weights', thus from 1 to 888.

Khāzinī will determine these systems for the two types, thus without and with counterbalancing. Here are his explanations:

[14] *If we wish to take the weights (...) using (for them) a single pan, where there is (thus) no need for compensation (= counterbalancing), we shall take them according to doubling. Thus we shall put as the first 1, as the second 2, as the third 4, as the fourth 8, as the fifth 16, as the sixth 32, as the seventh 64, as the eighth 128, as the ninth 256, as the tenth 512. It is (however) one more than the usual (set of) weights, but with greater convenience for determining the successive quantities (since) without use of subtraction (= counterbalancing).*

Indeed, with weights 1, 2, 4, 8, 16, ..., we see at once how to obtain the sequence of successive natural numbers: **1**, **2**, $2+1$, **4**, $4+1$, $4+2$, $4+2+1$, **8**, $8+1$, $8+2$, $8+2+1$, $8+4$, $8+4+1$, $8+4+2$, $8+4+2+1$, **16**, ..., and so on to 1023, their sum.

---

[33] See edition of the Arabic text, pp. 108–110, or pp. 136–138 of Ibel's study.

[34] A natural choice since balancing and counterbalancing could be used in the same way for each decimal order, e.g. for units with the sequence 1, 2, $1 + 2$, $5 - 1$, 5, $5 + 1$, $5 + 2$, $5 + (1 + 2)$, $10 - 1$ (the weights preceded by the sign $-$ are in the second pan). By the way, modern (19th-century) European scales also displayed weights of 1, 2, 5, 10, 20, 50, 100, 200, 500 grams.

[14′] *If we wish to establish the minimal number of weights when admitting the subtraction by way of compensation, we shall set their quantities from 1 on by successive multiplications by 3. Thus doing, the first (weight) will be 1, the second 3, the third 9, the fourth 27, the fifth 81, the sixth 243, the seventh 729. There is one weight less than the usual (set of) weights.*[35]

§3. Finding the solution

The initial step is the same in both cases. Supposing an ascending sequence of integral weights $p_1, p_2, \ldots, p_k$, the first must be 1 since otherwise we could not weigh the one immediately following p_2. Thus $p_1 = 1$. The difference between the two systems arises with the choice of p_2. So let us designate with $p_i^{(1)}$ the weights put in just one pan and with $p_i^{(2)}$ those of the second system, using counterbalancing (Fig. 41).

(1) Consider in the first system that we have determined a set of appropriate weights $p_1^{(1)}, p_2^{(1)}, \ldots, p_k^{(1)}$, with $p_1^{(1)} = 1$. The maximal weight we can measure is when all are together in one pan. Thus the next weight must necessarily be

$$p_{k+1}^{(1)} = \sum_1^k p_i^{(1)} + 1.$$

With this rule in mind we may determine the sequence of successive weights: the sum of all previous weights is each time increased by 1, they are thus, as seen before, 1, 2, 4, 8, ... Indeed, since the sum of these weights is of the form $1+2+2^2+2^3+\cdots+2^{k-1}$, which equals 2^k-1, our next weight will be $p_{k+1}^{(1)} = 2^k$. The quantities we may weigh are continuous since from the last possible weighing we pass to the immediately successive

---

[35] K͟hāzinī's text being of fundamental importance, we reproduce the original:

اذا اردنا ان نتخذ صنجات مرتبة توالى الاعداد الطبيعية على الترتيب فى كفّة واحدة ولا يحتاج فيها الى التقابل فانّا نتخذها على ترتيب التضعيف على ان نجعل الاولى واحدًا والثانية اثنين والثالثة اربعة والرابعة ثمانية والخامسة ستّة عشر والسادسة اثنين وثلاثين والسابعة اربعة وستّين والثامنة مائة وثمانية وعشرين والتاسعة مائتين وستّة وخمسين والعاشرة خمسمائة واثنى عشر زاد فيها على المعهود عدد واحد الّا انّه كثّرت الفائدة فى وجود اعداد الترتيب فى غير استثناء.

اذا اردنا اثبات قلّة عدد الصنجات وجواز الاستثناء من مقابلته فانّا نفرض اعدادها من الواحد مضروبة فى ثلاثة على الترتيب نحو ان تكون الاولى واحدًا والثانية ثلاثة والثالثة تسعة والرابعة سبعة وعشرين والخامسة احدًا وثمانين والسادسة مائتين وثلاثة واربعين والسابعة سبعمائة وتسعة وعشرين وهى اقلّ من المعهود بصنجتين.

one; it is also clear that the set of weights thus determined is minimal and their combination, for a given weight, unique.

weight	one pan	two pans
1	$\boxed{p_1^{(1)}}$	$\boxed{p_1^{(2)}}$
2	$\boxed{p_2^{(1)}}$	$p_2^{(2)} - p_1^{(2)}$
3	$p_1^{(1)} + p_2^{(1)}$	$\boxed{p_2^{(2)}}$
4	$\boxed{p_3^{(1)}}$	$p_2^{(2)} + p_1^{(2)}$
5	$p_3^{(1)} + p_1^{(1)}$	$p_3^{(2)} - p_2^{(2)} - p_1^{(2)}$
6	$p_3^{(1)} + p_2^{(1)}$	$p_3^{(2)} - p_2^{(2)}$
7	$p_3^{(1)} + p_1^{(1)} + p_2^{(1)}$	$p_3^{(2)} - p_2^{(2)} + p_1^{(2)}$
8	$\boxed{p_4^{(1)}}$	$p_3^{(2)} - p_1^{(2)}$
9	$p_4^{(1)} + p_1^{(1)}$	$\boxed{p_3^{(2)}}$
10	$p_4^{(1)} + p_2^{(1)}$	$p_3^{(2)} + p_1^{(2)}$
11	$p_4^{(1)} + p_1^{(1)} + p_2^{(1)}$	$p_3^{(2)} + p_2^{(2)} - p_1^{(2)}$
12	$p_4^{(1)} + p_3^{(1)}$	$p_3^{(2)} + p_2^{(2)}$
13	$p_4^{(1)} + p_3^{(1)} + p_1^{(1)}$	$p_3^{(2)} + p_2^{(2)} + p_1^{(2)}$
14	$p_4^{(1)} + p_3^{(1)} + p_2^{(1)}$	$p_4^{(2)} - p_3^{(2)} - p_2^{(2)} - p_1^{(2)}$
15	$p_4^{(1)} + p_3^{(1)} + p_2^{(1)} + p_1^{(1)}$	$p_4^{(2)} - p_3^{(2)} - p_2^{(2)}$
16	$\boxed{p_5^{(1)}}$	$p_4^{(2)} - p_3^{(2)} - p_2^{(2)} + p_1^{(2)}$

Fig. 41

(2) Consider now the case of the weights put in the two pans. Clearly, the number of required weights will be less than before since counterbalancing will increase the difference between two consecutive weights. Indeed, suppose we have determined a set of appropriate weights $p_1^{(2)}$, $p_2^{(2)}$, ..., $p_k^{(2)}$, with $p_1^{(2)} = 1$ and $S_k = \sum p_i^{(2)}$. If $p_{k+1}^{(2)}$ is the next weight, we shall be able to measure with it any integral weight from $p_{k+1}^{(2)} - S_k$ (putting all previous weights with the merchandise) to $p_{k+1}^{(2)} + S_k$ (merchandise alone, counterbalanced by all weights, including the new one). Now the largest weight we could determine before admitting $p_{k+1}^{(2)}$ was the sum of all the weights, thus S_k; so the next, $S_k + 1$, must coincide with the smallest

obtained by counterbalancing with the new weight the sum of the others, thus $p_{k+1}^{(2)} - S_k$. Setting equality gives $p_{k+1}^{(2)} = 2S_k + 1$. This enables us, starting with $p_1^{(2)} = 1$, to compute successively the $p_i^{(2)}$: 1, 3, 9, 27, 81, ..., that is, $p_{k+1}^{(2)} = 3^k$.

Thus, just as in the first case we had the successive powers of 2, we now have the successive powers of 3. Fig. 41 shows how in each case the succession of integral weights is obtained. As a matter of fact, the solution in the first case amounts to saying that each natural number may be expressed *in a single way* as the sum of consecutive powers of 2 with the coefficients 0 (weight not used) or 1. Again, in the second case, any natural number may be expressed *in a single way* as the sum of consecutive powers of 3 with the coefficients 0, 1, −1 (counterbalancing).

§4. Mediaeval examples

The French mathematician Nicolas Chuquet, already encountered, proposes a slightly different problem.[36]

[15] *A man has a pair of scales and three different pieces of weight, which altogether weigh 10 pounds, by means of which he may weigh from 1 pound to 10, neither more nor less,[37] as if he had all kinds of weight.[38] One wishes to know the weight of each piece.*

For performing such computations, one may[39] take two numbers in double proportion, which are 1 for the first and 2 for the second, add them together, which gives 3, which we double, and add 1 to the result, which gives 7; thus the first piece of weight weighs 1 pound, the second 2 pounds, and the third 7 pounds. If he wishes to weigh 1 pound, he can do it for he has the weight of one pound. If he wishes to weigh 2, he has the weight of 2 pounds. If he wishes to weigh 3, he has (the weights of) 1 pound and 2 pounds. If he wishes to weigh 4 pounds, he can put 7 pounds in one of the parts of the scales and 3 in the other. If he wants to do 5, he can put 7 pounds on one side and 2 on the other. If 6 pounds, he can put 7 pounds on one side and 1 pound on the other. If he wants to weigh 7 pounds, he can do it, for he has the weight of 7. If he wants to weigh 8 pounds, he has 7 pounds and 1. If 9 pounds, he has 7 and 2 pounds. If 10 pounds, he has 7 pounds, 2 pounds and 1 pound. It thus appears that he can easily, by means of these three pieces of weight, use the scales and each time weigh everything he wants to weigh from 1 pound to 10, inclusively.

---

[36] *Appendice*, pp. 451–452; MS., fol. 201^{v} – 202^{r}.

[37] The two limits thus fixed.

[38] One for each weight.

[39] He will mention below another possibility (with the successive powers of 2).

In the same manner as we have set up these three pieces of weight, we may set up 4 pieces or more by beginning with 1 and doubling to the penultimate, like 1, 2, 4, and so on, while the last must be twice the sum of all the previous, plus 1. Thus, the 4th will be 15 if the three preceding ones are 1, 2, 4.

Here we have a combination of the two types already seen: we take the sequence of the first type but determine the last weight in the manner of the second type; for, as we have seen, counterbalancing is allowed. Thus, suppose we take the weights $p_1^{(1)} = 1$, $p_2^{(1)} = 2$, $p_3^{(1)} = 4$, $p_4^{(1)} = 8$, ..., $p_k^{(1)} = 2^{k-1}$, which add up to $S_k^{(1)} = 2^k - 1$. The last weight p_{k+1} must then be $2S_k^{(1)} + 1 = 2^{k+1} - 1$. Then we may indeed measure all integral weights from 1 to $3S_k^{(1)} + 1$: those up to $S_k^{(1)} = 2^k - 1$ by putting a combination of the previous weights in the pan without the merchandise, the subsequent one by putting p_{k+1} on one side and $S_k^{(1)}$ on the other (which weighs $p_{k+1} - (2^k - 1) = 2^{k+1} - 1 - (2^k - 1) = 2^k$), and so on with the next weights, by first reducing the subtractive quantities and then putting them together with p_{k+1}. This way of proceeding combines thus the simplicity of the first system with the economy of terms of the second.

But this does not mean that Chuquet did not consider the two systems individually. About the first, he says in the same passage that it is the most suitable and easiest (*convenable*) method, for there is no counter-balancing (and therefore only straightforward computations).[40] But the second is even much better (*encores trop plus propices*), for with fewer pieces one may measure greater weights. This is best illustrated with the example of taking the first four weights: with the first system, we may weigh continuously from 1 to 15, whereas with the second from 1 to 40.

Some seventy years later Tartaglia asserts that with the weights 1, 2, 4, 8, 16, 32 we may weigh up to 63; he illustrates this with the first com-binations, but for weighing 3 he takes 4 counterbalanced with 1.[41] There seems at times to have been some confusion between the two systems.

Now this is surprising, for three and a half centuries earlier the second system is clearly explained by Leonardo Fibonacci in his *Liber abaci*: he considers the example of the four first powers of 3 weighing all integral weights from 1 to 40 (*a libra una usque in libris 40*); he adds that the system can be extended at will (*et sic eodem ordine possunt addi pesones in infinitum*), the next weight being obtained by either tripling the last or

---

[40] Kḥāzinī said just the same (above, [14]). Indeed, counterbalancing requires more reflection.

[41] *General trattato*, II, fol. 14ʳ (*falso*: 17ʳ).

adding 1 to twice the sum of the previous ones.[42] Again, in the fifteenth century, Benedetto da Firenze explicitly states that 40 is the limit with four weights (*chominciando a una libra adimandasi quanto possono pesare el più*).[43] Fig. 42 displays, from a contemporary text, the five weights (pounds, *libre*) by means of which one can weigh from 1 to 121.[44]

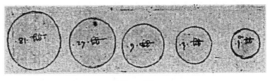

Fig. 42

Note incidentally the following strange (indeed impractical) application: we are asked to divide a piece of marble as above, the purpose being to be able to measure integral weights from 1 to 40.[45]

In Chuquet's problem [15], the last weight is determined, for it has the form $2S_k^{(1)} + 1$, and the limit of weighing is also determined, as $3S_k^{(1)} + 1$. But suppose now we wish not to go beyond a certain limit C, thus with $S_k^{(1)} < C < 3S_k^{(1)} + 1$. The additional weight will then be set as

$$p_{k+1} = C - S_k^{(1)} = C - 2^k + 1.$$

The main application is the payment of an untrustworthy worker one day at a time. This already occurs in Fibonacci's work: a worker, hired to work thirty days at the wage of one silver mark a day, will be paid using five silver cups, not to be broken up, weighing 1, 2, 4, 8, 15 marks.[46] Thus the payment of the 14th day, $2 + 4 + 8$, will have to be returned by the worker on the 15th day, and he will receive in exchange the cup of 15, while for the subsequent payments the employer will return to the use of the powers of 2. One sees why the powers of 2 are used: when he gives the cup of 2, the employer takes that of 1; were he to give that of 3, as in the second system, the worker would gladly do nothing on the third day.

Of course, if the number of days happens to be of the form $2^k - 1$, we shall use only the powers of 2, as in the case of this owner of a house renting it at one mark a day for 15 days with the (odd) requirement of

---

[42] *Liber abaci*, p. 297.

[43] MS. Siena BCI L.IV.21 (dated 1463), fol. 294^r.

[44] MS. Rome Accad. naz. dei Lincei Cors. 1875, fol. 52^v.

[45] *Rascioni d'algorismo*, No. 71.

[46] *Liber abaci*, p. 298. Taken over by Pacioli, *Summa*, fol. 97^v. The mark is a unit of weight, commonly (hardly here) 250 gr.

receiving each day exactly one mark, no more and no less;[47] four cups of 1, 2, 4, 8 silver marks will satisfy this condition.

Tartaglia, too, has the problem of an untrustworthy worker, where the situation is quite clearly explained:

[16] *A gentleman hires a craftsman to do a certain task for 60 days at the agreed wage of 1 mocenigo a day. But since the gentleman did not really trust the crafts- man, and thus would not pay him in advance, he had minted 6 silver coins worth altogether 60 mocenighi, but with values set in such a way that using the 6 coins he could pay the craftsman day by day, each evening for the day worked only.*[48]

For this purpose, he takes five coins worth 1, 2, 4, 8, 16, adding up to 31, and one last coin of $(60 - 31 =)$ 29, to be given the craftsman on the 29th day in exchange for $28 = 16 + 4 + 8$; the craftsman will end up on the evening of the 60th day with all six coins. (Note the absurdity of having these coins minted instead of paying day by day with 60 *mocenighi*.)

We just saw how the weights equal to the powers of 2 are supplemen- ted by another, not part of this sequence. The same is also encountered in the second system, thus with the sequence of the powers of 3. Cardan has two examples of it, just after asserting that the minimal number of weights is obtained by this sequence.[49] In his first example, we are told that, in order to be able to weigh continuously to $C = 100$, we shall choose 1, 3, 9, 27 (summing up to 40) and $(C - 40 =)$ 60. Next, to weigh to $C = 300$, we may take 1, 3, 9, 27, 81 (summing up to 121) and any weight p with $179 \leq p \leq 243 = 3^5$. Cardan finally observes that with the first ten weights in accordance with the sequence of the powers of 3 we could weigh continuously to 29 524. Note that this is precisely the problem of the ten weights encountered at the beginning of this chapter (p. 24), thus seen to have been considered half a millenium earlier.

§ 5. Generalization

The use of geometrical progressions for solving the weights problem is limited to the powers of 2 and 3. As remarked by Bachet after treating this problem, the further progressions would be of no use to solve it; thus, he says, with the weights 1, 4, $(4^2 =)$ 16 pounds we could not weigh 2, 6, 7, 8, 9, 10. There is, though, another kind of generalization, which will make the two previous ones particular cases.

---

[47] (...) *en esta manera que no lo diese plata de mas nin de menos sy non su marco cabal por cada día* (*Arte del alguarismo*, p. 155).

[48] *General trattato*, II, fol. 13ᵛ, No. 32. The *mocenigo* was a silver coin minted in Venice under the rule of Pietro Mocenigo (1474-1476).

[49] *Practica arithmetice*, Ch. LXV, No. 12.

Suppose that we have, of each of these new weights p_i $(i > 1)$, a fixed number of m pieces —and no longer a single one as previously. As before the first weight will be set as 1. Let us see what the next ones will be.

- If there is no counterbalancing, we shall take, as before,

$$p_{k+1}^{(1)} = S_k^{(1)} + 1 = m \sum_1^k p_i^{(1)} + 1,$$

thus, since $p_1^{(1)} = 1$,

$p_2^{(1)} = m + 1,$

$p_3^{(1)} = m(m + 2) + 1 = (m + 1)^2,$

$p_4^{(1)} = m[(m + 1)^2 + m + 2] + 1 = m^3 + 3m^2 + 3m + 1 = (m + 1)^3,$

and generally

$$p_{k+1}^{(1)} = (m + 1)^k.$$

- If there are counterbalancing weights, we shall put, as before,

$$p_{k+1}^{(2)} = 2\,S_k^{(2)} + 1 = 2m \sum_1^k p_i^{(2)} + 1,$$

thus, since $p_1^{(2)} = 1$,

$p_2^{(2)} = 2m + 1,$

$p_3^{(2)} = 2m(2m + 2) + 1 = (2m + 1)^2,$

$p_4^{(2)} = 2m[(2m + 1)^2 + 2m + 2] + 1 = (2m + 1)^3,$

and generally

$$p_{k+1}^{(2)} = (2m + 1)^k.$$

It thus appears that, with the assumption of just one piece of each weight $(m = 1)$, we shall have $p_{k+1}^{(1)} = 2^k$ and $p_{k+1}^{(2)} = 3^k$, which are the two cases seen above; for two pieces of each weight, we shall have $p_{k+1}^{(1)} = 3^k$ and $p_{k+1}^{(2)} = 5^k$, respectively; for three pieces, $p_{k+1}^{(1)} = 4^k$ and $p_{k+1}^{(2)} = 7^k$, respectively. That is, with the number of identical weights increasing we shall, if using a single pan, go through the powers of the successive natural numbers and, if the two pans are involved, through those of the successive odd natural numbers.

Chapter IV. Successive distributions

The coming problems are mostly concerned with the repeated reduction or increase in a certain amount of money. It may be a thief who is stopped several times and must each time leave a fraction of what is in his possession and something more; his initial loot, which is unknown, will be determined by knowing the final remainder and the fractions and additional amounts left successively. It may also be that an avaricious man enters various churches and each time prays for what he possesses to be doubled, in which case he will leave a certain sum to the church; however, as a reward for this display of cupidity, he will often end up with empty pockets. A third type of problem is that of a dying man who bequeaths to each of his heirs, taken in turn, the same given fraction of his estate and some additional money; since the parts are supposed to be equal while the estate is each time reduced, the (given) additional amount of money will change accordingly.

These problems must originate in antiquity, as suggested by the presence of the first two types in Anania S̲h̲irakatsi's 7th-century collection.[50] In one of these, he tells us that one of his students wanted to bring him apples, but that three groups of young people met successively left the student each time with a fourth of what he previously had, so that the gift to his teacher became five apples only.[51] In another problem, a man going through three churches sees each time his capital doubled, but his repeated gift of 25 will leave him with nothing at the end.[52]

§ 1. Doorkeepers

Here too, the object is mostly a quantity of apples, commonly stolen from an orchard, with part of it being each time used to bribe successive doorkeepers. The most common example is that of three doors and three keepers arresting a pilfering child, or just a lover wishing to bring a gift to his lady-love, both left in the end with hardly any apples (usually one, but also two: one for the lady and one for himself).[53]

In all these situations solving starts with the final situation. If one apple is left after the third door, and half the quantity remaining just

---

[50] This set of problems has been studied by Kokian.

[51] Problem 13 (the initial quantity of apples was 320). In fact, the earlier *Anthologia Graeca* also has such problems (XIV, 116–120, 138).

[52] Problem 19 (the initial capital was $21 + \frac{7}{8}$).

[53] *Livre de chiffres et de getz*, rules and problems, No. 10; MSS. Tours 399, fol. 128^v – 129^r; Paris BNF fr. 1339, fol. 76^v; Paris BNF fr. 2050, fol. 92^v – 93^r (two apples left) —these texts in *Récréations*, pp. 50–51. Jacopo da Firenze, ed. p. 271 (three oranges left).

before, plus one apple, was given at the door, the previous number of apples must have been four. In order for four apples to remain after the second door, the number before must have been ten, since half of it and one more were given. Likewise, before the first door, the thief must have had $2(10 + 1) = 22$, and such was the quantity stolen. Generally, if the final remainder is r_3, we find accordingly, reckoning backwards, $2(r_3 + 1) = r_2$, $2(r_2 + 1) = r_1$, $2(r_1 + 1) = S = 2\{2[2(r_3 + 1) + 1] + 1\}$. Fig. 43 is a contemporary illustration of the problem of three doors.[54]

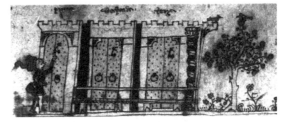

Fig. 43

Clearly, the situation can be extended at will. In the early 13th century, Fibonacci considers a single apple resulting from passing through seven doors, with each time half the apples plus one given; he finds accordingly, starting here too from the final situation,

$$\left(\left[\left(\left[\left((r_7 + 1)\, 2 + 1\right]\, 2 + 1\right)\, 2 + 1\right]\, 2 + 1\right)\, 2 + 1\right]\, 2 + 1\right)\, 2 = 382.$$

That is, he adds, how to solve such problems: reckoning backwards from the final situation.[55]

Consider now, more generally, that the thief has picked S apples and that the quantity left at the ith door $(i = 1, \ldots, n)$ is the proper fraction $\frac{p_i}{q_i}$ of the quantity possessed just before, and a supplement s_i. After passing through the successive doors, the following quantities r_i will remain:

$$r_1 = S - \frac{p_1}{q_1}S - s_1 = \left(1 - \frac{p_1}{q_1}\right)S - s_1$$

$$r_2 = \left(1 - \frac{p_2}{q_2}\right)r_1 - s_2 = \left(1 - \frac{p_1}{q_1}\right)\left(1 - \frac{p_2}{q_2}\right)S - \left(1 - \frac{p_2}{q_2}\right)s_1 - s_2$$

$$r_3 = \left(1 - \frac{p_3}{q_3}\right)r_2 - s_3 =$$

$$\left(1 - \frac{p_1}{q_1}\right)\left(1 - \frac{p_2}{q_2}\right)\left(1 - \frac{p_3}{q_3}\right)S - \left(1 - \frac{p_2}{q_2}\right)\left(1 - \frac{p_3}{q_3}\right)s_1 - \left(1 - \frac{p_3}{q_3}\right)s_2 - s_3$$

---

[54] MS. Rome Accad. naz. dei Lincei Cors. 1875, fol. 50^r (*prima, sechonda, terza*).

[55] *Et sic revertendo, secundum quod propositum fuerit, in ordinem retro poteris quamlibet similium positionum reperire* (*Liber abaci*, p. 278).

and finally

$$r_n = \prod_{i=1}^{n}\left(1-\frac{p_i}{q_i}\right)S - \prod_{i=2}^{n}\left(1-\frac{p_i}{q_i}\right)s_1 - \prod_{i=3}^{n}\left(1-\frac{p_i}{q_i}\right)s_2 - \cdots - \left(1-\frac{p_n}{q_n}\right)s_{n-1} - s_n,$$

whence

$$r_n = \prod_{i=1}^{n}\left(1-\frac{p_i}{q_i}\right)S - \sum_{k=1}^{n-1} s_k \prod_{l=k+1}^{n}\left(1-\frac{p_l}{q_l}\right) - s_n. \qquad (1)$$

The case of three doors is, as we have seen in the beginning, the most common situation, and various simplified situations occur. So consider now the first three equalities above. The reconstructed expression of S will be

$$S = \left\{\left[(r_3 + s_3)\,\frac{1}{1-\frac{p_3}{q_3}} + s_2\right]\frac{1}{1-\frac{p_2}{q_2}} + s_1\right\}\frac{1}{1-\frac{p_1}{q_1}}$$

$$= \left\{\left[(r_3 + s_3)\,\frac{q_3}{q_3-p_3} + s_2\right]\frac{q_2}{q_2-p_2} + s_1\right\}\frac{q_1}{q_1-p_1}. \qquad (1')$$

In the particular case where $q_i - p_i = 1$, that is, if in the fraction the numerator is less than the denominator by a unit, this takes the simpler form

$$S = \left\{\left[(r_3 + s_3)\,q_3 + s_2\right]q_2 + s_1\right\}q_1. \qquad (1'')$$

Such was the expression which enabled us to solve the problem of three doors seen above (and this was then extended to Fibonacci's larger number of terms).

In the case of the fraction added taking the same form as the part of the capital left, thus $s_i = \frac{p_i}{q_i}$, this last relation becomes

$$S = \left(\left[r_3 \cdot q_3 + p_3\right]q_2 + p_2\right)q_1 + p_1. \qquad (1''')$$

The general formula (1) may also take a simpler form in other situations. Thus, if $\frac{p_i}{q_i} = \frac{p}{q}$ and $s_i = s$ for all i, it becomes

$$r_n = \left(1-\frac{p}{q}\right)^n S - s\left[\left(1-\frac{p}{q}\right)^{n-1} + \left(1-\frac{p}{q}\right)^{n-2} + \cdots + \left(1-\frac{p}{q}\right) + 1\right]$$

$$= \left(1-\frac{p}{q}\right)^n S - s\,\frac{\left(1-\frac{p}{q}\right)^n - 1}{\left(1-\frac{p}{q}\right) - 1} = \left(1-\frac{p}{q}\right)^n S + s\,\frac{\left(1-\frac{p}{q}\right)^n - 1}{\frac{p}{q}}. \qquad (2)$$

Finally, if there are no supplements involved ($s = 0$), the expression for S becomes simply

$$S = \frac{r_n}{\left(1 - \frac{p}{q}\right)^n}. \qquad (2')$$

All these formulae correspond to actual computations in mediaeval and 16th-century examples, as will be illustrated now.

[17] *A man with bezants wanted to leave a town which had ten doors. At the first, he had to give $\frac{2}{3}$ of his bezants plus $\frac{2}{3}$ of a bezant; at the second, half the bezants he brought plus $\frac{1}{2}$ bezant; at the third, a third and $\frac{1}{3}$ of a bezant; at the fourth, a fourth and $\frac{1}{4}$ of a bezant; and so on to the tenth, where he gave a tenth of the bezants he brought and $\frac{1}{10}$ of a bezant. He was left with 1 bezant.*[56]

Fibonacci solved this in two ways, but in a rather lengthy manner. As a matter of fact, using our formula (1) and writing the numbers in figures simplify the computations considerably, for many numerators and denominators cancel one another out. Indeed, with the data $\frac{p_1}{q_1} = \frac{2}{3} = s_1$, $\frac{p_m}{q_m} = \frac{1}{m} = s_m$ for $m \geq 2$, we find

$$r_{10} = \frac{1\,1\,2\,3\,4\,5\,6\,7\,8\,9}{3\,2\,3\,4\,5\,6\,7\,8\,9\,10} S - \left(\frac{2}{3}\frac{1}{10} + \frac{1}{2}\frac{2}{10} + \frac{1}{3}\frac{3}{10} + \frac{1}{4}\frac{4}{10} + \frac{1}{5}\frac{5}{10} + \frac{1}{6}\frac{6}{10}\right.$$

$$\left. + \frac{1}{7}\frac{7}{10} + \frac{1}{8}\frac{8}{10} + \frac{1}{9}\frac{9}{10} + \frac{1}{10}\right) = \frac{1}{30} S - \left(\frac{2}{3}\frac{1}{10} + \frac{9}{10}\right) = \frac{1}{30} S - \frac{29}{30},$$

and therefore, since $r_{10} = 1$, $S = 59$.

[18] *While attempting to flee, a thief who had stolen gold coins in a palace was arrested by a guard to whom his appearance looked suspicious; the cunning thief threw him, like a meat ball to a growling dog, half of the money stolen and three gold coins more. Then, to a second guard, he gave a third of the remaining money and, in addition, four gold coins. He then escaped a third guard by handing him a fourth of the remainder less one gold coin, and finally walked out calmly with a hundred gold coins. Required the number of gold coins the thief had first laid hands on.*[57]

Using formula (1), we find that

$$100 = \left(1 - \frac{1}{2}\right)\left(1 - \frac{1}{3}\right)\left(1 - \frac{1}{4}\right) S - 3\left(1 - \frac{1}{3}\right)\left(1 - \frac{1}{4}\right) - 4\left(1 - \frac{1}{4}\right) + 1$$

$$= \frac{1}{2} \cdot \frac{2}{3} \cdot \frac{3}{4} S - 3 \cdot \frac{2}{3} \cdot \frac{3}{4} - 4 \cdot \frac{3}{4} + 1 = \frac{1}{4} S - \left(3 + \frac{1}{2}\right),$$

whence $S = 414$. The author reaches this answer using algebra.

[19] *A lord has a child whom he sends to a garden for 7 apples, telling him: You will find 3 doorkeepers, each of whom will tell you: I want half of the whole plus*

---

[56] *Liber abaci*, pp. 316–318. The 'bezant' (*bizantius*) was originally a Byzantine gold coin; obviously, in these texts that is not always the case (see pp. 41n, 161n).

[57] Buteo, *Logistica*, Ch. V, No. 6, pp. 334–335.

two of those remaining after halving. I ask how many he had initially picked, given that 7 are to remain to him.[58]

We may apply (1″) and thus follow the (mediaeval) reasoning of beginning with the final situation: in order to be left with 7, he must have had $(7+2)\,2 = 18$ before the last door, $(18+2)\,2 = 40$ before the second, and $(40+2)\,2 = 84$ initially.

The same book also has such a problem for a sum of denarii.[59] There are again three doors, but the fraction left is $\frac{1}{3}$, the supplements 6 and the final remainder 24. Since we are no longer in the situation of (1″), we shall use (1′) or (2). The successive sums are then 45, $76 + \frac{1}{2}$, $123 + \frac{3}{4}$.

Fig. 44

[20] *A robber, having stolen in a garden a certain quantity of apples, meets successively, whilst going out, three others who threaten to denounce him; in order to appease them, he gives $\frac{1}{2}$ to the first, who for kindness sake returns 12; then he gives again $\frac{1}{2}$ of the remainder to the second, who returns 7; next, he gives $\frac{1}{2}$ of the remainder to the third, who returns 4; in the end, he is left with 20. Required how many apples he had picked in said garden.*[60]

The supplements returned here vary, according to the 'kindness' of each of the other robbers. Using formula (1″) enables us (and the author) to quickly find the result:

$$ S = \Big(\big[\,(20-4)\,2 - 7\,\big]\,2 - 12\Big)\,2 = 76. $$

The author rightly says that we reach this by starting from the final situation. But he also asserts that one should proceed in the same way if the three fractions given were $\frac{1}{2}, \frac{1}{3}, \frac{1}{4}$, or 'other fractions'. This would be true if the fractions were $\frac{1}{2}, \frac{2}{3}, \frac{3}{4}$ (thus $q_i - p_i = 1$); otherwise, we are to use our (1′).

---

[58] Treatise attributed to Paolo dell'Abbaco, No. 47 (this illustration from MS. New York Columbia Plimpton 167, fol. 77ʳ).

[59] *Ibid.*, No. 71.

[60] Trenchant, *Aritmetique*, III, 'diverses questions', No. 6 (ed. 1561, p. 258; ed. 1643, p. 321).

[21] *A child entered a garden to steal apples. There were three guards and three doors. The first guard wanted to take away all his apples, but the child appeased him by giving him half of all his apples and, in addition, half the remainder. He did the same with the second and the third guard, and thus had only one apple upon leaving. Required the number of apples he had (initially). Answer: 64.*[61]

The computations are not performed. By (2′),

$$S = \frac{r_3}{\left(1 - \frac{p}{q}\right)^3} = \frac{1}{\left(1 - \frac{3}{4}\right)^3} = 64.$$

There exists also a more romantic form of these problems, with the thief becoming a suitor and the guards, girls being courted.

[22] *Someone must give apples to four girls; to the first, half of all and $\frac{1}{2}$ more, to the second, $\frac{2}{3}$ of the remainder and $\frac{2}{3}$ of an apple, to the third, $\frac{3}{4}$ of the remainder and $\frac{3}{4}$ of an apple; to the fourth, he must give $\frac{4}{5}$ of the remainder and in addition $\frac{4}{5}$ of an apple, and he must keep a single one for himself.*[62]

The text refers to a rule it formulates, corresponding to our above formula (1‴), and gives the answer: 239. Keeping the fractional terms, we shall have, as in the first expression of (1′),

$$S = \left(\left\{\left[\left(1 + \frac{4}{5}\right) \cdot 5 + \frac{3}{4}\right] \cdot 4 + \frac{2}{3}\right\} \cdot 3 + \frac{1}{2}\right) 2.$$

When the expression is written in this manner, it is evident that the result will be integral. It is less evident when the expression is described verbally, and some authors specify that the results, both intermediary and final, will be integral even though fractions of apples intervene:

[23] *A man shared apples between three girls in such a way that to the first he gave half of his apples and half of an apple, his apples remaining whole; to the second, he gave $\frac{2}{3}$ of his (remaining) apples and $\frac{2}{3}$ of an apple, his apples remaining whole; to the third he gave $\frac{3}{4}$ of his (remaining) apples and $\frac{3}{4}$ of an apple, his apples remaining whole; he will then be left with 2 whole apples. Required how many apples he had at the beginning.*[63]

The computation corresponds to that of our formula (1‴), thus

$$S = \left(\left[r_3 \cdot q_3 + p_3\right] q_2 + p_2\right) q_1 + p_1 = \left(\left[2 \cdot 4 + 3\right] 3 + 2\right) 2 + 1 = 71.$$

---

[61] MSS. Tours 399, fol. 129ʳ (text in *Récréations*, p. 155), and Paris BNF fr. 1339, fol. 76ᵛ.

[62] MS. Dijon 268 (cat. 447), fol. 120ʳ.

[63] MSS. BNF fr. 1339, fol. 75ᵛ – 76ʳ, or Dijon 268 (447), fol. 119ᵛ; see (text) *Récréations*, pp. 56–57.

§ 2. Giving to the church

Such is, from the time of (at least) late antiquity as seen at the beginning of this chapter, the usual object of this type of problem. But it may also take other forms, as we shall see.

[24] *A hermit entered a church where there were (statues of) three saints, namely Saint Peter, Saint Paul and Saint Lawrence. He came first to Saint Peter and prayed to him saying: 'Should it please you to double the coins I have in my purse, I shall give you 6 of them'. Which was done. Then he came to Saint Paul and said to him: 'May it please you to double the coins I have in my purse, and I shall give you 6 of them'. Which was done. Then he came to Saint Lawrence and said to him: 'May it please you to double the coins I have in my purse, and I shall give you 6 of them'. Which was done, and nothing remained to him. I ask how many coins he had in his purse.*

Answer. He had $5\frac{1}{4}$. In order to check it: double it, it makes $10\frac{1}{2}$, next you must give 6 to Saint Peter, there remain only $4\frac{1}{2}$; double them, it gives 9, then give 6 of them to Saint Paul, there remain 3; double them, it makes 6, which you give to Saint Lawrence, and nothing remains. This is why you may say that there were 5 coins and $\frac{1}{4}$.[64]

Keeping the same symbols as before, but with here q designating a multiple, we have $n = 3$, $r_3 = 0$, $q = 2$, $S = 5 + \frac{1}{4}$.[65]

Fig. 45

Let us suppose generally that the initial capital S is multiplied by factors q_i (> 1) and that a donation s_i is made each time. Then S becomes first $q_1 S - s_1 = r_1$, and the other capitals left successively will be $r_2 = q_2 r_1 - s_2 = q_2 (q_1 S - s_1) - s_2 = q_2 q_1 S - (q_2 s_1 + s_2)$, $r_3 = q_3 r_2 - s_3 = q_3 q_2 q_1 S - (q_3 q_2 s_1 + q_3 s_2 + s_3)$, generally

---

[64] *Livre de chiffres et de getz*, No. 12. Here the 'coins' are actually named: they are 'grans blans', minted in France during the second half of the 15th century.

[65] Illustration (Fig. 45): Christ with St Lawrence, St Peter and St Paul. Basilica di San Lorenzo fuori le mura, Rome (contains the tomb of St Lawrence).

$$r_n = S \prod_{i=1}^{n} q_i - \sum_{k=1}^{n-1} s_k \prod_{l=k+1}^{n} q_l - s_n$$

$$= \left\{ \cdots \left[\left\{ \left[(q_1 S - s_1) q_2 - s_2 \right] q_3 - s_3 \right\} q_4 - s_4 \right] \cdots \right\}. \qquad (1)$$

In the particular case with $q_i = q$, S will take the form

$$S = \frac{1}{q^n} \left[r_n + \sum_{i=1}^{n} q^{n-i} s_i \right]$$

which may also be written as

$$S = \frac{1}{q} \left(\frac{1}{q} \left[\frac{1}{q} \left\{ \frac{1}{q} \left[\cdots \frac{1}{q} (r_n + s_n) \cdots + s_4 \right] + s_3 \right\} + s_2 \right] + s_1 \right). \qquad (1')$$

If, in addition, $s_i = s$, we shall have

$$S = \frac{1}{q^n} \left[r_n + s \left(1 + q + q^2 + \cdots + q^{n-1} \right) \right] = \frac{1}{q^n} \left[r_n + s \frac{q^n - 1}{q - 1} \right]. \qquad (1'')$$

Finally, when the final remainder is zero, we shall have simply

$$S = s \left(\frac{1}{q} + \frac{1}{q^2} + \cdots + \frac{1}{q^{n-1}} + \frac{1}{q^n} \right). \qquad (1''')$$

An example of this last case is seen in the *Annales Stadenses*.[66] It is again Tirri (above, p. 1), on Christmas Eve, questioning his friend.

[25] *To honor this grand occasion, an offering must be made thrice.[67] Some wretch, wishing to make an offering this number of times, had not enough. So he asked God to give him as much as he already had in his purse. He received as much, and gave one denarius during mass, at first cock-crow. He asked once again that as much as remained be given to him. This was done, and he offered again one denarius. A third time he obtained as much as he had and, having given one denarius at the end of mass, he was left with nothing. Say, Firri, how much he brought to the church. Firri said: One obol, plus half an obol, plus half of half an obol.*

Since an obol is half a denarius, Firri's answer is

$$S = \frac{1}{2} + \frac{1}{2} \frac{1}{2} + \frac{1}{2} \frac{1}{2} \frac{1}{2} = \frac{7}{8},$$

which is our formula $(1''')$ with $n = 3$, $s = 1$, $q = 2$.

Since, as in the case of crossing a river (p. 2), Firri gives the answer immediately, it must have been a well-known formula in or before the 13th century. In any case, this rule is explained in the 14th and in the 15th

---

[66] Ed. Pertz, p. 334.
[67] Namely at matins, high mass and vespers.

century.[68] As the 15th-century sources explain, if there is just one altar and the believer leaves with nothing after his money has been doubled, he must have entered the church with $\frac{1}{2}s$, s being the amount of the donation; if there are two altars, he must have come with $(\frac{1}{2} + \frac{1}{2}\frac{1}{2})s$, and so on, generally

$$S = \tfrac{1}{2}s + \tfrac{1}{4}s + \tfrac{1}{8}s + \tfrac{1}{16}s + \tfrac{1}{32}s + \dots,$$

that is, as said in the text, the initial amount will be the sum of as many such terms, each half the preceding, as there are altars.

A variant of this problem involves the earnings and expenditures of a commercial traveller, as in the following one by the Pisan Fibonacci.

[26] *Someone went to Lucca to trade, then to Florence, and came back to Pisa. In each town he made the double, and in each he spent 12 denarii. In the end, he was left with nothing. Required how much he had at the beginning.*[69]

The answer is $S = 12 \cdot \frac{7}{8} = 10 + \frac{1}{2}$ (by 1′′′, or just above).

Such are the easiest cases. But the sums earned and spent each time may take different values:

[27] *Someone traded with a certain sum, doubled this sum, and spent from it one drachma; then he traded with the remainder, doubled it, and spent from it two drachmas; after that, he traded with the remainder, doubled it, and spent from it three drachmas. He was left with ten. What was the initial sum?.*[70]

Answer (by 1′): $2 + \frac{5}{8}$.

[28] *Someone went to four markets. At the first he doubled, plus 8, and spent 8; he doubled again, plus 8, and spent 9; he doubled again, plus 7, and spent 6; he doubled again, plus 6, and spent 5. Then he counted his denarii and found just 5. I ask how many denarii he took from home.*[71]

Using formula (1′), which is how the text computes, thus

$$S = \frac{1}{2}\left[\frac{1}{2}\left(\frac{1}{2}\left[\frac{1}{2}(5+5-6)+6-7\right]+9-8\right)+8-8\right],$$

we find $S = \frac{3}{8}$.

---

[68] *Subtilitates*, No. 5; MSS. Tours 399, fol. 128^r–128^v, and Paris BNF fr. 1339, fol. 75^r–75^v (text in *Récréations*, p. 59).

[69] *Liber abaci*, pp. 258 & 329.

[70] *Liber augmenti et diminutionis*, pp. 323–324. The Greek 'drachma' became the Latin 'dragma' and the Arabic 'dirham'. Like the *denarius* or the bezant, it often designates a coin of unspecified value.

[71] *Rascioni d'algorismo*, No. 4.

[29] *A merchant went to three fairs. At the first fair he had with him an unknown quantity of denarii; of these, he made 9 out of 8, and spent 6 denarii. At the second, he made 3 out of 2 and spent 12 denarii. At the third fair he made 2 out of 1 and spent 24 denarii. He found that the money taken to the first fair was doubled. Required how many denarii he had then.*[72]

According to the enunciation, which is also the general formula (1), we have

$$\left[\left(S\cdot\frac{9}{8}-6\right)\frac{3}{2}-12\right]2-24=2\,S,\quad\text{whence}\ \ S=48.$$

Fibonacci also has less simple examples than the above [26]. In one of them, $n=3$, $q_1=\frac{3}{2}$, $q_2=\frac{5}{4}$, $q_3=\frac{7}{6}$, $s=15$, $r_3=0$, whence $S=24+\frac{6}{7}$. In another one, $n=4$, $q_1=2\ (=\frac{2}{1})$, $q_2=\frac{3}{2}$, $q_3=\frac{4}{3}$, $q_4=\frac{5}{4}$, $s_1=13$, $s_2=16$, $s_3=18$, $s_4=20$, $r_4=0$, leading to $S=20+\frac{1}{3}$.[73] A third problem, the data of which relate it to the one before, is more elaborate; here there are two unknowns, S and s_1:

[30] *Someone made on a first voyage the double, on a second three out of two, on a third four out of three, on a fourth five out of four. What he spent on the first voyage I do not know, on the second 3 more than on the first, on the third 2 more than on the second, on the fourth 2 more than on the third; it is said that he was left with nothing at the end. Both his expense and his capital must be integral numbers.*[74]

Thus the expense s_1 ('*res*') and the capital S ('*summa*') are unknown, while $s_2=s_1+3$, $s_3=s_2+2=s_1+5$, $s_4=s_3+2=s_1+7$. This leads to the equation[75]

$$5\,S-\left(6+\frac{5}{12}\right)s_1-\left(18+\frac{1}{4}\right)=0.\quad(*)$$

Since the solutions must be integral, Fibonacci has first s_1 taking the form $9+12\cdot t$, with t natural; whereby not only will the s_i be integers but the factor 9 will make the fraction $\frac{3}{4}$ appear in the second term, so that the constant term will become an integer. We indeed obtain

$$5S=54+72t+\tfrac{45}{12}+5t+18+\tfrac{1}{4}=77t+76.$$

---

[72] Chuquet, *Appendice*, p. 423; MS., fol. 156ʳ.

[73] *Liber abaci*, pp. 261, 262–263.

[74] *Liber abaci*, pp. 264–265.

[75] It is quite exceptional to have equality to zero (*que equantur 0*, writes Fibonacci); we have such an occurrence in Abū Kāmil's *Algebra* (fol. 80ᵛ, p. 160 of the printed reproduction), but this time to express an impossible situation (see *Récréations*, p. 61n, and below, p. 250). Note the Latin designation of the two unknowns (below, p. 252).

In order for S also to be an integer, the right-hand side must be divisible by 5. The choice $t = 2$ already fulfills this, since then $S = 46$, with $s_1 = 33$.

We might also, observes Fibonacci, add to this value of s_1 multiples of 60, that is, take $s_1 = 33 + k \cdot 60$ with $S = 46 + k \cdot 77$; by varying k, we could thus find what is required in an infinite number of ways.[76]

As seen here, Fibonacci has calculated a particular integral solution (as a matter of fact, the lowest) of the indeterminate equation $(*)$ —which may also be written $60S - 77s_1 = 219$— and has then given its general solution in natural numbers.

In another example of his, we know the initial capital, the multiplying factor, the expenses and the remainder (here too, 0) but not the number of journeys:

[31] *Someone has 13 bezants. With them, he made an unknown number of journeys, in each of which he doubled (his amount of money) and spent 14 bezants. Required the number of journeys.*[77]

Our formula $(1'')$, which is $q^n S = r_n + s \frac{q^n - 1}{q - 1}$ becomes (with $q = 2$, $S = 13$, $r_n = 0$ and $s = 14$) $2^n \cdot 13 = 14 \left(2^n - 1\right)$, thus $2^n = 14$. Since 14 is not a (rational) power of 2, Fibonacci proceeds by trial and error, taking for n positive successive integers. After the first journey there remains $2 \cdot 13 - 14 = 12$; after the second, $2 \cdot 12 - 14 = 10$; after the third, $2 \cdot 10 - 14 = 6$. But the fourth journey will not be completed: unlike the previous remainders, this one will be negative $(2 \cdot 6 - 14 = -2)$. Thus Fibonacci seeks the fraction of the fourth journey which will lead to a remainder 0. In our terms, with x the fraction of the fourth journey, we must have $(2 - x)\, 6 = (1 - x)\, 14$, whence $8x = 2$. This $x = \frac{1}{4}$ means that the traveller will find himself short of money after $3 + \frac{3}{4}$ journeys (the true value would be closer to $3 + \frac{4}{5}$). However, observes Fibonacci, *to say that someone has made $\frac{3}{4}$ of a journey seems inappropriate,* which implies that some modification of the initial hypotheses is needed; for the fourth journey he thus suggests changing, in accordance with the above equality, the multiplying factor 2 into $\frac{7}{4}$ and the expense 14 into $\frac{3}{4}$ of 14. This reminds us of the subterfuges he adopted to avoid ending up with negative solutions in his linear systems.[78]

---

[76] *Totiens vis adde 60 super primum expendium, scilicet super 33, et totiens adde 77 super inventum capitale, scilicet super 46; et habebis quesitum infinitis modis* (p. 265).

[77] *Liber abaci,* p. 266.

[78] See our *Introduction to the history of algebra,* pp. 107–115 (pp. 111–120 in the original French edition). See also below, pp. 56–59.

Here he has kept the value zero of the final remainder. But we might also modify its value, and Fibonacci mentions a few cases. We have completed his reasoning by setting out a table of the different possibilities, with the final remainder r varying in function of the number of journeys n (Fig. 46).

r	$13 < r < 12$	12	$12 < r < 10$	10	$10 < r < 6$	6	$6 < r < 0$	0
n	$0 < n < 1$	1	$1 < n < 2$	2	$2 < n < 3$	3	$3 < n < 3 + \frac{3}{4}$	$3 + \frac{3}{4}$

Fig. 46

Let us conclude these problems of commercial travellers with an example by Pacioli, treated in the same way.

[**32**] *Someone made a certain number of journeys, and carries (at the outset) as much money as he is to make journeys. He doubles his money on each journey. At the end, he happens to have 30 ducats altogether. Required how many journeys he made and how many ducats he took with him.*[79]

If n is the number of journeys, then $n \cdot 2^n = 30$. Taking $n = 2$, Pacioli finds for the left-hand side 8, then $n = 3$ gives him 24, and $n = 4$, 64. Therefore, the traveller makes three journeys and part of a fourth. Putting this part x (*poni che facesse 3 viaggi più 1 co(sa)*, see p. 252), his capital, initially $3 + x$, becomes at the end of the first journey $6 + 2x$, after the second $12 + 4x$, after the third $24 + 8x$. Since he will double this capital on the fourth journey, he would earn during this single journey $24 + 8x$ (*24 p̄ 8 co*). Let us apply, as does Pacioli, the rule of three: if the whole journey brings him $24 + x$, what will the fraction x of it bring him? This will be $24x + 8x^2$. The equation is then $24 + 8x + 24x + 8x^2 = 30$, thus $4x^2 + 16x = 3$, the positive solution of which is $x = \sqrt{4 + \frac{3}{4}} - 2$, which is indeed Pacioli's result (*harai la cosa valere* $\mathcal{R}$ *4 3/4 m̄ 2*, see Fig. 47). The initial sum was thus $\sqrt{4 + \frac{3}{4}} + 1$.[80]

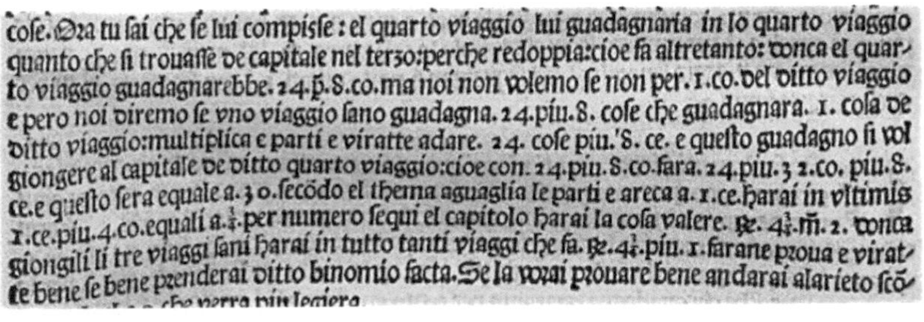
coſe. Ora tu ſai che ſe lui compiſſe: el quarto viaggio lui guadagnaria in lo quarto viaggio
quanto che ſi trouaſſe de capitale nel terzo: perche redoppia: cioe fa altretanto: donca el quarto viaggio guadagnarebbe. 24. p̄. 8. co. ma noi non volemo ſe non per. 1. co. del ditto viaggio
e pero noi diremo ſe vno viaggio ſano guadagna. 24. piu. 8. coſe che guadagnara. 1. coſa de
ditto viaggio: multiplica e parti e viratte adare. 24. coſe piu. 8. ce. e queſto guadagno ſi vol
giongere al capitale de ditto quarto viaggio: cioe con. 24. piu. 8. co. fara. 24. piu. 32. co. piu. 8.
ce. e queſto ſera equale a. 30. ſecōdo el thema aguaglia le parti e areca a. 1. ce. harai in vltimis
1. ce. piu. 4. co. equali a. 3/4. per numero ſequi el capitolo harai la coſa valere. R. 4 3/4. m̄. 2. donca ←
giongili li tre viaggi ſani harai in tutto tanti viaggi che fa. R. 4 3/4. piu. 1. farane proua e viratte bene ſe bene prenderai ditto binomio facta. Se la vorai prouare bene andarai alarieto ſcō-

Fig. 47

[79] *Summa*, fol. 187r, No. 8.

[80] On the symbols $\mathcal{R}^2$ and $\bar{m}$, see p. 251.

§3. The last will

The following example is from a sequence of problems found in the mathematical notes of a Florentine merchant living in southern France (Avignon).[81]

[33] *A count is close to death. He does not know how many children he has (!) or the value of his estate. He says: 'I bequeath to my oldest son one ounce of gold and a seventh of the remaining gold; to the second, 2 ounces of gold and $\frac{1}{7}$ of the remaining gold; to the third, 3 ounces of gold and $\frac{1}{7}$ of the remaining gold'. He goes on likewise to the last. I wish to know how many children he has, what his estate is, and the part of each one.*

Do it in this way. Since he says 'a seventh', take away 1 (from 7), there remains 6, and such is the number of children he has. In order to know how many ounces of gold he has, multiply 6 by 6 ounces, this makes 36; he has 36 ounces of gold, and 6 children, the part of each one is 6 ounces.[82]

¶ Ein Teſtament

Item es lygt ein vatter am todtbet vnd ſtirbt auch vñ er leſt kinder vñ ſagt nicht wieuil/ vñ leſt gelt vñ ſagt auch nit wie vil/ vnnd beſtelt ſeine letſte wille alſo/daſ man eine Kind ſo vil

Eyn geſchefft

¶ Itm Eſ ligt eyn Uater am todtpet vñ ſtirbt auch vñ er leſt kinder vñ ſagt nicht wy vil. vnnd leſt gelt vnnd ſagt auch nicht wie v lvñ beſtelt ſeynē lecztñ willen alſo das man eynem kind ſzo vol ſol gebñ als dē andern Uñ dē erſtñ gibt man 1 fl vñ $\frac{1}{10}$ des vberigñ gelcz. Uñ dē andern 2 fl vnd auch $\frac{1}{10}$ des vberi~

Fig. 48   Fig. 49

The same problem as the above with the same data, is solved by Fibonacci in the same way —except that the dying person is an ordinary man (*quidam ad finem veniens*).[83] The Byzantine monk Planudes, in the second half of the 13th century, has the same problem, but the father dies without finishing his will —a sad situation, certainly, but mathematically irrelevant (see below):

---

[81] No. 8 in F. Bartoli's († 1425) collection. See below, p. 263.

[82] The left-hand illustration (*Item es lygt ein vatter am todtbet und stirbt auch*) is taken from J. Widman's book, ed. 1508, fol. 97ᵛ. In the original, 1489 edition, the illustration is much less elaborate and bears the title *eyn geschefft* (good business for the children?). The first part is '1 fl(orin) and $\frac{1}{10}$ of the remaining money'.

[83] *Liber abaci*, p. 279.

[34] *Someone dying called his sons, asked for his coffer of gold to be brought, and ordered as follows. I want my gold to be distributed among my sons equally (ἐξ ἴσου): the first is to take one gold coin and a seventh of the remainder; the second, two gold coins and a seventh of the remainder; the third, three (gold coins) and a seventh of the remainder. With these words he passed away, without finishing with all the sons and the gold in its entirety. I wish to know how many sons and how much gold he had.*[84]

The solution is found by the same rule as before, without any justification.[85] The main difference between this and the previous problem is that here equality of the parts is explicitly mentioned. The individual parts are then computed as a verification of the answer.

In Byzantium again, taking a different form:

[35] *Apples are offered for breakfast. To the first are given 1 apple and $\frac{1}{7}$ of the remaining apples, to the second two and $\frac{1}{7}$ of the remainder, to the third 3 and $\frac{1}{7}$ of the remaining apples, to the fourth 4 and $\frac{1}{7}$ of the remainder, and to the others likewise according to this stipulation. It must be said how many had breakfast and how many apples were given.*[86]

Here the same rule is followed as in the other problems. Again (as in all cases but one) it is not specified that the parts shall be equal.

Let us suppose generally that the first is to receive s_1 and the kth part of the remainder, the second s_2 and the kth part of the remainder, and so on to the last, whereupon no capital is left. The first is thus to receive

$$s_1 + \frac{1}{k}\left(S - s_1\right)$$

and the second

$$s_2 + \frac{1}{k}\left[S - \left(s_1 + \frac{1}{k}\left(S - s_1\right)\right) - s_2\right]$$

$$= s_2 + \frac{1}{k}S - \frac{1}{k}s_1 - \frac{1}{k^2}S + \frac{1}{k^2}s_1 - \frac{1}{k}s_2.$$

We do not need to express the other parts: since they must be identical, we shall just equate these two expressions (the dying father thus needs only to mention the first two parts). After removing the common terms, we are left with

---

[84] Ed. Gerhardt, p. 46.

[85] Λαμβάνω τὸν μονάδι ἐλάττονα τοῦ ὁμωνύμου τῷ ἑβδόμῳ ἀριθμῷ τοῦ ζ τὸν ⸱ καὶ πολλαπλασιάζω τοῦτον ἐφ᾽ ἑαυτὸν καὶ γίνεται λϚ· τοῦτόν φημι εἶναι τὸν ἀριθμὸν τοῦ χρυσίου, τὴν δὲ πλευρὰν ('the side' = the root) αὐτοῦ ἤτοι τὸν ⸱ τὸν ἀριθμὸν τῶν παίδων.

[86] Vogel, *Byz. Rechenbuch*, No. 84.

$$\frac{1}{k^2} S = s_2 + \frac{1}{k^2} s_1 - \frac{1}{k} s_2 - s_1,$$

whence $S = k^2 s_2 + s_1 - k s_2 - k^2 s_1$ and therefore

$$S = k \left[k(s_2 - s_1) - s_2 \right] + s_1. \qquad (1)$$

In the particular case with $s_2 = s_1 + d$ (arithmetical progression with common difference d), this reduces to

$$S = k \left[kd - s_1 - d \right] + s_1 = k \left[d(k-1) - s_1 \right] + s_1. \qquad (2)$$

The part of the first being $s_1 + \frac{1}{k}(S - s_1) = d(k-1)$, dividing S by that gives $k - \frac{s_1}{d}$ as the number of children.

If, in addition, the first term of the progression equals the common difference, thus $s_1 = d$ —as in the previous examples— the expression for S reduces once again, and becomes

$$S = kd(k-2) + d = d(k-1)^2, \qquad (3)$$

with this time $k - 1$ children.

In the above examples we had $k = 7$, $s_1 = d = 1$, thus $S = 36$ and six children. Chuquet has a similar example, of donations, with $k = 10$ and $s_1 = d = 1$ (so also Widman, see p. 45n); then the number of children is $k - 1 = 9$, which is also the amount of each part. Exactly the same problem is found in a contemporary Arabic manuscript, with bequests.[87] Another variant in the situation is found in contemporary German and French texts, with a father sending his sons one after the other to a moneychanger.[88] An example of this case is the following.

Fig. 50

[36] *A lawyer had given money to a changer but had forgotten how much. In order to know it, and taking back all his money, he found the following trick. He told one of his sons, for he had many: 'Go to the changer so-and-so and bring me one*

---

[87] MS. Paris BNF ar. 4441, fol. 44ʳ. See (for both) *Récréations*, p. 67.

[88] *Algorismus Ratisbonensis*, No. 114; MS. Paris BNF fr. 1339, fol. 75ᵛ (text of the latter in *Récréations*, p. 68). The illustration (changer in Jewish, not moneychanger's, dress) is from J. Widman's book (ed. 1508, fol. 105ᵛ).

florin and the tenth part of the money I have given him'. This was done. Another time he told another son: 'Go to the changer and bring me two florins and the tenth part of the remainder'. He told likewise the others, to the last of whom he said: 'Go to the changer and bring me all the money remaining'. This was done. And they brought so much the one as the other. I ask you how much money there was, how many sons, and how much everyone brought.

Answer. For (answering) these three questions, take the number they all bring —that is, the tenth part (thus)— 10. From 10 subtract 1, there remains 9. Therefore you may say that he had 9 sons, and that each brought 9 florins. In order to know how much he had given, multiply 9 by itself, which is 81. Therefore he had given 81 to the changer. In order to verify it for the first son, take 81, subtract 1, and add (to 1) the tenth part of the remainder.[89]

Chuquet has a numerical variant, with the children taking, respectively, 2 denarii, 5, 8, and so on, and, in addition, $\frac{1}{8}$ of the remainder.[90] To find the solution he applies formulae he has just described (called by him 'general rule for all such problems', which are our (2) above), namely for the part of each, the capital, the number of children, respectively,

$$d\,(k-1), \quad k\,[\,d\,(k-1)-s_1\,]+s_1, \quad k-\frac{s_1}{d},$$

and thus finds, since $k=8$, $s_1=2$, $d=3$, the capital $S=8[3\cdot7-2]+2=154$ and the parts $d(k-1)=21$ for each of the $7+\frac{1}{3}$ children. The presence of a fractional child is not commented, neither here nor in his next two examples, with, respectively, $\frac{1}{k}=\frac{2}{11}$, $s_1=2$, $d=3$ and $\frac{1}{k}=\frac{3}{13}$, $s_1=3$, $d=2$, giving $4+\frac{5}{6}$ and $2+\frac{5}{6}$ as numbers of children.

In the previous examples, each received first a sum and then a fraction of the remainder. But one may also suppose that the fraction of the capital is received before the supplementary sum. The first two parts will then be

$$\frac{1}{k}\,S+s_1, \quad \frac{1}{k}\left[S-\frac{1}{k}S-s_1\right]+s_2,$$

the equating of which leads to, successively,

$$\frac{1}{k^2}\,S=s_2-s_1-\frac{1}{k}s_1, \quad S=k^2(s_2-s_1)-ks_1, \quad \text{thus}$$

$$S=k\,[\,k\,(s_2-s_1)-s_1\,]. \quad (1')$$

With an arithmetical progression of constant difference $d=s_2-s_1$, this reduces to

---

[89] *Livre de chiffres et de getz*, No. 21.
[90] *Appendice*, p. 449 or MS., fol. 197^r–197^v (text: *Récréations*, p. 69).

$$S = k\,(kd - s_1) \qquad (2')$$

so that each is to receive, like the first, kd, and there will be $k - \frac{s_1}{d}$ children. This will be reduced once more if $d = s_1$; for then

$$S = kd\,(k - 1), \qquad (3')$$

with the parts kd for the $k - 1$ children. Here are examples of this simplified case, first by Chuquet.

Fig. 51

[37] *A man has children in unknown number, and he has in his coffer denarii the number of which is not said. This man says to the first of his children: Go to my coffer and take the $\frac{1}{7}$ part of the denarii there, plus 2. To the second, he says: Go to the coffer and, after your brother has taken his part, take the $\frac{1}{7}$ part of the denarii there, plus 4. The third he tells to take the $\frac{1}{7}$ part and 6 more. And so on to the others up to the last, whom he tells to take what his brothers have left. Thus doing, they have as many denarii the one as the other. Required the number of children, how many denarii each will have, and also how many denarii were in the coffer.*[91]

Since $s_1 = 2$, $d = 2$, $k = 7$, and thus $s_1 = d$, the rule given by Chuquet —which corresponds to our $(3')$— leads him to the result $S = 84$, with six children receiving 14 each. In his next example, with $s_1 = 3 = d$ and the fraction $\frac{2}{11}$ (thus $k = \frac{11}{2} = 5 + \frac{1}{2}$), he finds $S = 74 + \frac{1}{4}$, the $4 + \frac{1}{2}$ children receiving $16 + \frac{1}{2}$ each (if complete). In his next problems, Chuquet considers the case with $d \neq s_1$. His formulae correspond then to $(2')$ above.

He is not the only one to have the fraction of the capital taken first. As a matter of fact, examples of that are found both earlier and later, as in the following problem by Cardan.[92]

[38] *A dying man left sons, at that time abroad, and gold coins of which he did not know the number; he stipulated that the first on his return should receive*

---

[91] *Appendice*, p. 450; MS. fol. 198ʳ. Illustration from Muscarello's *Algorismus*, MS. Pennsylvania LJS 27, fol. 85ᵛ (but usual case of taking first cash then a fraction of the remainder, with $k = 9$, $s_1 = 1 = d$; ed., II, pp. 204–205).

[92] *Practica arithmetice*, Ch. LXVI, No. 65.

$\frac{1}{7}$ *of the whole, plus 100, the second $\frac{1}{7}$ of the remainder, plus 200, the third $\frac{1}{7}$ of the remainder, plus 300, and so forth for the others. After sharing the coins, they found them to be in equal number. Required the number of coins and of sons.*

Since here $k = 7$ and $s_1 = 100 = d$, calculating according to (3'), as does Cardan, gives him the number of coins ($7 \cdot 100 \cdot 6 = 4200$) and that of sons ($7 - 1 = 6$).

Before, in the 15th century, Benedetto da Firenze has these two cases successively, thus with the fraction of the capital taken either after or before taking the supplementary sum; with $k = 10$, $s_1 = d = 1000$, he finds $S = 81000$, then $S = 90000$, to be divided up among nine children.[93]

---

[93] See Arrighi, *Scritti scelti*, pp. 346–347 (from MS. Florence BNC Magl. XI 76). In fact, such a pair of cases is already found in Fibonacci's *Liber abaci*, with $k = 7$ and $d = 1$ (above, p. 45n), and the results $S = 36$ then $S = 42$.

Chapter V. Mutual borrowing

§ 1. Moschos' problem

Being both merchant and mathematician, Fibonacci used his travel opportunities around the Mediterranean to collect information and meet local mathematicians. Among the numerous problems gathered by him there is the following one, which, as he reports, was communicated to him 'in Constantinople by the most learned Constantinopolitan master Muscus' —obviously the Greek Μόσχος.[94] It is about five men wishing individually to buy a ship, each not having enough money for it but able to pay the price by borrowing from one another a certain fraction of their money. In our terms, with x_i the individual amounts, y the cost of the ship and the given data, it takes the form

$$
\begin{cases}
x_1 + \left(\frac{2}{3} + \frac{1}{5}\right)\left(x_2 + x_3 + x_4 + x_5\right) = y \\
x_2 + \left(\frac{2}{3} + \frac{1}{6} + \frac{1}{480}\right)\left(x_3 + x_4 + x_5 + x_1\right) = y \\
x_3 + \left(\frac{2}{3} + \frac{1}{6} + \frac{1}{638}\right)\left(x_4 + x_5 + x_1 + x_2\right) = y \\
x_4 + \left(\frac{2}{3} + \frac{1}{7} + \frac{1}{420}\right)\left(x_5 + x_1 + x_2 + x_3\right) = y \\
x_5 + \left(\frac{2}{3} + \frac{1}{10} + \frac{1}{27} + \frac{1}{810}\right)\left(x_1 + x_2 + x_3 + x_4\right) = y,
\end{cases}
$$

thus also

$$
\begin{cases}
x_1 + \frac{13}{15}\left(x_2 + x_3 + x_4 + x_5\right) = y \\
x_2 + \frac{401}{480}\left(x_3 + x_4 + x_5 + x_1\right) = y \\
x_3 + \frac{799}{957}\left(x_4 + x_5 + x_1 + x_2\right) = y \\
x_4 + \frac{341}{420}\left(x_5 + x_1 + x_2 + x_3\right) = y \\
x_5 + \frac{326}{405}\left(x_1 + x_2 + x_3 + x_4\right) = y.
\end{cases}
$$

Putting $S = x_1 + x_2 + x_3 + x_4 + x_5$ and adding to both sides the complement of each fraction, for instance, for the first equation, $\frac{2}{15}\left(x_2 + x_3 + x_4 + x_5\right)$, we shall obtain successively, as does Fibonacci,

---

[94] *Questio nobis proposita a peritissimo magistro musco constantinopolitano in constantinopoli* (*Liber abaci*, p. 249). We do not know anything about this mathematician. In the recently published (2002) English translation of the *Liber abaci*, we are told about 'a most learned master of a Constantinople mosque'; there was of course no mosque there before the Turkish occupation (1453).

$$\begin{cases} S = y + \frac{2}{15}\left(x_2 + x_3 + x_4 + x_5\right) \\ S = y + \frac{79}{480}\left(x_3 + x_4 + x_5 + x_1\right) \\ S = y + \frac{158}{957}\left(x_4 + x_5 + x_1 + x_2\right) \\ S = y + \frac{79}{420}\left(x_5 + x_1 + x_2 + x_3\right) \\ S = y + \frac{79}{405}\left(x_1 + x_2 + x_3 + x_4\right), \end{cases} \qquad \begin{cases} \frac{2}{15}\left(x_2 + x_3 + x_4 + x_5\right) = S - y \\ \frac{79}{480}\left(x_3 + x_4 + x_5 + x_1\right) = S - y \\ \frac{158}{957}\left(x_4 + x_5 + x_1 + x_2\right) = S - y \\ \frac{79}{420}\left(x_5 + x_1 + x_2 + x_3\right) = S - y \\ \frac{79}{405}\left(x_1 + x_2 + x_3 + x_4\right) = S - y, \end{cases}$$

and therefore

$$\begin{cases} x_2 + x_3 + x_4 + x_5 = \frac{15}{2}\left(S - y\right) \\ x_3 + x_4 + x_5 + x_1 = \frac{480}{79}\left(S - y\right) \\ x_4 + x_5 + x_1 + x_2 = \frac{957}{158}\left(S - y\right) \\ x_5 + x_1 + x_2 + x_3 = \frac{420}{79}\left(S - y\right) \\ x_1 + x_2 + x_3 + x_4 = \frac{405}{79}\left(S - y\right). \end{cases}$$

This problem being indeterminate, we may choose a value for $S - y$; most appropriately, Fibonacci (perhaps following Moschos) takes the smallest common multiple of all the denominators, namely 158. Then

$$\begin{cases} x_2 + x_3 + x_4 + x_5 \equiv S - x_1 = \frac{15}{2} \cdot 158 = 1185, \quad \text{and, likewise,} \\ x_3 + x_4 + x_5 + x_1 \equiv S - x_2 = 960 \\ x_4 + x_5 + x_1 + x_2 \equiv S - x_3 = 957 \\ x_5 + x_1 + x_2 + x_3 \equiv S - x_4 = 840 \\ x_1 + x_2 + x_3 + x_4 \equiv S - x_5 = 810. \end{cases}$$

Since in each of the five left-hand sides just one of the x_i is missing, adding them will give $4S = 4752$, whence $S = 1188$; then $y = S - (S - y) = 1030$, $x_1 = S - 1185 = 3$, $x_2 = 228$, $x_3 = 231$, $x_4 = 348$, $x_5 = 378$.

The way of solving used by Moschos —as well as, by the way, the form of this linear system— does not differ from what is found in antiquity, and solving such systems by adding up their equations is even attributed to a member of the Pythagorean School, Thymaridas (4th century BC).[95] The (throughout verbal) reasoning found in Fibonacci's text corresponds exactly to that. The similar systems we shall now see are solved in the same way. Note too that, although these systems may be considered to be

---

[95] See e.g. Heath, *History of Greek Math.*, I, pp. 94–96 or Bulmer-Thomas, *Greek math. works*, I, pp. 138–141 (from Iamblichos, *In Nicom. Arithm. introd.*, ed. Pistelli, pp. 62, 18 – 63, 2).

applicable in the context of real life, they are nevertheless of recreational nature, with the data and/or the situation being rather fanciful. This will have a mathematical consequence: in some of them one of the unknowns takes a negative value and Fibonacci, who, *more temporum*, does not recognize negative numbers, will feel free to provide an interpretation of his own in order to give it some meaning. And, as we shall see, this was to be the first step towards the recognition of negative numbers.

There are three main subjects involving such linear systems: a group of people finds a purse; or they wish to buy a horse (similar to Moschos' example); or they put in common individual sums of which each then steals a part.[96] We shall first examine the solving of these systems and then examples of them.

§ 2. Finding a purse

A group of n persons, each of which possesses a certain (unknown) sum, finds a purse with a certain (unknown) amount; this amount, being added to the sum each one possesses, is said to equal a (given) multiple of the sum held by the $n - 1$ others.

Such problems occur first in ancient Greece, namely, as said, in Iamblichos' Commentary to Nicomachos' *Arithmetic*, and then in the first book of Diophant's *Arithmetica* (I, 18–19; in numbers). Later, we meet them in Byzantium, in India and in Moslem countries.[97]

Fig. 52

The proposed system is thus of the form

$$x_j + y = m_j \sum_{i \neq j} x_i \qquad (j = 1, \ldots, n)$$

---

[96] These systems in their most general forms are solved in our *Appearance of negative solutions*, and examples are given; see also Vogel's *Zur Geschichte der linearen Gleichungen*.

[97] Tropfke, *Geschichte der Elementarmathematik*, p. 607. Illustration from Muscarellos' *Algorismus*, MS. Pennsylvania LJS 27, fol. 65ᵛ, three finding a purse (ed., II (text), pp. 172–173).

with y the contents of the purse, m_j the given multiples, and x_i (taken in cyclical order) the individual sums.

Adding on both sides $\sum_{i \neq j} x_i$ and putting $S = \sum_1^n x_i$, we obtain

$$S + y = (m_j + 1) \sum_{i \neq j} x_i,$$

whence

$$\sum_{i \neq j} x_i = (S + y)\, \frac{1}{m_j + 1}. \qquad (*)$$

Summing now these equations on the index j we obtain, since the indices are covered cyclically and each left-hand sum contains $n - 1$ of the x_i's, and thus omits each time one,

$$(n - 1)\, S = (S + y) \sum_1^n \frac{1}{m_i + 1}$$

and therefore

$$S = \frac{S + y}{n - 1} \sum_1^n \frac{1}{m_i + 1}. \qquad (**)$$

Now since

$$x_j \equiv S - \sum_{i \neq j} x_i,$$

we find, by $(*)$ and $(**)$

$$x_j = (S + y) \left[\frac{1}{n - 1} \sum_1^n \frac{1}{m_i + 1} - \frac{1}{m_j + 1} \right]$$

so that the condition for a positive solution x_j will be

$$\frac{1}{n - 1} \sum_1^n \frac{1}{m_i + 1} > \frac{1}{m_j + 1}.$$

This condition is fulfilled in most of the examples solved by Fibonacci.

§3. Buying a horse

Each person in a group of n people possesses some (unknown) sum, less than the (unknown) price of a horse he wishes to acquire; he asks the $n - 1$ others for a (known) fraction of what they possess to reach the price of the horse. Required their individual sums and that price.

Here too, such systems are found earlier, in Greece (Diophant's *Arithmetica*, I, 24–25; in numbers) and (with concrete quantities) in Byzantium, in Moslem countries and in China.[98]

---

[98] See Tropfke, pp. 608–609; early examples in Islam are Abū Kāmil, *Algebra*, fol.

Fig. 53

Let us thus consider the system, with y and the x_i unknown and the m_j given (proper) fractions,

$$x_j + m_j \sum_{i \neq j} x_i = y \quad (j = 1, \dots, n).$$

By simple comparison with the previous case, where the terms differ only by their place, we may deduct the solving formula:

$$x_j = (S - y) \left[\frac{1}{n-1} \sum_{1}^{n} \frac{1}{1 - m_i} - \frac{1}{1 - m_j} \right].$$

Since the m_i are proper fractions and $S > y$ (otherwise buying is not possible), the condition for a positive solution x_j will be

$$\frac{1}{n-1} \sum_{1}^{n} \frac{1}{1 - m_i} > \frac{1}{1 - m_j}.$$

§4. The dishonest partners

An unknown sum belongs in common to a group with each member having a given fraction of it. For reasons of security, they decided to put it in a chest, of which each had a key. This was unwise of them, for each in turn went once to steal some money, with the result that they emptied the chest completely. With none of them admitting to being guilty, they finally agreed that each would return a given fraction of what he had taken and that the resulting sum would then be shared equally between them. It turns out that each obtains, between what he has kept of his theft and received from the sharing, exactly what he had initially. Required the parts stolen (from which the parts initially possessed are inferred).

This problem is certainly of ancient origin. Around 880, the Egyptian mathematician Abū Kāmil solves it in two ways, one of which, he says, is used by '(the other) mathematicians'.[99]

95[r] (p. 189 of the reproduction), then al-Karajī's Fakhrī (Woepcke, III.26–27) and Kāfī (Hochheim, III, pp. 16–17). The illustration is taken from P. M. Calandri's *Tractato d'abbacho*, MS. Florence BML Acq. e doni 154, fol. 188[r] (ed., p. 169).

[99] *Algebra*, fol. 100[r] – 101[r] (reproduction of the manuscript: pp. 199–201).

Let x_j be the sum stolen by the jth partner, p_j his (given) part of the whole capital S, m_j the (given) fraction he keeps of his theft, and thus $(1 - m_j)x_j$ the part he returns. We are therefore to solve the system

$$m_j x_j + \frac{1}{n} \sum_1^n (1 - m_i)\, x_i = p_j S \quad (j = 1, \dots, n),$$

with $S = \sum_1^n x_i$. After a few transformations,[100] we arrive at

$$x_j = \frac{S}{m_j \sum \frac{1}{m_l}} \left[\sum_{i \neq j} \frac{p_j - p_i}{m_i} + 1 \right].$$

Such is the part stolen by the jth partner; note that, in order for it to be positive, we must have

$$\sum_{i \neq j} \frac{p_j - p_i}{m_i} > -1.$$

§5. Examples

Fig. 54

[39] Among his problems of finding a purse, Fibonacci has the following.[101]

$$\begin{cases} x_1 + y = \left(2 + \frac{1}{2}\right)\left(x_2 + x_3 + x_4 + x_5\right) \\ x_2 + y = \left(3 + \frac{1}{3}\right)\left(x_3 + x_4 + x_5 + x_1\right) \\ x_3 + y = \left(4 + \frac{1}{4}\right)\left(x_4 + x_5 + x_1 + x_2\right) \\ x_4 + y = \left(5 + \frac{1}{5}\right)\left(x_5 + x_1 + x_2 + x_3\right) \\ x_5 + y = \left(6 + \frac{1}{6}\right)\left(x_1 + x_2 + x_3 + x_4\right). \end{cases}$$

He first transforms all his equations in the same manner. Consider, for instance, the first one. In $x_1 + y = \frac{5}{2}\left(x_2 + x_3 + x_4 + x_5\right)$ we complete the left-hand side to $S + y$; we shall have $S + y = \frac{7}{2}\left(x_2 + x_3 + x_4 + x_5\right)$, whence $x_2 + x_3 + x_4 + x_5 = \frac{2}{7}\left(S + y\right)$ which, inserted into the right-hand

---

[100] Details in our *Appearance of negative solutions*, pp. 115–116.

[101] *Liber abaci*, pp. 215–216. Illustration from MS. Florence BNC Magl. XI 86 (Paolo dell'Abbaco's treatise, ed. p. 99: *tre uomini truovano una borsa di danari*).

side of the original equation, gives $x_1 + y = \frac{5}{7}(S + y)$. The same being done for all equations, the system takes the form

$$
\begin{cases}
x_1 + y = \frac{5}{7}(S + y) & = 259\,935 \cdot 4 \\
x_2 + y = \frac{10}{13}(S + y) & = 279\,930 \cdot 4 \\
x_3 + y = \frac{17}{21}(S + y) & = 294\,593 \cdot 4 \\
x_4 + y = \frac{26}{31}(S + y) & = 305\,214 \cdot 4 \\
x_5 + y = \frac{37}{43}(S + y) & = 313\,131 \cdot 4.
\end{cases}
$$

The numerical quantities on the right-hand side are obtained as follows. The problem being indeterminate, Fibonacci has first put for $S + y$ the lowest common denominator of the fractions, thus 363 909, and has then calculated the individual right-hand terms. When it turns out that y (and therefore S) will not be integral, he introduces the above factor 4. Adding next the resulting equations, he obtains $(S+y)+4y = 1\,452\,803 \cdot 4$; thus, $4y = (1\,452\,803 - 363\,909) \cdot 4$, and therefore $y = 1\,088\,894$ and $S = 366\,742$.

Now the first equation above gives $x_1 + y = 1\,039\,740$ whereas $y = 1\,088\,894$. Since, writes Fibonacci, there is more in the purse than between the purse and the first man, *either the problem as stated will be unsolvable, or the first man will have a debt, namely that which is missing from the sum of his (amount of) denarii and the purse to (make up) the amount of denarii in the purse, that is, what there is from 1 039 740 to 1 088 894, which is 49 154.*[102] Computing then the remaining unknowns, Fibonacci finds $x_2 = 30\,826$, $x_3 = 89\,478$, $x_4 = 131\,962$, $x_5 = 163\,630$. The *negative* asset of the first man has thus become a *positive* debt, to be subtracted in the equations. That means modifying the sign of x_1 in the given system, thus also changing the conditions of the original problem. But this was the only way to get around the issue of a negative solution without rejecting the problem as impossible. That no doubt paved the way to the acceptance of negative results, as we shall see below.[103]

[40] Among the few problems of the dishonest partners type, one leads to a negative result.[104] Here too, Fibonacci will attempt to find an

---

[102] *Sed quia superius magis in bursa repertum est quam id quod inter bursam et primum hominem habent, aut positio huius questionis indissolubilis erit, aut primus homo debitum habebit, illud videlicet quod deest a summa denariorum ipsius et burse usque ad summam denariorum burse, scilicet id quod est a 1039740 usque in 1088894, quod est 49154.*

[103] All problems of Fibonacci leading to a negative result, as well as later mediaeval ones, are listed in our *Appearance of negative solutions*.

[104] *Liber abaci*, pp. 296–297.

interpretation making some sense. As before, that means modifying the conditions of the problem so as to make the negative result positive. It should be noted that Fibonacci has again considered a way to avoid rejecting *a priori* the negative solution and thus the problem itself.

Three men are said to have put in a chest a sum, of which half belongs to the first, two-fifths to the second and a tenth to the last. After grabbing each a handful of this money, they must put back in the empty chest a given fraction of what they took, namely $\frac{1}{2}$ for the first, $\frac{1}{3}$ for the second, $\frac{1}{6}$ for the third. We already know that, by receiving a third of the sum returned, each will recover his initial part. The system to solve is thus the following:

$$\begin{cases} \frac{1}{2}x_1 + \frac{1}{3}\left(\frac{1}{2}x_1 + \frac{1}{3}x_2 + \frac{1}{6}x_3\right) = \frac{1}{2}S \\ \frac{2}{3}x_2 + \frac{1}{3}\left(\frac{1}{2}x_1 + \frac{1}{3}x_2 + \frac{1}{6}x_3\right) = \frac{2}{5}S \\ \frac{5}{6}x_3 + \frac{1}{3}\left(\frac{1}{2}x_1 + \frac{1}{3}x_2 + \frac{1}{6}x_3\right) = \frac{1}{10}S. \end{cases}$$

Let us put with Fibonacci $S = 470$ for the whole capital (for convenience, to avoid fractional results). Calculating the unknown quantities gives the following results:

	initial capitals	amounts stolen	amounts returned	final amounts
1st	$\frac{1}{2}S = 235$	$x_1 = 326$	$\frac{1}{2}x_1 = 163$	$\frac{1}{2}x_1 + 72 = 235$
2nd	$\frac{2}{5}S = 188$	$x_2 = 174$	$\frac{1}{3}x_2 = 58$	$\frac{2}{3}x_2 + 72 = 188$
3rd	$\frac{1}{10}S = 47$	$x_3 = -30$	$\frac{1}{6}x_3 = -5$	$5 + 72 = 77 = 47 + 30$
	$S = 470$		total : 216	total : $470 + 30$

Fig. 55

Fibonacci first computes x_1 and x_2, and notes that their sum is larger than 470. *Thus,* he says, *this problem cannot be solved unless it is solved with some other money belonging to the third man.* The interpretation he gives is the following. The common capital is 470, but the third man had in addition a sum of 30 of his own, which he entrusted to the chest. Now these 500 are stolen by the two others, the first taking $x_1 = 326$ and the second, $x_2 = 174$; therefore not only was all the capital removed, leaving nothing to be taken by the third, but in addition that made him lose his own money ($x_3 = -30$). Their gain is thus opposed to his loss.

Next, the first man returns half of what he has taken, thus 163, the second, a third of his theft, thus 58; the third, who according to the computations should return a sixth of -30, in fact takes the opportunity to recover 5 from the money just given back by the two others. Sharing among the three the money returned by the first two, less 5, each will receive 72. With that and the amount they kept, the first two will recover their initial capital; as to the third, he will have $72 + 5 = 77$, that is, his initial capital plus his own sum.[105]

All these intellectual contortions in order to explain the situation arising from a negative solution had no purpose but to show that, in a system of equations, one should not *a priori* reject a negative result when it occurs among a set of positive ones. This approach unblocked the hitherto prevailing situation, even if it did not mean acceptance of a negative solution, thus a recognition of negative numbers; indeed, the interpretation changes the problem's conditions: *the negative number is made a positive one by inverting the concept it represents*, a sum possessed becoming a debt, or a gift a theft, or a gain a loss, *and the corresponding quantity is then subtracted* whereas, according to the initial conditions, it was to be added. This, however, was a necessary stage on the way to acceptance of negative numbers. Note too that the problems considered were to involve concrete quantities; for only then could the concepts they represented be inverted. Observe finally that Fibonacci always leaves the reader faced with an alternative: either the problem must be regarded as impossible, or its results are to be interpreted according to the above mode of inversion.

[41] Let us now examine the two ancient examples of 'buying a horse' (in numbers, with no concrete magnitudes involved).

$$\begin{cases} x_1 + \frac{1}{3}(x_2 + x_3) = y \\ x_2 + \frac{1}{4}(x_3 + x_1) = y \\ x_3 + \frac{1}{5}(x_1 + x_2) = y \end{cases} \qquad \begin{cases} x_1 + \frac{1}{3}(x_2 + x_3 + x_4) = y \\ x_2 + \frac{1}{4}(x_3 + x_4 + x_1) = y \\ x_3 + \frac{1}{5}(x_4 + x_1 + x_2) = y \\ x_4 + \frac{1}{6}(x_1 + x_2 + x_3) = y. \end{cases}$$

As said, they are found in the first book of Diophant's *Arithmetica* (I, 24–25), thus among well-known school problems presented to the reader

---

[105] (...) *erunt 470 pro summa pecunie eorum; quam cum diviseris ordine demonstrato, invenies primum sumpsisse 326, secundum 174; que cum insimul iunguntur, faciunt plus de 470. Quare hec questio non potest solvi, nisi solvatur cum aliqua propria pecunia tertii hominis. Et tunc erit talis questio quod pecuniam, quam ipsi tres habebant comunem, nec non et pecuniam propriam tertii hominis, sumpserunt fortuitu inter primum et secundum hominem; post hec primus posuit in comune $\frac{1}{2}$ ex hoc quod ceperat, secundus $\frac{1}{3}$; ex quibus positionibus tertius homo cepit $\frac{1}{6}$ de ipsa pecunia propria, quam socii habuerant. Post hec, de residuo unusquisque sumpsit tertiam partem; et sic habuit quilibet ipsorum id quod suum erat (...).*

in order to familiarize him with algebraic formalism and computations.[106]

As may be observed, the sequence of numbers in the fractions given is regular, beginning with $\frac{1}{3}$. But why not with $\frac{1}{2}$? The answer is given right away by our previous solving formula, namely

$$x_j = (S - y) \left[\frac{1}{n-1} \sum_{1}^{n} \frac{1}{1 - m_i} - \frac{1}{1 - m_j} \right];$$

because if we were to begin with $\frac{1}{2}$, we would obtain positive solutions only for the two cases $n = 3$ and $n = 4$: for $n = 5$ we would obtain

$$\frac{1}{4} \sum_{1}^{5} \frac{1}{1 - m_i} = \frac{437}{240} < \frac{1}{1 - m_1} = 2,$$

whereas starting with $\frac{1}{3}$ we may go on to $n = 6$. Since Diophant's reader is expected to solve by himself further similar examples,[107] it would not be fair to trap him with a negative solution in the very next one.

Two 15th-century Provençal arithmetics played a pioneering rôle in the history of negative numbers. The first, an anonymous treatise written around 1435 in the town of Pamiers, in southern France, admits for the first time a negative result in a problem, without any attempt to interpret it; the other, a book by Frances Pellos published in Turin in 1492, merely reproduces this problem but is nonetheless the first printed text to display an accepted negative solution.[108]

[42] In the Pamiers arithmetic (and thus in Pellos' text) three problems of the horse-type are solved successively. The first two ($n = 3$, $n = 4$), with $\{m_i\} = \frac{1}{2}, \frac{1}{3}, \frac{1}{4}$, and $\{m_i\} = \frac{1}{2}, \frac{1}{3}, \frac{1}{4}, \frac{1}{5}$, respectively, deal with the purchase of a horse and both end with positive solutions. The next example, this time on purchasing a piece of cloth, is precisely the one alluded to above, namely

$$\begin{cases} x_1 + \frac{1}{2} (x_2 + x_3 + x_4 + x_5) = y \\ x_2 + \frac{1}{3} (x_3 + x_4 + x_5 + x_1) = y \\ x_3 + \frac{1}{4} (x_4 + x_5 + x_1 + x_2) = y \\ x_4 + \frac{1}{5} (x_5 + x_1 + x_2 + x_3) = y \\ x_5 + \frac{1}{6} (x_1 + x_2 + x_3 + x_4) = y. \end{cases}$$

---

[106] Such is indeed the purpose of Book I; see our *Introduction to the history of algebra*, p. 32 (French edition, p. 33).

[107] As implied by his introduction to Book IV, see the edition of the part preserved in Arabic, p. 87.

[108] See our *Arithmétique de Pamiers*, problem C.78, and Pellos, *Compendion*, fol. 64^v – 65^r.

Before treating these three problems the author explains the solving rule. We are to choose a number C and find numbers $C_i = \frac{1}{1-m_i} C$; then $x_j = \frac{1}{n-1} \sum C_k - C_j$ and $y = \frac{1}{n-1} \sum C_k - C$.[109] Comparing with the above formula shows that the number set, C, is $S - y$.

Thus, in the last example, putting $C = 60$, gives $\{C_i\} = 120, 90, 80,$ 75, 72, and therefore $x_j = 109 + \frac{1}{4} - C_j$. Consequently, $x_1 = -(10 + \frac{3}{4})$ (*the first possesses 10 and $\frac{3}{4}$ 'less than nothing'*, mens de non res), $x_2 = 19 + \frac{1}{4}$, $x_3 = 29 + \frac{1}{4}$, $x_4 = 34 + \frac{1}{4}$, $x_5 = 37 + \frac{1}{4}$ (Fig. 365).[110] There is no attempt to explain the negative result; the only hint as to its particularity is that in this problem, and in this problem only (by the way, the only one leading to a negative solution), there is a verification: the values found are substituted in the given equations, the term containing x_1 being each time subtracted.

ar as per lo pruiñier leua en.1 2 o. reſtan.i o. be tres quarts
mes de nó res. Jte p lo fegót leua en.90.reſtá.19. be vn quart.
Jte per lo ters eua en.80.rcſtá.2 9.be vn qnart. Jte per lo quart
leua en.7. 5.reſtá.34..be vn quart. Jtem per lo quint leua en.7 2,
reſtá .3 7.be vn quart.et enfins aues trobat la rafon.   Fig. 56

This verification, instead of an explanation or an interpretation, is important. It means that the negative solution is accepted as such *because it satisfies the given system.* Such is the first (known) acceptance of a negative number as the solution of a problem. As a matter of fact, like many important discoveries in mathematics, it had to come because the time was ripe for it, with all the necessary conditions for this extension of the number domain being present. In any event, the algebraic solution of the equations of third and fourth degrees a few decennia later would have made recourse to negative numbers unavoidable.

Other problems with negative solutions appear in the second half of the 15th century, in Italy and France, sometimes with concrete quantities, but sometimes also with pure numbers, so that the question of interpretation no longer arises, as in Chuquet's following example.[111]

---

[109] *Pausa 1 nombre qual que tu vulhas, sobre lo qual tu atrobaras autres nombres dels quals, levadas que ne fossan las partidas prepausadas, aquell nombre resta entierament; los quals nombres sobre ells trobas totz ensemps sens ell ajustatz se partiscan per 1 mentz que non son aquels en los quals se deu fer la raso. Et de so que vendra del partiment sustray los nombres atrobatz, car so que restara sera lo nombre d'aquell que demanda tal partida; et quant tu sustrayras lo nombre de ta positio, so que restara sera lo pretz que tu voles saber (Arithmétique de Pamiers, p. 316).*

[110] Illustration from Pellos' printed text (computation of the x_i); no fraction bars for typographical reasons.

[111] *Triparty*, p. 642 (p. 90 in the separate print, now available online).

[**43**] *I want to find five numbers such that all of them without the first make 120, without the second 180, without the third 240, without the fourth 300, and without the fifth 360.   To find them I add together all these five numbers, and this makes 1200; which I divide by 4, and I obtain 300; from which I subtract the five aforesaid numbers, namely 120, 180, 240, 300, and 360, and there remain to me 180, 120, 60, 0, and minus 60, which are the five numbers that I desired.*

Consider thus

$$\begin{cases} x_2 + x_3 + x_4 + x_5 = 120 \\ x_3 + x_4 + x_5 + x_1 = 180 \\ x_4 + x_5 + x_1 + x_2 = 240 \\ x_5 + x_1 + x_2 + x_3 = 300 \\ x_1 + x_2 + x_3 + x_4 = 360. \end{cases}$$

Compared with some of Fibonacci's systems, this is a simple case, and it is solved directly with Thymaridas' rule, that is, by adding the equations; indeed, since each of n such equations is of the form $S - x_i = a_i$, we obtain, by addition, $(n-1)S = \sum a_i$, whence

$$x_j = S - a_j = \frac{1}{n-1} \sum a_i - a_j.$$

There is no comment here about the negative number; what is significant, though, is that this problem follows a few instructions for computing with subtracted numbers.[112] Note too that one of the solutions found is zero, a situation which was avoided before (p. 16). In any event we may consider the number domain in problem solving to have been thus extended by the close of the 15th century.

Nevertheless, even if we encounter more and more problems involving negative solutions from the second half of the 15th century on, these results will still be considered with some embarrassment in textbooks: any justification of their presence always seemed somewhat forced. Only later was this constraint overcome. In the second half of the 19th century, Hermann Hankel wrote that the existence of negative numbers need not be justified by occasional applications to real phenomena: they must be introduced in order for the operation of subtraction to be performed without restriction.[113]

---

[112] Already in Diophant (has nothing to do with acceptance of negative numbers).

[113] *Theorie der complexen Zahlensysteme*, I, §2 (where he writes: *als unmöglich gilt dem Mathematiker streng genommen nur das, was logisch unmöglich ist, d. h. sich selbst widerspricht*) and III, §10.

Chapter VI. Filling and emptying cisterns

One of the most well-known examples of such a problem is that of the lion in the *Anthologia Graeca*:

[44] *I am a brass lion. Two jets come out from my eyes, another from my mouth, another from my foot. In two days my right eye fills the basin, my left eye in three, my foot in four days; to fill it, six hours suffice for the water jet of my mouth. If all the jets, from my eyes, from my mouth and from my foot flow together, in how many hours will the basin be full?*[114]

With such antique traces it is not surprising to find examples in the Middle Ages, both in the Christian and in the Islamic world. In the latter case, in the 10th century, Abū'l-Wafā' Būzjānī has a few examples in his *Arithmetic*.[115]

The solution is easy to find: if the kth pipe fills a cistern or a basin in t_k days, it will fill in one day the fraction $\frac{1}{t_k}$ of it; all of them will fill in one day $\sum \frac{1}{t_k}$ of the cistern, which therefore will be completely filled in

$$t = \frac{1}{\sum \frac{1}{t_k}}$$

days.

§ 1. Mediaeval examples

[45] *There are above a cistern three pipes the first of which fills it in two days, the second in three, and the third in four. In how many days will the three running together fill it?*[116]

In one day they fill the fractions $\frac{1}{2}$, $\frac{1}{3}$, $\frac{1}{4}$ of the cistern, respectively, thus altogether the fraction $\frac{1}{2} + \frac{1}{3} + \frac{1}{4} = \frac{13}{12}$; that is, they fill slightly more than the whole cistern in one day. The cistern itself will be filled completely after the fraction $\frac{12}{13}$ of a day has passed.

Sometimes it is the duration of *emptying* the full receptacle which is required. We may then be given the individual times needed, as in

---

[114] Book XIV, No. 7; the answer is $3 + \frac{33}{37}$ or $4 + \frac{44}{61}$, depending on whether we consider a day of 12 or 24 hours; other, simpler examples are Nos. 130–133, 135. See also Nos. 20–21 of the late treatise (by *pseudo* Heron) *De mensuris* (*Opera*, V, pp. 176–177), or Anania Shirakatsi's 24th problem.

[115] Thus one with three pipes filling a basin in, respectively, 2, 3, 10 days: فان كانت البركة ينصب اليها ثلاثة انهار وهى تمتلىء من احدها فى يومين ومن الاخر فى ثلاثة ايام ومن الاخر وهو الثالث فى عشرة ايام وفتحت الثلاثة الانهار واردنا ان نعلم المدّة التى تمتلىء فيها البركة (ed., pp. 366–367); answer: $1 + \frac{1}{14}$ of a day.

[116] *Liber mahameleth*, B.333; the same problem is found in Abū Kāmil's *Algebra*, fol. 105^v (p. 210 of the printed reproduction).

this example of a barrel with three outlets, the running of which singly would empty it in one, two, three days, respectively; with all three open, $1 + \frac{1}{2} + \frac{1}{3} = \frac{11}{6}$ barrels are emptied in a day, thus (considering the full day with 24 hours) the barrel itself in $13 + \frac{1}{11}$ hours.[117] We may also just be given the outflowing quantities in a certain unit of time, as with this barrel containing 56 measures (*charges*) with 4, 6, 8 flowing in one hour, thus 18 in one hour and the whole in $3 + \frac{1}{9}$ hour.[118]

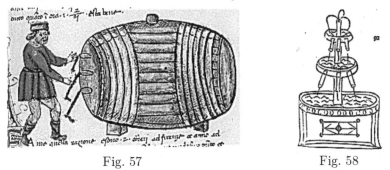

Fig. 57   Fig. 58

Some pipes above the basin may fill it while, at the same time, others below empty it, as in the following example by the painter Piero della Francesca, with one basin filling another (Fig. 58).

[46] *A fountain has two basins, one above and the other below. Each has three pipes. The first pipe above fills the lower basin in two hours, the other ones being closed, the second in three hours, the third fills it in 4 hours. All (these pipes) being closed, opening the first below empties it in three hours, opening the second empties it in four hours, opening the third empties it in five hours. Now, at a same time, I open the pipes above and those below. I ask in how much time the lower basin will be filled.*[119]

The procedure is just the same as before, except that the sum of the fractions for the emptying pipes are subtracted from the sum of the fractions for the filling pipes, all reduced to one hour; since here $(\frac{1}{2} + \frac{1}{3} + \frac{1}{4}) - (\frac{1}{3} + \frac{1}{4} + \frac{1}{5})$ reduces to $\frac{1}{2} - \frac{1}{5} = \frac{3}{10}$, the lower basin will be filled in $3 + \frac{1}{3}$ hours.

[47] A 15th-century Byzantine text, written in (non-classical) Greek dur-

---

[117] MS. Tours 399, fol. 136^r – 136^v; see (text) *Récréations*, p. 74. The illustration of Fig. 57 is taken from Muscarello's *Algorismus*, MS. Pennsylvania LJS 27, fol. 81^r (*lo vino ne uscerà tutto quanto in ora 1$\frac{2}{21}$, e sta bene*; ed., II, p. 196).

[118] MS. Paris BNF fr. 2050, fol. 77^r – 77^v.

[119] Edtion, p. 61. The illustration is taken from Pacioli's *Summa de arithmetica*, fol. 66^v in the part on geometry; the numerical quantities he chooses are 1, 2, 3 and 2, 3, 4, respectively.

ing the Turkish occupation, displays a source of confusion.[120] Three outlets are placed vertically at three different levels on a barrel containing 120 'measures' of wine, namely at those of 90, 50, 0 measures (with the last thus on the ground). Taken individually, we are told, they would empty their part of the barrel, thus 30, 40, 50 measures, in 4, 3, 2 hours, respectively, that is, in one hour they would empty $7 + \frac{1}{2}$, $13 + \frac{1}{3}$, 25 measures, respectively. Since this gives $45 + \frac{5}{6}$ altogether, the whole barrel, we are told, will be emptied in $120 : (45 + \frac{5}{6}) = 2 + \frac{34}{55}$ hours. Now this result would make some sense if the three outlets were on the bottom. Here, for instance, the top outlet will no longer contribute after 30 measures have run out.

The situation of different outlets —with, this time an equal vertical distance between them— had already been considered three centuries before by Fibonacci, but more adequately.[121]

[48] *A barrel has four openings one above the other, at the height of each quarter.*[122] *When the first above is opened, a quarter of the barrel will be emptied in one day. This being emptied, and the second being opened, the part of the barrel between the first and the second, thus the second quarter, will be emptied in two days. Likewise, the first two quarters being emptied, if one opens the third, the next quarter of the barrel, between the second opening and this one, will be emptied in three days. Likewise, if one opens the fourth, the next fourth of the barrel will be emptied in four days. Required in how much time the whole barrel will be emptied if all four are opened simultaneously.*

Since, when the upper opening has finished emptying, it no longer takes part in the emptying, according to the hypotheses, it is necessary to determine the emptying by each opening individually.

We are thus to consider the emptying, as the author does, quarter by quarter, taking into account how many outlets are still in use.

With all four outlets open, there will flow out $1 + \frac{1}{2} + \frac{1}{3} + \frac{1}{4} = \frac{25}{12}$ of a quarter in one day; the first quarter will thus be emptied in $\frac{12}{25}$ of a day. With the last three open, there will flow out $\frac{1}{2} + \frac{1}{3} + \frac{1}{4} = \frac{13}{12}$ of a quarter in one day; the second quarter will thus be emptied in $\frac{12}{13}$ of a day. With the last two open, there will flow out $\frac{1}{3} + \frac{1}{4} = \frac{7}{12}$ of a quarter in one day; the third quarter will thus be emptied in $\frac{12}{7}$ of a day. With the last open, there will flow out $\frac{1}{4}$ of a quarter in one day; the last quarter will thus be emptied in four days. Therefore, the full barrel

---

[120] Hunger-Vogel, No. 82.
[121] *Liber abaci*, pp. 183–184.
[122] On its bottom part.

will be empty in $4 + \frac{12}{7} + \frac{12}{13} + \frac{12}{25} = 4 + \frac{7092}{2275} = 7 + \frac{267}{2275}$ days. Fibonacci, who considers a day of twelve hours (daylight), obtains 7 days, one hour and $\frac{929}{2275}$ of an hour.

Even if his variation law is fanciful, Chuquet deserves credit for having taken into account the inequality of flow according to the level of the liquid.[123]

[49] *In a vessel full of wine or water containing 10 $\frac{2}{3}$ barrels there is an opening of a size such that when it is open the wine will at first gush out impetuously and very strongly, and then all the more quietly and slowly that the flow will progress, in such a way that the first barrel comes out in one hour, the second barrel in two hours, the third in 3 hours, and so forth increasing by 1 and continuing till the 10 $\frac{2}{3}$ barrels are emptied. Required in how many hours this vessel will be empty.*

Answer. Since the number of terms of this progression is 10 $\frac{2}{3}$ and the progression begins with 1 and increases by 1, we are to add 1 to 10 $\frac{2}{3}$, making 11 $\frac{2}{3}$, to be multiplied by half of 10 $\frac{2}{3}$, which is 5 $\frac{1}{3}$; the product makes 62 $\frac{2}{9}$, and in this number of hours will the vessel be empty. [124]

Fig. 59   Fig. 60

Chuquet's next problem is similar, with the same data, but in this case the diminution of the flow corresponds to the sequence of odd numbers for the hours (the use of such different laws of flowing confirms that Chuquet's interest in fluid dynamics was limited). Therefore, the tenth barrel will come out in 19 hours. The sum from 1 to $2n - 1$ being n^2, he computes $(10 + \frac{2}{3})^2 = 113 + \frac{7}{9}$.

Remark. In both cases Chuquet applies the summation of arithmetical progressions to $10 + \frac{2}{3}$, thus to an integer with a fraction. He could in fact have proceeded in two steps, first adding up to $n = 10$ and

---

[123] *Appendice*, p. 441; MS., fol. 184^v – 185^r.

[124] A similar problem, solved in the same way, with the data *10 $\frac{1}{2}$ barili* and the answer $60 + \frac{3}{8}$ hours, is found in Pacioli's *Summa*, fol. 44^v, No. 32. His result is changed into $60 + \frac{1}{2}$ by Tartaglia (*General trattato*, II, fol. 12^v, No. 24); see below, *Remark*. Illustrations taken from Tagliente's *Componimento*, ed. 1547 and 1554 (each vessel with two outlets —and both with a rat on top).

then increasing the result by $\frac{2}{3}$ of the flow in the eleventh hour, which would give $62 + \frac{1}{3}$ and 114 hours, respectively. There is in the *Liber mahameleth* the problem of the cost of $12 + \frac{1}{4}$ bushels of corn with the prices increasing in arithmetical progression; in this case it is perfectly clear that to the price of the 12 bushels will be added a quarter of the price of the thirteenth.[125]

§ 2. Related problems

The cistern problem takes various other forms, which may describe practical situations in daily life, although their actual application may be rather limited.

1. Sails

[50] *A ship must travel from one place to another. With one sail, it will do the journey in 2 days, with the second sail it will do the journey in 3 days, with the third sail it will do the journey in 4 days. We are to say in how much time this ship will do the journey when all the sails are up.*[126]

The solution is the same as in the first cistern problem ([45]). These computations of time are, however, hardly realistic: the speed will not increase proportionally to the sails' total surface, since in any event they partly hide one another.[127]

Fig. 61

Fig. 62

2. Escaping from a jail

[51] *Three men are in a jail and wish to demolish it. The first says that he will demolish it in 6 hours, the second says that he will demolish it in 12 hours, the third*

---

[125] Problem B.53.

[126] *Arte del alguarismo*, pp. 156–157. Fig. 61 is from Muscarello's *Algorismus*, MS. Pennsylvania LJS 27, fol. 78ʳ (ed., II, p. 193).

[127] Other example: MS. Paris BNF fr. 2050, fol. 88ᵛ – 89ᵛ (odd way of solving, see *Récréations*, p. 79).

says that he will demolish it in 18 hours. Required, if all three work together, in how much time they will demolish the jail.[128]

Since all three demolish in one hour $\frac{1}{6} + \frac{1}{12} + \frac{1}{18} = \frac{11}{36}$ of their jail, they will be free in $3 + \frac{3}{11}$ hours.

3. Animals feasting

[52] *A lion ate a sheep in 4 hours, a leopard, in 5 hours, a bear, in 6 hours. Required, if a sheep is thrown among them, in how many hours they will have eaten it.*[129]

Fig. 63

[53] *A wolf, a bear, a dog and a fox eat all four a gelding. The bear alone would eat it in one day, the wolf alone would eat it entirely in two days, the dog alone would eat it entirely in three days, and the fox in four. Required in how much time all four together will eat it.*[130]

4. Building

[54] *A master wants to build a house; but he decided that it should be done rather quickly. A building contractor promised him that he would construct it in 32 days, another contractor promised to do it, more rapidly, in 16 days, still another in 8 days. Required, if all three are working together, in how many days they will do it.*[131]

Fig. 64   Fig. 65

---

[128] From F. Calandri's manuscript, with the illustration of Fig. 62 (MS. Florence Biblioteca Riccardiana 2669, fol. 91^v; text ed. p. 183: *tre huomini in una prigione*).

[129] Fibonacci, *Liber abaci*, p. 182; they need $1 + \frac{23}{37}$ hour. In F. Calandri's printed *Opusculum* (with picture) lion, leopard and wolf eat the sheep in 1, 2, 3 days, respectively, together in $\frac{6}{11}$ of a day.

[130] Tartaglia, *General trattato*, II, fol. 13^r ['17'], No. 28; they need $11 + \frac{13}{25}$ hours, in a day of 24 hours.

[131] They need $4 + \frac{4}{7}$ days; see the Byzantine text, ed. Hunger-Vogel, No. 6. See also *Anthologia Graeca*, XIV, No. 136. Fig. 64 is from Calandri's *Opusculum*.

[55] *Someone wants to construct a wall. He finds a master builder to do it in 6 days; he finds another to do it in 5 days, another in 4 days, another in 3 days, another in 2 days, and another in 1 day. I want to know, if all these builders work together, in how many days they will have done it.*[132]

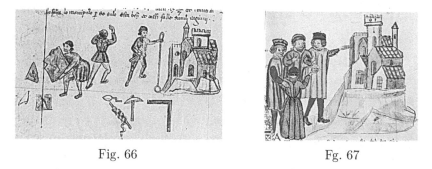

Fig. 66   Fg. 67

5. Drinking bout

[56] *An amphora of wine having been set, three good drinkers, but with different capacities, met together. The first alone would empty the amphora in twenty-four hours, the second in twelve, the third in eight. Required in how many hours these three drinkers, drinking simultaneously, would empty the amphora.*[133]

Fig. 68

---

[132] Bartoli, No. 33; they need $\frac{20}{49}$ of a day. In the example corresponding to the illustration of Fig. 65, taken from F. Calandri's *Opusculum*, three contractors need 10, 12, 15 days, respectively, and, together, 4; then Calandri calculates how many bricks of given size are required for a wall of given size. Fig. 66 & 67 (three master builders, *maystri*; four wishing to acquire a house) are from Muscarello's *Algorismus*, MS. Pennsylvania LJS 27, fol. 58^r–58^v (ed., II, pp. 158–162).

[133] Buteo, *Logistica*, Ch. IV, No. 6, pp. 205–206; in 4 hours. Note, above the drinkers in Fig. 68, the words *Terribilis est locus* (to be followed by *iste, hic domus Dei est et porta caeli, et vocabitur aula Dei*, Gen. 28, 17), thus from the MS. of a Gregorian chant for dedicating a Church; strange contrast. The subject of the discussion might be the 14th-century saying (MS. Reims G. 531 (cat. 695), fol. 1^r)
Vinum subtile creat in sene cor iuvenile (Fine wine makes young the old man's heart
Sed vinum vile reddit iuvenile senile.   Cheap wine renders a young one senile.)

[57] *A drinker by himself empties a cask of wine in 20 days, while with the help of his wife, each one drinking in proportion* (to ability), *they consume as much wine in 14 days. In how much time would the wife by herself empty the full vessel?*[134]

Here one of the times of consumption is unknown. From $\frac{1}{20} + \frac{1}{x} = \frac{1}{14}$ it follows that the wife will empty the cask in $46 + \frac{2}{3}$ days.

6. Grinding

[58] *Working day and night, one mill grinds twenty cafiz, another thirty, yet another forty. A man wants to grind ten cafiz with these three working simultaneously. How much will he put in each one?*[135]

Here the interest is not in the common working time ($\frac{1}{9}$ of a day) but in the respective parts of the ten cafiz they should receive in order to be working simultaneously. Since their productivity is in the proportion $2 : 3 : 4$, they are to receive $\frac{2}{9} \cdot 10$, $\frac{3}{9} \cdot 10$, $\frac{4}{9} \cdot 10$.[136]

Fig. 69

Fig. 70

---

[134] Gemma Frisius, *Arithmetica practica*, fol. 37^v – 38^r (ed. 1561, pp. 74–75). Illustration (Fig. 69, a couple encouraged to enter a tavern) from MS. Paris Bibliothèque de l'Arsenal fr. 5107 rés., fol. 61^v (Fr. Jehan de Vignay French translation of Jacobus de Cessolis' *Liber de moribus*, see below, p. 256).

[135] *Liber mahameleth*, B.169. The *qafīz* (قفیز) is an Arabic measure of capacity, which became in mediaeval Spain *caficius*, and in modern times *cafiz* and *cahiz*.

[136] Similar problem in the MS. Paris BNF fr. 2050, fol. 76^v, with three mills grinding daily, respectively, 10, 7, 5 bushels, and 100 bushels to be ground with the three mills working simultaneously; the parts will be $45 + \frac{10}{22}$, $31 + \frac{18}{22}$, $22 + \frac{16}{22}$, with the work completed in $4 + \frac{6}{11}$ days. In Muscarello's *Algorismus*, MS. Pennsylvania LJS 27, fol. 77^v (see Fig. 70), the three mills grind, respectively, 9, 8, 5 *some* ('load') and the quantity brought is 6 ($= 2 + \frac{5}{11} + 2 + \frac{2}{11} + 1 + \frac{4}{11}$; ed., II, pp. 192–193).

Chapter VII. Messengers

§ 1. Introduction

In the simplest case (§ 2), a single person covers a distance d at a uniform daily speed a. The duration of his journey will thus be $t = \frac{d}{a}$ days. But one may also suppose that his daily speed changes, with the most simple case of a regular change being that his daily speed follows an arithmetical progression, or even a geometrical one. That is certainly unrealistic, but some mediaeval authors, undeterred, considered that with training daily speed might be increased.[137]

Such problems of movement may be extended to two persons. In the case of pursuit (§ 3), they move in the same direction and one must catch up with the other. Here again, the two daily speeds may be uniform, but the second person —who starts then either with a delay τ or is initially behind by a distance δ— progresses with a greater daily speed $a_2 > a_1$; the first will be caught up after a time t determined from $a_1\tau + a_1 t = a_2 t$. The situation may become more intricate by considering that the daily speeds are not uniform, but change according to arithmetical or geometrical progressions.

In problems of meeting (§ 4), two persons starting from two places distant by d move in opposite directions and will meet after t days, with $a_1 t + a_2 t = d$ in the case of uniform speeds and simultaneous departure. Here again, variations of one of the daily speeds or both are possible.

The case of pursuit with uniform speeds may be varied by supposing that one animal, behind another at a given distance, is to catch it up (§ 5); here the daily movement is replaced by the number of jumps in a certain time or over a certain distance, with these jumps, though remaining constant for each animal, differing between them in length and frequency.

A further intricacy in the case of advance, pursuit and meeting is to suppose that the progress made during the day is reduced by a nightly retreat (§ 6). Some authors attempt to justify this: thus a ship, making regular progress by day, is blown slightly back each night by the wind; or a squirrel on a tree, trying to escape a cat, keeps losing each night a little of the distance gained, thinking that 'the cat had gone away'.[138] On

---

[137] As does F. Calandri for the master who, pursuing his servant after a theft, covered three miles the first day, six the second, and so on, thus progressing *perché e' non era uso a caminare* (*Opusculum*).

[138] *Liber mahameleth*, B.370 (see below [96], p. 99); F. Calandri, *Opusculum* (*lo scoiactolo, credendo che la gatta si fussi partita, voleva scendere del decto albero*).

the contrary some authors, instead of searching for a justification, add to the absurdity by supposing that the distance separating the two is itself variable, increasing by day and decreasing by night.

§2. Progress of a single person

According to whether the daily progress is uniform, or in arithmetical or geometrical progression with daily additive or multiplicative increment r, the distance covered after an integral number t of days will be, if the progress in the first day is a_0,

$$d = t \cdot a_0, \quad d = t \cdot a_0 + r \cdot \frac{(t-1)t}{2}, \quad d = a_0 \cdot \frac{r^t - 1}{r - 1}.$$

1. Uniform movement

The very first example of Alcuin's Collection is a movement problem.

[59] *A snail was invited to eat by a swallow at a distance of one league. But it could not advance more than one inch (uncia) of a foot each day. May whoever wishes to do so say in how many days the snail will reach the place of the meal.*

Solution. There is in a league one thousand five hundred paces, 7500 feet, 90 000 inches.[139] *There have been as many days as inches, which corresponds to 246 years and 210 days* (with Alcuin rounding off the year to 365 days).

2. Movement in arithmetical progression

$$d = t \cdot a_0 + r \cdot \frac{(t-1)t}{2}.$$

[60] *A man goes the first day 1 league, the second day 2 leagues, the third day 3 leagues, and so on in increasing by 1. Required in how many days he will have covered 18 leagues.*[140]

As Chuquet puts it, *we are to find a number which, when added to 1 and this addition being multiplied by half this number, the multiplication gives 18.* (...) *You will find (si trouveras)* $\mathcal{R}^2\ 36\tfrac{1}{4}\ \bar{m}\ \tfrac{1}{2}$ (Fig. 71). Indeed, with $a_0 = 1 = r$ and $d = 18$, the above relation becomes

$$\frac{(t+1)t}{2} = 18, \quad \text{whence} \quad t = \sqrt{36 + \frac{1}{4}} - \frac{1}{2}.$$

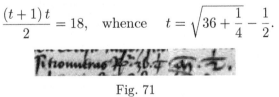

Fig. 71

---

[139] For one league = 1500 paces, one pace = 5 feet, one foot = 12 inches. The mediaeval value of the league (from the time of Charlemagne) is about 4 km, but with wide variations.

[140] Chuquet, *Appendice*, p. 440; MS., fol. 184^{v}.

A similar problem is that of a man digging a well when the price increases with the depth according to an arithmetical progression.

[61] *A workman undertakes to dig a well 11 cubits deep at a cost of 11 livres; but since the deeper the well the higher the cost, it is agreed that whatever the first cubit costs, the second will cost twice the first, the third thrice the first, the fourth the quadruple of the first, and so on always increasing by 1. It happens that the workman has dug 6 cubits, and both the master and he agree that he does not have to dig any more. Required how much is due to the workman according to the initial agreement.*[141]

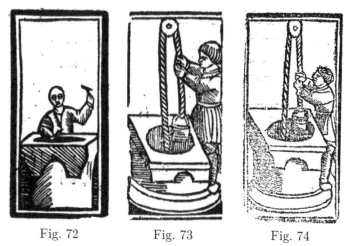

Fig. 72   Fig. 73   Fig. 74

Since $1 + 2 + 3 + \cdots + 11 = 66$ while $1 + 2 + 3 + \cdots + 6 = 21$, the amount of the wage p will be determine by $p : 21 = 11 : 66$, and will thus be three and a half livres.

Pacioli's *Summa* being the first substantial arithmetic published in Italy, it had considerable influence. Because or in spite of that, it was heavily criticized fifty years later by both Cardan and Tartaglia, who liked to point out its errors. Sometimes even the situation described in the problem was criticized, as by Tartaglia here, as follows: *The deeper the workman digs, the greater his effort; one may well believe that, but it is now a matter of dispute that the second cubit will cost him an effort twice the first, the third an effort thrice the first, and the same may be argued for the other deeper cubits. Now I say that things which are debatable are not known, in fact, indeed, are uncertain, and the mathematician should not care about uncertain things, but*

---

[141] Pacioli, *Summa*, fol. 40ᵛ, No. 9; in French version by de la Roche, *Arismethique*, fol. 217ᵛ, with *toises* instead of Pacioli's *braccia* ($\approx$ 60 cm; the teis (French *toise*) is some six feet long, thus about 2 m). Illustrations of digging a well in Tagliente's *Componimento*, ed. 1525, 1547 and 1554 (data different).

only about those certain. Therefore such a problem is neither mathematical nor belongs to the mathematician's domain.[142] Here Tartaglia is being unfair; and just after he will solve an even more absurd problem (see below, p. 109).

Fig. 75   Fig. 76

Pacioli's *Summa* has numerous other examples of this kind, two of which are less banal, namely the following ones.

[62] *Someone covers 300 miles. The first day, he covers 25 plus (as many miles as) the number of days he will take to finish his journey; the second day, he doubles the miles of the first day; and so on, he always doubles the previous (day's) progress. Required in how many days he will reach his destination and how many miles he has covered the first day.*[143]

We must first, the author says, determine by trial and error (*a tastoni*) the integral number of days. The successive daily progresses being (in our terms) $25 + t$, $50 + 2t$, $100 + 4t$, $200 + 8t$, and thus the miles covered in three and four days, respectively, $175 + 7t$ and $375 + 15t$, Pacioli may already say that the duration of the journey will be between three and four days. Putting then this duration equal, in days, to (in our terms) $3 + x$, with x the fraction of the fourth day (*donca poni che'l andasse in 3 dì, più 1 co(sa) de dì per lo 4°*), the traveller will have successively covered, during the first three days, $28 + x$, $56 + 2x$, $112 + 4x$. Since he would, during this fourth day, cover (at a constant speed) $224 + 8x$ miles, we infer, by the rule of three, that in the fraction x of the fourth day he will cover $224x + 8x^2$ miles. With the sum of the miles covered in the first three days, $196 + 7x$, Pacioli obtains as the equation and its solution

$$8x^2 + 231x + 196 = 300, \quad \text{whence} \quad x = \sqrt{221 + \tfrac{113}{256}} - \left(14 + \tfrac{7}{16}\right),$$

from which he infers the two required quantities:

$$t = \sqrt{221 + \tfrac{113}{256}} - \left(11 + \tfrac{7}{16}\right), \quad 25 + t = 13 + \tfrac{9}{16} + \sqrt{221 + \tfrac{113}{256}}.$$

This may remind us of another problem by Pacioli (above, p. 44).

---

[142] *General trattato*, II, fol. 10ʳ, No. 16.

[143] *Summa*, fol. 41ʳ, No. 13. Illustrations from Tagliente's *Componimento*, ed. 1525 and 1547.

[63] *There is from Florence to Rome 100 miles, and four companions leave Florence to go to Rome at different speeds. The first covers the first day 1 mile, the second day, 2, the third, 3, and continues always likewise, with an increment of one mile a day. The second companion goes the first day 1 mile, the second, 3, the third, 5, and increases always in this manner, by 2 miles each day. The third companion goes the first day 2 miles, the second, 4, the third, 6, and increases always in this manner, by 2. The fourth companion goes the first day 4, the second, 8, the third, 12, and increases always in this manner, by 4 each day. Since they wish to arrive in Rome at the same time, we ask after how many days they are successively to leave.*[144]

Since $a_0,\ a_0 + r,\ a_0 + 2r, \ldots, a_0 + (t_i - 1)\, r$ add up to

$$t_i \cdot a_0 + r\, \frac{(t_i - 1)\, t_i}{2} = d,$$

the equations for the various numbers of days t_i will be:

$$\frac{r}{2}\, t_i^2 + \left(a_0 - \frac{r}{2}\right) t_i = d.$$

In Pacioli's problem, this gives the equations and the solutions

$(a_0 = 1,\ r = 1)$	$t_1^2 + t_1 = 200,$	$t_1 = \sqrt{200 + \frac{1}{4}} - \frac{1}{2},$
$(a_0 = 1,\ r = 2)$	$t_2^2 = 100,$	$t_2 = 10,$
$(a_0 = 2,\ r = 2)$	$t_3^2 + t_3 = 100,$	$t_3 = \sqrt{100 + \frac{1}{4}} - \frac{1}{2},$
$(a_0 = 4,\ r = 4)$	$t_4^2 + t_4 = 50,$	$t_4 = \sqrt{50 + \frac{1}{4}} - \frac{1}{2},$

with $t_1 > t_2 > t_3 > t_4$, as already suggested by the hypotheses. Thus, concludes Pacioli, the second will start $t_1 - t_2 = \sqrt{200 + \frac{1}{4}} - (10 + \frac{1}{2})$ days after the first, the third $t_2 - t_3 = (10 + \frac{1}{2}) - \sqrt{100 + \frac{1}{4}}$ days after the second, the fourth $t_3 - t_4 = \sqrt{100 + \frac{1}{4}} - \sqrt{50 + \frac{1}{4}}$ days after the third.

3. Movement in geometrical progression

$$d = a_0 \cdot \frac{r^t - 1}{r - 1}.$$

[64] *A man travels the first day one league, the second 3 leagues, the third day 9 leagues, and so on continuing and tripling his (daily) journeys. Required the number of leagues he has walked in 5 days $\frac{1}{2}$.*

Answer. To do it, I put 1, 3, 9, 27, 81, 243, which are six (daily) journeys, of which the fifth is 81 leagues and the sixth, 243. And since there are only 5 days $\frac{1}{2}$, we are to search for a middle proportional between 81 and 243; this is R^2 19683. Which must be tripled; this gives R^2 177147. From this triple we must take away 1, there remains R^2 177147 m̄ 1. From this remainder we must take the half, which

---

[144] *Summa*, fol. 41^r, No. 14.

is $\mathcal{R}^2$ 44286 $\frac{3}{4}$ $\bar{m}$ $\frac{1}{2}$. *This is the number of leagues he has travelled in 5 days and* $\frac{1}{2}$. *Therefore he has travelled in this* $\frac{1}{2}$ *day* $\mathcal{R}^2$ 44286 $\frac{3}{4}$ $\bar{m}$ 121 $\frac{1}{2}$*, for in the 5 previous days he had travelled 121 leagues.*[145]

Chuquet's problem requires, given a_0, r, t, to find d for an intermediary time between t and $t+1$, namely here $t+\frac{1}{2}$ with $t=5$. The formula used by him is that for the sum of a geometrical progression, but extended to a fractional exponent (see above, p. 66, *Remark*). Let us indeed consider the distance covered after t days (with a single league covered in the first day, thus $a_0 = 1$):

$$d_t = 1 + r + r^2 + \cdots + r^{t-1} = \frac{r^t - 1}{r - 1}.$$

Applying this by analogy to the journey after $t+\frac{1}{2}$ days, we obtain

$$d_{t+\frac{1}{2}} = 1 + r + r^2 + \cdots + r^{t-1} + r^{t-\frac{1}{2}} = \frac{r^{t+\frac{1}{2}} - 1}{r - 1}.$$

Now

$$r^{t+\frac{1}{2}} = r \cdot r^{t-\frac{1}{2}} = r \cdot r^{\frac{2t-1}{2}} = r \cdot \sqrt{r^{t-1} \cdot r^t}.$$

This gives us the formula used by Chuquet and (for $t = 5$, $r = 3$) its numerical value

$$d_{t+\frac{1}{2}} = \frac{r \cdot \sqrt{r^{t-1} \cdot r^t} - 1}{r - 1} = \frac{3 \sqrt{81 \cdot 243} - 1}{2}.$$

§3. Pursuit

A first messenger is sent to a certain place, and a second is sent to catch up with him.

1. Uniform movements

The daily progresses a_1, a_2 of the two messengers are constant (with $a_2 > a_1$), but the second starts with either an initial distance to be made up δ or a delay of τ days —which comes to the same since we may immediately change one into the other ($\delta = \tau \cdot a_1$). We shall then calculate the number of days t_2 taken by the second to catch up with the first by equating the journeys made during the respective numbers of days, since the distance they have covered must be the same. Thus $t_1 \cdot a_1 = t_2 \cdot a_2$, with $t_1 = t_2 + \tau$, whence $(t_2 + \tau)\, a_1 = t_2 \cdot a_2$, and

$$t_2 = \frac{\tau \cdot a_1}{a_2 - a_1} = \frac{\delta}{a_2 - a_1}.$$

---

[145] Chuquet, *Appendice*, p. 440, and MS., fol. 183^{v}.

The time thus calculated being that of the pursuer, we shall add to it τ if we wish to know the number of days the first has travelled.

(i) *The given difference is a time.*

$$t_2 = \frac{\tau \cdot a_1}{a_2 - a_1}.$$

[65] *A messenger is sent to a town and progresses daily by twenty miles. In how many days will another messenger, sent five days later and progressing daily by thirty miles, overtake him?*[146] $(a_1 = 20,\ \tau = 5,\ a_2 = 30,\ \text{thus } t_2 = 10)$.

[66] *Two are on the way to Rome. The first progresses daily by 6 miles, and he has started 3 days before the other. The second progresses daily by 8 miles. The question is: in how many days will they meet?*[147]

The author puts x (*1r. = 1 res*, see p. 252) as the common number of days (thus x is our t_2). Equating the journeys gives $8x = (x + 3)\,6 = 6x + 18$, whence $x = 9$, and the first was on his way 12 days. See Fig. 77: *8 r. gleych 6 r. + 18.*

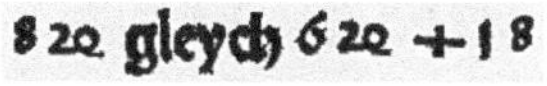

$$\text{8 2\varrho\ \textbf{gleych}\ 6 2\varrho\ \textbf{+18}}$$

Fig. 77

(ii) *The given difference is a distance.*

$$t_2 = \frac{\delta}{a_2 - a_1}.$$

[67] *A priest goes from Paris to Rome where there is a church vacancy, and progresses daily by ten leagues. Another, wanting the same appointment, follows him, and progresses 15 leagues daily; but the first has already walked 60 leagues. I ask in how many days he will reach him.*

Answer. You are to consider how many leagues the second walks more than the first; he walks 5 leagues each day. Thus, divide the 60, which the first has walked more, by 5; it gives 12. Therefore, you can answer that he will meet him in 12 days. You can prove it as follows. The second walks 15 leagues daily; therefore multiply 12 by 15, it gives 180. The first walks 10 leagues daily; multiply 12 by 10, it gives 120, and add the 60 he has walked more; this also gives 180 leagues, like the other. (Thus the) answer: he will meet him in 12 days, 180 leagues away from Paris.[148]

2. Movements in arithmetical progression

(i) *Only one of the movements is a progression.*

---

[146] *Liber mahameleth*, B.368. Same problem in Abū Kāmil's *Algebra*, fol. 103ʳ (p. 205 of the printed reproduction).

[147] Rudolff, *Die Coß*, fol. 305ᵛ, No. 185.

[148] *Livre de chiffres et de getz*, No. 16.

[68] *Two pilgrims start at Montpellier to go to Santiago (de Compostella); both start at the same time, and there is from Montpellier to Santiago 250 leagues.*[149] *They start together, but one walks every day 10 leagues, no more and no less, and the other walks the first day 1 league, the second day 2 leagues, the third day 3 leagues, and walks every day one league more, till he has joined his companion. I ask you in how many days he will have joined his companion and how many leagues they will have covered when he joins him.*[150] The text finds, correctly, 19 days.

Indeed, let a_1 be the daily progress of the first, a_2 the initial progress of the second and r the daily increment, therefore $a_2 + r(i-1)$ is his progress on the ith day. Since the departure is simultaneous, they will meet after t days with

$$t \cdot a_1 = t \cdot a_2 + r \frac{(t-1)\,t}{2}, \qquad (1)$$

whence $2(a_1 - a_2) = r(t-1)$, which gives

$$t = \frac{2(a_1 - a_2) + r}{r}. \qquad (1')$$

With $a_1 = 10$, $a_2 = 1 = r$, we find indeed $t = 19$.[151]

Fig. 78   Fig. 79

But if in formula $(1')$, the divisor does not divide the dividend exactly, we shall have to consider, for the additional fraction, the distances covered at the beginning and at the end of the day $t + 1$ on which the meeting will take place. Let thus $d_t^{(1)}$, $d_t^{(2)}$ be the respective distances covered by the end of the day t, with $d_t^{(1)} > d_t^{(2)}$, and $d_{t+1}^{(1)}$, $d_{t+1}^{(2)}$ the distances which would be covered at the end of the day $t+1$, with, this time, $d_{t+1}^{(1)} < d_{t+1}^{(2)}$. The fraction of said day will thus be given by

$$\frac{d_t^{(1)} - d_t^{(2)}}{\left(d_{t+1}^{(2)} - d_t^{(2)}\right) - \left(d_{t+1}^{(1)} - d_t^{(1)}\right)},$$

---

[149] This is anecdotal, for in pursuits the distance of the destination is irrelevant.

[150] MS. Paris BNF fr. 2050, fol. 90$^\text{v}$.

[151] Illustrations from Tagliente's *Componimento*, ed. 1525 and 1547 (same problem, but with $a_1 = 20$ and $a_1 = 30$, respectively, thus $t = 39$ and $t = 59$).

and the above formula $(1')$ may now be written as

$$t = \left[\frac{2\,(a_1 - a_2) + r}{r}\right] + \frac{d_t^{(1)} - d_t^{(2)}}{\left(d_{t+1}^{(2)} - d_t^{(2)}\right) - \left(d_{t+1}^{(1)} - d_t^{(1)}\right)}, \qquad (1^*)$$

with the square brackets in the first term meaning that the integral part of the fraction is to be taken. This formula is valid as long as the daily speed is considered to remain unchanged during the day.

If in the progression the first term is equal to the constant difference, and thus the daily progress of the second on the ith day is $r \cdot i = a_2 \cdot i$, formulae (1), $(1')$ become

$$t \cdot a_1 = a_2 \,\frac{t(t+1)}{2} \qquad \text{and} \qquad t = \frac{2a_1 - a_2}{a_2} = 2 \cdot \frac{a_1}{a_2} - 1. \qquad (1'')$$

This last relation is enunciated by Fibonacci: *Divide the number of miles covered daily by the first by that of the other, double the result and take 1 away from the quantity which has been doubled; the remainder will be the number of days after which he will catch up with him.*[152] But this, Fibonacci adds, is valid only if a_2 divides a_1; if there is a fraction left, we shall have to consider the daily progress for the next day. Thus, in an example of his with $a_1 = 10$, $a_2 = 3 = r$, he finds, with the formula, $5 + \frac{2}{3}$, corresponding therefore to $56 + \frac{2}{3}$ miles as the distance covered. Now after five days the first will have covered 50 miles and the second, 45 miles. At the end of the sixth day, the distance covered by the second must be 18 miles, that is, 8 miles more than what the first will have covered. The remaining distance of 5 miles will be covered after $\frac{5}{8}$ of the sixth day, as Fibonacci indeed finds. It is precisely the result given by formula (1^*) with $d_t^{(1)} = 50$, $d_t^{(2)} = 45$, $d_{t+1}^{(1)} = 60$, $d_{t+1}^{(2)} = 63$. Accordingly, the meeting will take place $56 + \frac{1}{4}$ miles from the point of departure.

As a matter of fact, our first example, with two pilgrims, was particularly easy to solve, for in the arithmetical progression the first term a_2 and the constant difference r were both equal to 1. Therefore formula $(1'')$ simply became $t = 2a_1 - 1$. Whence the rule given in a 15th-century manuscript: *Double the number of leagues of the one who moves each day an equal number of leagues, and subtract from this doubled number a unit, and the remainder will show you, without fail, on which day they will meet.*[153]

---

[152] *Numerum miliariorum, que primus cotidie vadit, per ascensionem alterius divide, et quod inde exierit duplica, et de duplicata summa 1 abice; residuum erit quantitas dierum in quibus eum consecutus erit* (*Liber abaci*, p. 168).

[153] *Doublez le nombre dez lieuez de celluy qui va chascun jour egal nombre de lieuez, et de ce nombre doublé ostez une unité, et le remenant vous monstrera à quel jour ilz s'entretrouveront sans faillir* (MS. Tours 399, fol. 135ᵛ). This rule is already found

Elsewhere the constant difference is no longer 1; as for example in the following:

[69] *Two set off together. The first progresses daily by 6 miles, the second 1 mile the first day, 3 miles the second day, 5 miles the third day, and so on, each day 2 miles more. The question is: when will they meet again?* Applying our formula (1′) we find, like the author, that they meet after 6 days.[154]

Fig. 80

[70] *A man was progressing by 30 miles a day, going from Milan to Naples. Another, starting at the same time, progressed the first day by 3 miles, the second by 8, the third by 13, the fourth by 18, and so on. Required when they will meet.*[155]

This problem, with thus $a_1 = 30$, $a_2 = 3$, $r = 5$, had been proposed before Cardan by G. Sfortunati in his arithmetical treatise; he calculated

$$t = \frac{2a_1 - a_2}{r} = \frac{2 \cdot 30 - 3}{5} = \frac{57}{5} = 11 + \frac{2}{5}$$

and informs us that with this the reader will have a general, infallible rule (*generale, laquale infallibil ti sarà*).[156] He is severely criticized for this assertion by Cardan, according to whom with *this problem one may see such great stupidity in said man that this problem is no more suitable for him than for the dullest schoolboy* (*in qua questione licet videre hominis stuporem tam magnum, ut non sit digna viro tali questio illa, ymo nec minimo discipulo*). And indeed Cardan computes the answer as in formula (1*) above, giving

$$t = \left[\frac{2(30 - 3) + 5}{5} \right] + \frac{330 - 308}{(366 - 308) - (360 - 330)} = 11 + \frac{11}{14}.$$

in the 14th century (see *Subtilitates*, Nos. 8 & 20) and was used by Tagliente in his two problems (above, p. 78*n*). For other 15th-century examples, see *Récréations*, pp. 94–95.

[154] Rudolff, *Die Coß*, fol. 345^v–346^r, No. 232. Similar problem in Muscarello's *Algorismus*, MS. Pennsylvania LJS 27, fol. 81^r–81^v (and illustration Fig. 80: *miglia 15, e sta bene*, that is, with the first progressing 15 miles a day and thus the meeting after 15 days; see ed., II, pp. 196–197).

[155] Cardan, *Practica arithmetice*, Ch. LXVI, No. 66.

[156] *Nuovo lume*, fol. 87^r–87^v (ed. 1544).

(*ii*) *The two movements are in arithmetical progression.*

The following example is by Chuquet.[157]

[**71**] *A man runs away and covers the first day 3 leagues, the second day 6 leagues, the third day 9 leagues, and so on increasing by 3. Another wants to follow him and starts the fourth day after the departure of the other, and goes the first day 5 leagues, the second day 10 leagues, the third day 15 leagues, and so on increasing by 5. Required after how many days he will catch up with his man.*

Chuquet describes how to solve it as follows.[158] *In order to do this computation, I put 1^1* (that is, $1 \cdot x$, here $1 \cdot t$) *as the number of days after which the one running away is reached by the one who progresses by 5; to which I add 1, which gives 1^1 plus 1. Which I multiply by two and a half (times) 1^1, which is $2^1 \frac{1}{2}$* (meaning $2 \cdot t + \frac{1}{2} \cdot t$); *the multiplication gives $2^2 \frac{1}{2}$ plus $2^1 \frac{1}{2}$, which I keep apart.*[159] *Next, for the one fleeing and progressing by 3, I put 1^1 plus 3 days, for he has three days more than the one following him. I add to this number 1, which gives 1^1 plus 4, which I multiply by one and a half (times) 1^1 plus 3, which is $1^1 \frac{1}{2}$ $\bar{p}$* ($=$ plus)[160] *$4 \frac{1}{2}$, which gives $1^2 \frac{1}{2}$ plus $10^1 \frac{1}{2}$ plus 18* (Fig. 81, but with 10 instead of 10^1), *equal to $2^2 \frac{1}{2}$ $\bar{p}$ $2^1 \frac{1}{2}$ which I kept apart. Now reduce the two sides, you will have 1^2 on the one and 8^1 $\bar{p}$ 18 on the other. Complete the remainder of this problem according to the rule for solving quadratic equations; you will have $\mathcal{R}^2$ 34 $\bar{p}$ 4* ($= \sqrt[2]{34} + 4$). *This is the number of days after which he will be caught up with.*

Chuquet verifies the answer: each has covered $135 + \sqrt{17212 + \frac{1}{2}}$. *Thus each has walked as much as the other. Therefore this computation is correct.*

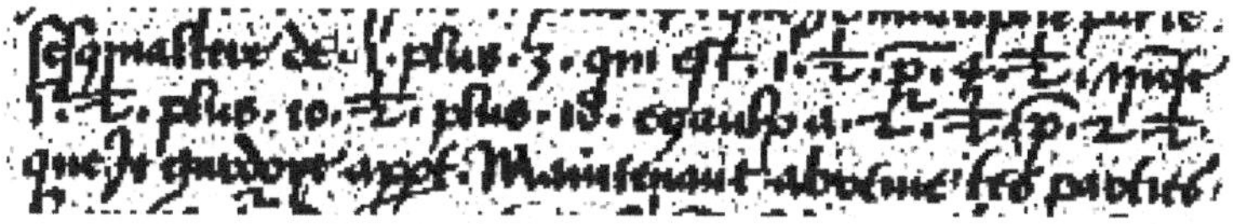

Fig. 81

Generally, let us put that the initial term of the arithmetical progression for the first is a_1, its constant difference r_1, the corresponding quantities for the second a_2 and r_2, with the delay of his departure τ days, and finally that t is the duration of the pursuer's journey. Since the distances they cover are, respectively,

---

[157] *Appendice*, p. 443; MS. fol. $187^v - 188^r$.

[158] On his algebraic symbolism, see p. 251.

[159] This is characteristic of verbal algebra and means that an intermediary result has been obtained, to be reverted to later; see our *Introduction to the history of algebra*, pp. 5 (*n.* 12) and 74 (*n.* 65).

[160] $\bar{p}$ is analogous to $\bar{m}$, already seen (pp. 44, 72). See p. 251.

$$d_1 = (t + \tau)\, a_1 + r_1 \frac{(t + \tau - 1)(t + \tau)}{2}, \qquad d_2 = t \cdot a_2 + r_2 \frac{(t - 1)\, t}{2}$$

and they must be equal, we obtain as the equation for t

$$t^2 (r_2 - r_1) = t \left[(r_2 - r_1) + 2\tau r_1 - 2(a_2 - a_1) \right] + \tau \left[r_1 (\tau - 1) + 2a_1 \right].$$

In the particular case where $a_1 = r_1$, $a_2 = r_2$, as in the previous problem, we shall have

$$d_1 = a_1 \frac{(t + \tau)(t + \tau + 1)}{2}, \qquad d_2 = a_2 \frac{t\,(t + 1)}{2},$$

$$t^2 (a_2 - a_1) = t \left[2\tau a_1 - (a_2 - a_1) \right] + \tau \left[a_1 (\tau + 1) \right].$$

With $a_1 = 3 \,(= r_1)$, $a_2 = 5 \,(= r_2)$, $\tau = 3$, as in Chuquet's problem, we indeed find $t^2 = 8t + 18$, of which he determines the (rather inconvenient) positive solution for t seen above. Simpler choices are considered many centuries earlier by Abū Kāmil: with $a_1 = 1 = r_1$, $a_2 = 1$, $r_2 = 3$, $\tau = 8$, he finds $t = 12$; with $a_1 = 1 = r_1$, $a_2 = 1$, $r_2 = 2$, $\tau = 84$, it will be $t = 204$.[161]

In the example below, one of the two progressions is decreasing.

[72] *A servant, after robbing his master, runs away, the first day by twenty-four miles, the second by 23, the third by 22, and so on: feeling safer, he reduced his flight each day by one mile. But the master, after a denunciation, started pursuing the thief on the same day, covering ten miles, the next day twelve, ensuring that the journey of one day always exceeded that of the previous day by one and an additional mile. I ask on which day the master will catch the fugitive and after how many miles.*[162]

The conditions give

$$t \cdot 24 - \left[1 + 2 + \cdots + (t - 1) \right] = t \cdot 10 + 2 \left[1 + 2 + \cdots + (t - 1) \right],$$

whence

$$t \cdot 14 = 3\, \frac{(t - 1)t}{2}\, ;$$

therefore $3t = 31$, and t equals ten days and a fraction. Since after ten days the servant will have covered 195 miles and the master 190, whilst on the eleventh day the first would progress by 14 and the second by 30, the difference of 5 miles will be cancelled after the fraction $\frac{5}{30-14} = \frac{5}{16}$ of the

---

[161] *Algebra*, fol. 103ᵛ (p. 206 of the manuscript reproduction).

[162] *Servus, expilato domino, fugit, primo die milliaria quatuor & viginti, altero 23, tertio 22; et sic in dies factus securior, de fuga remittebat milliare. Dominus autem, indicio facto, eodem die furem rectà prosequitur, ad milliaria decem, postridie verò duodecim milliaria pergit; et ita semper curam intendens, iter diei præcedentis uno, atque altero milliari superabat. Quæro & ad quem diem & ad quot milliaria fugitivum dominus apprehendit* (Buteo, Ch. IV, No. 31, pp. 231–233).

eleventh day has elapsed (above, p. 79, formula 1*), thus at the distance of $199 + \frac{3}{8}$ miles. This is indeed Buteo's result, obtained by summing the progressions and, for the fractional part, by a rule of three.

3. One movement is in geometrical progression

Let a_1 be the daily progress of the first and $a_2 \cdot r^{i-1}$ that of the second on the ith day. They will meet after t days if

$$t \cdot a_1 = a_2 \frac{r^t - 1}{r - 1},$$

and thus $a_2 \cdot r^t = a_1 t \,(r - 1) + a_2$. The first two examples below give tentative determinations of t (in the third, t is known and a_2 required).

[73] *A thief, fleeing after his misdeed to escape the hands of justice, rides each day regularly 30 leagues. The provost marshal, together with his sergeants, follows him. He rides the first day only one league, the second 2, the third four, the fourth 8, and so on in pursuit by increasing each day by half.*[163] *Required in how many days and (after) how many leagues said provost will reach said thief.*[164]

The answer is not computed. We are merely told that it will be on the 8th day, for in eight days the thief rides 240 leagues and the provost 255 (thus more). This appears by setting out a table of the distances covered, with t integral and $a_1 = 30$, $a_2^{(i)} = 2^{i-1}$ (Fig. 82).

t	1	2	3	4	5	6	7	8
$30 \cdot t$	30	60	90	120	150	180	210	240
$2^t - 1$	1	3	7	15	31	63	127	255.

Fig. 82

A contemporary text by Piero della Francesca does give a more precise answer in a similar case (assuming, though, uniformity of speed during the day itself).

[74] *Two wish to walk. The first is (initially) ahead by 25 miles and walks each day 25 miles. The second follows him and walks the first day 1 mile, the second 2, the third 4, the fourth 8, and continues walking in this manner, doubling (the distance) each day, until he meets the other. I ask in how many days that will be.*[165]

We may again show in a table the distances after each day up to the meeting (Fig. 83).

---

[163] The journey on any given day is half that of the subsequent one.
[164] MS. Nantes 456, fol. 81^v.
[165] Edition, p. 102.

t	1	2	3	4	5	6	7	8
$25(t+1)$	50	75	100	125	150	175	200	225
$2^t - 1$	1	3	7	15	31	63	127	255.

Fig. 83

Since this gives the eighth day as that of meeting, the author then considers the situation at both the beginning and the end of it. As in some other similar cases ([62], p. 74, [72], p. 82), the fraction of the last day is then determined, here algebraically by putting $200 + 25x = 127 + 128x$, whence $x = \frac{73}{103}$.

A more intricate form of such a problem is solved by Cardan.[166]

[75] *Two were walking, the first 10 miles a day, the second a certain number of miles the first day, the second (day) $\frac{1}{3}$ more, the third day $\frac{1}{5}$ more than in the second, the fourth day $\frac{1}{3}$ more than in the third, the fifth day $\frac{1}{5}$ more, and so on in the proportions one and a third and one and a fifth, alternately. They met in 19 days, at the very end of the 19th day. Required the journey of the second on the first day, and also on the last.*

Let us solve it in our way. Since we know that they will meet at the end of the 19th day, when the first will have covered $19a_1 = 190$ miles, we must find for which value of a_2 the sum of these nineteen terms

$$a_2, \ \tfrac{4}{3}a_2, \ \tfrac{4}{3}\tfrac{6}{5}a_2, \ \left(\tfrac{4}{3}\right)^2\tfrac{6}{5}a_2, \ \left(\tfrac{4}{3}\right)^2\left(\tfrac{6}{5}\right)^2 a_2, \ \left(\tfrac{4}{3}\right)^3\left(\tfrac{6}{5}\right)^2 a_2, \ \ldots, \ \left(\tfrac{4}{3}\right)^9\left(\tfrac{6}{5}\right)^9 a_2$$

will equal that number. Now these terms may be represented as two geometrical progressions, namely

$$a_2, \ \left(\tfrac{4}{3}\tfrac{6}{5}\right)a_2, \ \left(\tfrac{4}{3}\tfrac{6}{5}\right)^2 a_2, \ \ldots, \ \left(\tfrac{4}{3}\tfrac{6}{5}\right)^9 a_2,$$

$$\tfrac{4}{3}a_2, \ \left(\tfrac{4}{3}\tfrac{6}{5}\right)\tfrac{4}{3}a_2, \ \left(\tfrac{4}{3}\tfrac{6}{5}\right)^2\tfrac{4}{3}a_2, \ \ldots, \ \left(\tfrac{4}{3}\tfrac{6}{5}\right)^8\tfrac{4}{3}a_2,$$

with the respective sums

$$a_2\frac{\left(\tfrac{4}{3}\tfrac{6}{5}\right)^{10}-1}{\tfrac{4}{3}\tfrac{6}{5}-1} = a_2\frac{\left(\tfrac{8}{5}\right)^{10}-1}{\tfrac{3}{5}} = a_2\frac{5}{3}\left[\left(\tfrac{8}{5}\right)^{10}-1\right],$$

$$\frac{4}{3}a_2\frac{\left(\tfrac{4}{3}\tfrac{6}{5}\right)^9-1}{\tfrac{4}{3}\tfrac{6}{5}-1} = a_2\frac{4}{3}\frac{\left(\tfrac{8}{5}\right)^9-1}{\tfrac{3}{5}} = a_2\frac{20}{9}\left[\left(\tfrac{8}{5}\right)^9-1\right],$$

the addition of which leads to setting the equality

$$a_2\frac{760\,054\,773\,046\,875}{2\,288\,818\,359\,375} = 190,$$

whence

$$a_2 = \frac{371\,093\,750}{648\,580\,073} = 0.572163354146096\ldots.$$

---

[166] *Practica arithmetice*, Ch. LXVI, No. 7.

Cardan himself obtains the fraction

$$a_2 = \frac{16\,328\,125\,000}{28\,537\,523\,207},$$

which, expressed in the decimal writing (then not in use, see p. 247), is indeed very close to our result ($\underline{0.572163354246344\ldots}$).

§4. Meeting

Since the messengers walk towards one another, it is the sum of their journeys which will equal the initial distance. In all subsequent examples the departures are simultaneous.[167]

Fig. 84

1. Uniform movements

With a_1, a_2 the daily progresses of the two messengers and d the initial distance between them, we shall have

$$t \cdot a_1 + t \cdot a_2 = d, \quad \text{whence} \quad t = \frac{d}{a_1 + a_2}.$$

The first problem below is solved in this way. But we may be given, instead of the daily progresses and the distance d, the durations τ_1 and τ_2 the journeys over the whole distance d would take; indeed, since

$$\tau_1 = \frac{d}{a_1}, \qquad \tau_2 = \frac{d}{a_2},$$

we infer from above that

$$t\frac{d}{\tau_1} + t\frac{d}{\tau_2} = d, \quad \text{or} \quad t = \frac{\tau_1 \cdot \tau_2}{\tau_1 + \tau_2}.$$

The second problem below may be solved in that latter way.

[76] *In one day and at the same time 2 men begin to walk towards one another. One goes from Paris to Lyons and walks each day 7 leagues. The other goes from Lyons to Paris and walks every day 9 leagues. There is from Paris to Lyons 80 leagues. I ask in how much time they will meet.*

Answer. Add the leagues they walk in one day, thus 7 and 9, it gives 16. Form the rule (of three): if 16 comes from one day, from how many will come 80, which they have to walk? Multiply 80 by 1, it is 80, divide it by 16, it gives 5. Therefore

---

[167] Illustration from the MS. Florence BNC Magl. XI 86 (Paolo dell'Abbaco's treatise, ed. p. 90).

they will meet in 5 days. The proof is as follows: the one going from Paris to Lyons walks in 5 days 35 leagues, and the other one 45, which (together) make 80 leagues.[168]

[77] *There are two men one of whom is in Lyons and the other in Paris, who set off at the same time, one from Lyons to go to Paris and the other from Paris to go to Lyons. The one going from Paris to Lyons walks in such a way that he completes his journey in 7 days, and the one who goes from Lyons to Paris makes his way in 8 days. Required in how many days they will meet.*[169]

Answer. The one doing the journey in 7 days covers $14\frac{2}{7}$ leagues a day, and the one doing it in 8 days covers $12\frac{1}{2}$ leagues a day. Thus both travel $26\frac{11}{14}$ leagues a day. Divide now 100 by $26\frac{11}{14}$, you will have $3\frac{11}{15}$, and in that many days will they meet.[170]

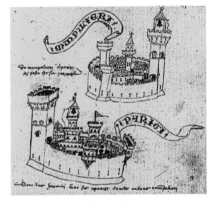

Fig. 85

[78] Similar problem in an Italian manuscript (Fig. 85), with the towns being Montpellier (*Munpalieri*) and Paris (*Parigi*), 900 miles apart, and the one from Paris covering the distance in 35 days and the other in 49 days. The number of days after which they will meet is calculated (see above) as

$$\frac{35 \cdot 49}{35 + 49} = \frac{1715}{84} = 20 + \frac{5}{12},$$

with their meeting taking place 375 miles away from Paris.[171]

---

[168] *Livre de chiffres et de getz*, No. 25.

[169] The distance between the two towns (here superfluous) is given further on ([80]) as 100. In the contemporary text above, it was 80. No importance should be attached to such variations: many authors borrow such problems from others and just adapt the names of towns to their region without bothering about distances.

[170] Chuquet, *Appendice*, p. 448; MS., fol. $195^{v} - 196^{r}$.

[171] MS. Rome Accad. naz. dei Lincei Cors. 1875, fol. $49^{r} - 49^{v}$, with the illustration

The following elementary Byzantine example is just to show that problems without any specifying of towns are found as well:[172]

[79] *Two left two towns, in a direct line one from the other and distant by 105 miles. One walked each day 31 $\frac{1}{2}$ miles, the other 21. Required when they will meet* (after two days).

2. Movements in arithmetical progression

This problem is again by Chuquet and again involves two men meeting midway between Lyons and Paris.[173]

[80] *Two men, one in Lyons and the other in Paris, at a distance of 100 leagues, leave at the same time, one to go from Lyons to Paris and the other to go from Paris to Lyons. The one going to Paris walks the first day 1 league, the second day two leagues, the third day 3 leagues, and so on, while the one going to Lyons walks the first day 2 leagues, the second day 4 leagues, the third day 6 leagues, and so on. Required the number of days after which they will meet and also how many leagues each one will have covered.*

With t the number of days to the meeting, the sum of the two progressions must equal the total distance. We then obtain

$$\frac{t(t+1)}{2} + 2\frac{t(t+1)}{2} = 100, \quad \text{thus} \quad t^2 + t = \frac{200}{3},$$

with the positive solution

$$\sqrt{\frac{1}{4} + \frac{200}{3}} - \frac{1}{2} = \sqrt{\frac{803}{12}} - \frac{1}{2} = \sqrt{66 + \frac{11}{12}} - \frac{1}{2}.$$

During this time, the first will progress by

$$\frac{1}{2}\left(\sqrt{66 + \frac{11}{12}} - \frac{1}{2}\right)\left(\sqrt{66 + \frac{11}{12}} + \frac{1}{2}\right) = \frac{1}{2}\left(66 + \frac{11}{12} - \frac{1}{4}\right) = 33 + \frac{1}{3},$$

while the second will have covered twice that distance.

3. Sequence of the natural numbers and their cubes

Cardan has the following problem involving the meeting of two birds.[174]

[81] *Two birds were on the same tree. One began to fly towards the East, (covering) the first day 1 mile, the second day 2, the third day 3, the fourth day 4, and so on. The other left towards the West, following the same line,[175] (covering) the first day*

(49^r). Same computation in a 14th-century text, with the same two towns, but distances covered in 11 and 9 days (Jacopo da Firenze's treatise, ed. p. 262).

[172] Vogel, *Byz. Rechenbuch*, No. 46.

[173] *Appendice*, p. 443; MS., fol. 188^r. Like many of Chuquet's problems, this one reappears in de la Roche's *Arismethique* (fol. 218).

[174] *Practica arithmetice*, Ch. LXVI, No. 14.

[175] They are to follow a great circle.

1 mile, the second day 8 miles, the third day 27 miles, and so on according to the cubes. The earth's circumference is, according to the belief of many, 44310 miles.[176] *Required in how many days the birds will meet.*

Let t be the time until they meet. The miles covered by the first, which increase daily in arithmetical progression, equal

$$\frac{t(t+1)}{2}.$$

The miles of the second, thus the sum of the cubes of the consecutive natural numbers, equal

$$\left(\frac{t(t+1)}{2}\right)^2.$$

We shall therefore put

$$\left(\frac{t(t+1)}{2}\right)^2 + \frac{t(t+1)}{2} = 44\,310,$$

whence, successively,

$$t^4 + 2t^3 + t^2 + 2t^2 + 2t = 177\,240,$$

$$t^4 + 2t^3 + 3t^2 + 2t + 1 = 177\,241,$$

$$t^2 + t + 1 = 421, \quad t^2 + t = 420,$$

and therefore $t = 20$. One may indeed verify that in 20 full days the first has flown 210 miles and the second, 44 100 (210 times more!), the sum of which makes the earth's circumference as set.

This problem must be taken from Pacioli, who formulates it as follows.[177]

[82] *Let us put that the terrestrial circumference is 20400 miles long.*[178] *On the equinox* (= equinoctial circle), *from the same place and at the same time, two mobile points proceed, one towards the East, the first day one mile, the second 2, the third 3, and so on, the other towards the West, the first day one mile, the second 8, the third 27, and so on. I ask in how many days the two points will be found in the same place.*

The equation here is, successively,

$$t^4 + 2t^3 + t^2 + 2t^2 + 2t = 81\,600,$$

$$t^4 + 2t^3 + 3t^2 + 2t + 1 = 81\,601,$$

$$t^2 + t + 1 = \sqrt{81\,601}, \quad t^2 + t = \sqrt{81\,601} - 1,$$

---

[176] This value has in fact been chosen merely for convenience, as it simplifies the computations. Cardan later uses another value, about half this one (below, p. 130).

[177] *Summa*, fol. 44$^\mathrm{r}$, No. 30.

[178] About this figure, notably lower than the previous one, see below, p. 125.

thus $t = \sqrt{\sqrt{81\,601} - \frac{3}{4}} - \frac{1}{2}$ ($\mathcal{R}$ *81601 m̄ $\frac{3}{4}$, la $\mathcal{R}$ del rimanente m̄ $\frac{1}{2}$, e in tanti giorni si trovaran li ditti ponti in un ponto*).[179]

As a matter of fact, this problem is found even earlier: in a manuscript a few decennaries older, the question concerns two men going to meet one another; it even tells us that its inventor, or transmitter, proposed it to a colleague in 1372.[180]

§ 5. Dog and hare

1. General problem

The problem of the dog catching up with a hare is one of the most common pursuit problems in the Middle Ages. Although of a simple kind, it takes a complex form because of the unusual units of measurement adopted. Indeed, the quantities involved are now:

l_d, l_h length of a jump of the dog and hare, respectively

n_d, n_h number of their jumps in a certain time τ

N_d, N_h number of their jumps until the meeting time t

s initial distance of separation

s_d, s_h initial distance expressed in dog's and hare's jumps (thus $s_d \cdot l_d = s = s_h \cdot l_h$).

The reduction of the initial distance of separation after the time τ is then equal to

$$\delta_\tau = n_d l_d - n_h l_h,$$

whence, since t is the time to the meeting (counted from the beginning of the dog's progress),

$$\frac{t}{\tau} = \frac{s}{\delta_\tau} = \frac{s}{n_d l_d - n_h l_h}. \qquad (1)$$

But since

$$\frac{N_d}{n_d} = \frac{t}{\tau}, \qquad \frac{N_h}{n_h} = \frac{t}{\tau},$$

we find as the number of jumps to the meeting

$$N_d = \frac{n_d \cdot s}{n_d l_d - n_h l_h}, \qquad N_h = \frac{n_h \cdot s}{n_d l_d - n_h l_h}. \qquad (2, 2')$$

Or else, since $s = s_d \cdot l_d = s_h \cdot l_h$,

---

[179] Note the verbal formulation, in the absence of brackets to enclose a mathematical expression (see p. 251).

[180] MS. Florence BNC Palat. 573, fol. 138^r–138^v; Arrighi, *Scritti scelti*, p. 165.

$$N_d = \frac{n_d \cdot s_d \cdot l_d}{n_d l_d - n_h l_h} = \frac{s_d}{1 - \frac{n_h}{n_d} \frac{l_h}{l_d}}, \qquad (3)$$

$$N_h = \frac{n_h \cdot s_h \cdot l_h}{n_d l_d - n_h l_h} = \frac{s_h}{\frac{n_d}{n_h} \frac{l_d}{l_h} - 1}. \qquad (3')$$

In such problems the given quantities are the initial distance, expressed as s, s_d, or s_h, then the length of the jumps, or their ratio, finally their number in a given time (our τ), or their ratio.

As said, the complexity of these problems is formal. Indeed, formula (1) just corresponds to the simplest case we have seen for pursuit (p. 77): τ was then one day while $n_d l_d$ and $n_h l_h$, which are the distances covered during τ, were the respective daily progresses. As to the pair of formulae (2, 2′), they give the number of jumps of each animal during the whole time t, and correspond thus, in jumps, to the distance covered by each animal, that of the dog being greater than that of the hare by the initial distance of separation. Is this last distance expressed in jumps, either dog's or hare's, we shall use the pair of formulae (3, 3′); remember too that s_d may easily be converted to s_h since $s_d : s_h = l_h : l_d$. Note finally that formulae (2, 2′) and (3, 3′) are not found as such in mediaeval and Renaissance texts, though they are *de facto* applied after their individual components have been computed.

2. Jumps of unequal length but equal in frequency

Since the jumps occur simultaneously, and thus the same number of jumps is made by each animal during the pursuit time t ($N_d = N_h = N$), it is their difference in length $l_d - l_h$ which will cancel the initial separating distance s:

$$N = \frac{s}{l_d - l_h};$$

this is a particular case of formulae (2), with $n_d = n_h$. With the initial distance expressed in jumps, we have, since $s = s_d l_d = s_h l_h$,

$$N = \frac{s_d}{1 - \frac{l_h}{l_d}}, \qquad N = \frac{s_h}{\frac{l_d}{l_h} - 1},$$

which is a particular case of formulae (3).

(*i*) *Distance s given.*

Alcuin's Collection contains an example of such a problem, with the initial distance expressed in feet:[181]

[83] *There is a field having a length of 150 feet, with a dog at one end and a hare at the other. The dog began to run after the hare. Now while the dog made in one*

---
[181] *Propositiones*, No. 26.

jump 9 feet, the hare covered 7. May whoever wishes it say how many feet or how many jumps the pursuing dog made or the hare running away until it was caught.

Here, $s = 150$, $l_d = 9$, $l_h = 7$, thus $N = 75$ jumps, corresponding to $9 \cdot 75 = 675$ feet for the dog and $7 \cdot 75 = 525$ for the hare, which indeed display a difference of 150. This problem is still found at the end of the Middle Ages.[182]

(*ii*) Distance s_h given.

[84] *Required, for a fox preceding a dog by 50 jumps, with 9 jumps of the fox running away being 6 jumps of the pursuing dog, after how many it will be caught.*[183]

Since $s_h = 50$, $9l_h = 6l_d$, thus $l_d : l_h = 9 : 6$, then

$$N = \frac{s_h}{\frac{l_d}{l_h} - 1} = \frac{50}{\frac{1}{2}} = 100.$$

A similar problem, but with a dog chasing a roe-deer (*capriolo*), with $s_h = 50$, $l_d : l_h = 7 : 5$, thus $N = 125$, is found in Tagliente's *Componimento*, with the illustrations of Fig. 86 (ed. 1547) & 87 (ed. 1554).

Fig. 86   Fig. 87

[85] *A dog has been unleashed to catch a hare which was ahead by 40 jumps, said dog making $\frac{1}{12}$ $\frac{1}{22}$ $\frac{1}{33}$ $\frac{1}{44}$ of a jump more than the hare. In how many jumps will the dog reach the hare?*[184]

Since here $s_h = 40$ and the hare's lead reduces, with each of the (simultaneous) jumps, by $\frac{1}{12} + \frac{1}{22} + \frac{1}{33} + \frac{1}{44} = \frac{24}{132} = \frac{2}{11}$ of one hare's jump, the dog will catch it after

$$N = \frac{40}{\frac{2}{11}} = 220 \text{ jumps.}$$

---

[182] MS. Tours 399, fol. $131^v - 132^r$. See (full text) *Récréations*, pp. 106–107. A similar problem, with $s = 355$ pace (*passus*), $l_d = 11$, $l_h = 9$, thus $N = 177 + \frac{1}{2}$, and (as also required) the distance covered by the dog $N \cdot l_d = 1952 + \frac{1}{2}$, is found in the Latin MS. Lyons 59 (127), fol. 64^r (see *Récréations*, p. 107).

[183] Fibonacci, *Liber abaci*, pp. 179–180.

[184] Vogel, *Byz. Rechenbuch*, No. 88. Fractions placed side by side are to be added.

Fig. 88

(*iii*) *Distance* s_h *or* s_d *given.*

Pacioli has a problem of the same kind.[185] But he deserves credit for pointing out the ambiguity inherent in the formulation.

[86] *A hare has a lead of 60 jumps over a dog, and for each 5 jumps of the dog the hare makes 7 of them. In the end, the dog catches it. I ask in how many jumps the dog will catch up with the hare.*

You are to know that there are in this problem many awkward and debatable aspects. First of all, it must be clarified whether the distance of 60 jumps between the hare and the dog is counted in dog's or hare's jumps. Next, it must be clarified whether the hare indeed makes these 7 jumps whilst the dog makes 5.[186] *Further, whether they begin to walk, or to run, simultaneously. For it is according to the hypotheses that the problem must be solved and the answer given.*

Here, s_h or s_d is 60, and $l_d : l_h = 7 : 5$. Thus, according to whether the initial distance is expressed in jumps of the hare or the dog,

$$N = \frac{s_h}{\frac{l_d}{l_h} - 1} = \frac{300}{2} = 150, \quad N = \frac{s_d}{1 - \frac{l_h}{l_d}} = \frac{420}{2} = 210 \ \text{jumps.}$$

About this type of problem, Pacioli adds that it would become somewhat more intricate if the groups of jumps also displayed a difference in time.[187]

Fig. 89

---

[185] *Summa*, fol. $42^v - 43^r$, No. 27. The picture of Fig. 88 is from F. Calandri's manuscript (MS. Florence Bibl. Riccardiana 2669, fol. 92^v; ed. p. 185).

[186] Considering here the possibility of some small difference in length (a possible difference in time will be alluded to at the end of the problem).

[187] Illustration of Fig. 89 from Muscarello's *Algorismus*, MS. Pennsylvania LJS 27, fol. 75^r, with the data $s_h = 70$, $l_d : l_h = 7 : 5$, thus $N = 175$ (ed., II, p. 189).

3. Jumps equal in length and unequal in frequency

Since $l_d = l_h$, we must necessarily have $n_d > n_h$ (in τ) and $N_d > N_h$ (in t). Furthermore, since $s_d l_d = s = s_h l_h$, $s_d = s_h$. Formulae (3) and (3') then become

$$N_d = \frac{s_d}{1 - \frac{n_h}{n_d}}, \qquad N_h = \frac{s_h}{\frac{n_d}{n_h} - 1}.$$

[87] *A hare precedes a dog by 90 jumps, and each time the hare makes 12 jumps, the dog makes 15, and the hare invariably jumps as far as the dog. The question is: how many times must the dog make 15 jumps so as to reach the hare?*[188]

Since $s_d = 90$, $n_h : n_d = 12 : 15$, we have

$$N_d = \frac{s_d}{1 - \frac{n_h}{n_d}} = \frac{90 \cdot 15}{3} = 30 \cdot 15 = 450.$$

The dog must then make 30 times 15 jumps. Rudolff finds the answer algebraically: putting (in accordance with the question) the multiple of 15 jumps equal to x (*1r.*, that is, 1 *res*, in the text), the animals will make $15x$ and $12x$ jumps, respectively, and will meet when $15x = 12x + 90$ (*15 r. gleych 12 r. + 90*), whence $x = 30$. The proof follows ($12x = 360$, indeed equal to $450 - 90$).

Fig. 90

4. Jumps unequal in length and unequal in frequency

Rudolff's next problem is about the general case.

[88] *A hunter pursues a fox. The fox is 60 jumps ahead. Every time the fox makes 9 jumps, the dog makes 6 jumps, but 3 jumps of the dog make 7 jumps of the fox. The question is: how many jumps will the dog make until it catches the fox?*[189]

Since $s_h = 60$, $n_d : n_h = 6 : 9$, $l_d : l_h = 7 : 3$, then, by (3'),

$$N_h = \frac{s_h}{\frac{n_d}{n_h} \frac{l_d}{l_h} - 1} = \frac{60}{\frac{6}{9} \frac{7}{3} - 1} = \frac{60}{\frac{14}{9} - 1} = \frac{60 \cdot 9}{5} = 108,$$

and thus $N_d = \frac{6}{9} \cdot 108 = 72$.

---

[188] Rudolff, *Die Coß*, fol. 305^v – 306^r, No. 186.

[189] Rudolff, *Die Coß*, fol. 306^r – 307^r, No. 187. The picture is from MS. Florence BNC Magl. XI 86 (Paolo dell'Abbaco's treatise, ed. p. 79).

Rudolff's solution proceeds as follows. Since 6 jumps of the dog cover the same distance as 14 of the fox while, during the time the dog makes 6 jumps, the fox makes 9, the difference, $14 - 9 = 5$ jumps of the fox, will be the reduction of the initial distance 60, which are jumps of the fox, whenever the dog makes 6 jumps. Since $\frac{60}{5} = 12$, after making $12 \cdot 6$ jumps the dog will have progressed as much as $12 \cdot 9 + 60$ jumps of the fox, that is, as much as the jumps of the initial distance plus those made during the chase.

In his reedition of Rudolff's *Coß*, Stifel remarks that, in comparison with really useful problems, such 'ridiculous' ones (*solliche spötliche Exempla*) require a disproportionate amount of explaining; he felt obliged, though, to keep them in his reedition, along with all the others.

Cardan has a similar problem, the idea of which, he says, was originally Pacioli's (our [86] above).[190]

[89] *A dog was pursuing a hare. The hare preceded the dog by 60 dog-jumps, and for each 3 dog-jumps the hare made 5 jumps, but with a delay of $\frac{1}{20}$ of the time to complete them —so that if, say, the dog makes 3 jumps in 20 moments (*momentum*, a short period of time), the hare will make 5 of his in 21 moments. And 3 jumps of the dog are exceeded by 7 jumps of the hare by $\frac{1}{20}$ of a dog-jump. Required when the dog will catch the hare, that is, at what distance.*

Friar Luca has a similar problem, much easier but, nevertheless, altogether faulty, leaving the reader quite bewildered by that man.

Using our formula (3), with $s_{\mathrm{d}} = 60$, $n_{\mathrm{d}} : n_{\mathrm{h}} = 3\left(1 + \frac{1}{20}\right) : 5 = 63 : 100$, $l_{\mathrm{h}} : l_{\mathrm{d}} = \left(3 + \frac{1}{20}\right) : 7 = 61 : 140$, the number of jumps made by the dog altogether is

$$N_{\mathrm{d}} = \frac{s_{\mathrm{d}}}{1 - \frac{n_{\mathrm{h}}}{n_{\mathrm{d}}}\frac{l_{\mathrm{h}}}{l_{\mathrm{d}}}} = \frac{60}{1 - \frac{100}{63}\frac{61}{140}} = \frac{60}{1 - \frac{305}{441}} = \frac{6615}{34} = 194 + \frac{19}{34}.$$

As to the number N_{h} of jumps made by the hare during the time of pursuit, it can be calculated either from

$$N_{\mathrm{h}} = \frac{s_{\mathrm{h}}}{\frac{n_{\mathrm{d}}}{n_{\mathrm{h}}}\frac{l_{\mathrm{d}}}{l_{\mathrm{h}}} - 1} = \frac{60 \cdot \frac{140}{61}}{\frac{63}{100}\frac{140}{61} - 1} = \frac{\frac{8400}{61}}{\frac{441}{305} - 1} = \frac{5250}{17} = 308 + \frac{14}{17},$$

since $s_{\mathrm{h}} = s_{\mathrm{d}} \cdot \frac{140}{61}$, or simply from $N_{\mathrm{h}} = N_{\mathrm{d}} \cdot \frac{100}{63}$.

Cardan's solving, expressed in our writing, proceeds as follows.

(a) First, express the initial distance in hare-jumps ($s_{\mathrm{h}} \cdot l_{\mathrm{h}} = s_{\mathrm{d}} \cdot l_{\mathrm{d}}$):

$$s_{\mathrm{h}} = s_{\mathrm{d}} \cdot \frac{7}{3 + \frac{1}{20}} = \frac{60 \cdot 140}{61} = 137 + \frac{43}{61}.$$

---

[190] *Practica arithmetice*, Ch. LXVI, No. 11. Cardan's criticism is unjustified here.

(b) Second, find the number $\nu_{\text{h}}^{(t)}$ of hare-jumps made during the same time as (say) 3 dog-jumps $(\nu_{\text{h}}^{(t)} : 3_{\text{d}}^{(t)} = n_{\text{h}} : n_{\text{d}})$:

$$\nu_{\text{h}}^{(t)} = \frac{3 \cdot 100}{63} = 4 + \frac{48}{63} = 4 + \frac{16}{21}.$$

(c) Third, find the number $\nu_{\text{h}}^{(s)}$ of hare-jumps covering the same distance as 3 dog-jumps $(\nu_{\text{h}}^{(s)} : 3_{\text{d}}^{(s)} = l_{\text{d}} : l_{\text{h}})$:

$$\nu_{\text{h}}^{(s)} = \frac{3 \cdot 140}{61} = \frac{420}{61} = 6 + \frac{54}{61}.$$

Thus, during 3 dog-jumps, that is, over the same distance as $6 + \frac{54}{61}$ hare-jumps, the hare makes $4 + \frac{16}{21}$ jumps. This means that, during 3 dog-jumps the distance separating them decreases, in hare-jumps, by

$$\nu_{\text{h}}^{(s)} - \nu_{\text{h}}^{(t)} = 6 + \frac{54}{61} - \left(4 + \frac{16}{21}\right) = 2 + \frac{54 \cdot 21 - 16 \cdot 61}{61 \cdot 21} = 2 + \frac{158}{1281}.$$

Consequently, there will be, until they meet, as many times 3 dog-jumps as is

$$\frac{s_{\text{h}}}{\nu_{\text{h}}^{(s)} - \nu_{\text{h}}^{(t)}} = \frac{137 + \frac{43}{61}}{2 + \frac{158}{1281}} = 64 + \frac{29}{34},$$

and these groups of three jumps equal altogether, in hare-jumps,

$$\left(64 + \frac{29}{34}\right)\left(4 + \frac{16}{21}\right) = 308 + \frac{14}{17}.$$

The reasoning of Cardan is thus the same as in Rudolff's *Coß*, the latter's data just being simpler: both consider an integral number α of dog-jumps, calculate the number of hare-jumps made in the same time and the number of hare-jumps covering the same distance; dividing s_{h} by their difference will then give after how many groups of α dog-jumps the dog will catch the hare.

§6. Moving back and forth

Such movement problems were highly regarded in the Middle Ages. The main difference between them and those of simple progress is that the backward movement must stop once the goal has been reached; curiously, however, many authors, including Fibonacci, failed to realize that. Simple cases, though, enable us to grasp the situation. Thus, Pacioli presents the problem of a person who has to cover n miles by walking m miles a day, but goes back $m - 1$ miles during the night (which already shows how absurd such problems were). It is clear that after $n - m$ days he will have covered that many miles, and that a single day will suffice for him to reach the goal.[191] A similar simple case is proposed a little earlier by

---

[191] *Summa*, fol. 42ᵛ, No. 24.

F. Calandri in his *Opusculum*: a man must walk from Florence to Pistoia, 20 miles away, and goes forward each day 4 miles whilst returning 3 miles during the night; the answer, 17 days, is correct.

1. Single subject

[90] *A cat is at the bottom of a tree 300 feet high, and climbs each day 17 feet and goes down nightly 12 feet. I ask in how much time it will be at the top.*

Answer. Remove, by subtraction, the night from the day, that is, 12 from 17, which leaves 5. Thus, it climbs 5 feet a day. Divide now 300 by 5, it gives 60. Therefore, it will be at the top in 60 days.[192]

As a matter of fact, it will have covered in 57 days, taking into account the back-and-forth movement, 285 feet; it will thus reach the top on the 58th day, during daylight.[193]

Fig. 91

[91] *A lion is in a well 50 palms deep. It climbs daily $\frac{1}{7}$ of a palm and goes down by $\frac{1}{9}$. Required in how many days it will be out of the well.*[194]

Fibonacci's computation corresponds to

$$\frac{50}{\frac{1}{7} - \frac{1}{9}} = \frac{50}{\frac{2}{63}} = 25 \cdot 63 = 1575.$$

Here again, it will be out already on the 1572nd day (!). Same problem and same answer in F. Calandri's *Opusculum*, but with a snake and *braccia* instead of *palmi*.

Problems involving a snake climbing up a tower or coming out of a well are indeed quite common. Examples of the first case are: height of the tower 60 *cannes* and fractions $\frac{3}{4}$ and $\frac{1}{3}$ (answer given: 144 full days; correct: 144th day, during daylight);[195] height 30 cubits and fractions $\frac{1}{3}$

---

[192] *Livre de chiffres et de getz*, No. 26. Note the impressive height of the tree.

[193] In Muscarello's *Algorismus*, MS. Pennsylvania LJS 27, fol. 73^v (Fig. 91), a tower is 30 *canne* high and a cat climbs each day $\frac{1}{2}$ *canna* and goes down $\frac{1}{3}$; the answer is calculated as 180 days (ed., II, pp. 187–188). A *canna* is a measurement of length about 1.80 m.

[194] Fibonacci, *Liber abaci*, p. 177. A palm is a fourth of a foot, thus about 7.5 cm (lazy lion).

[195] MS. Paris BNF fr. 2050, fol. 86^r (text: see *Récréations*, p. 114).

and $\frac{1}{4}$ (in 360 days, instead of 356, end of daylight);[196] depth of the well 100 teises and fractions $\frac{1}{15}$ and $\frac{1}{16}$ (in 24 000 days instead of 23 985, end of daylight).[197]

[92] *A tower is 10 cubits high and a pigeon flies down it each day $\frac{2}{3}$ of a cubit then flies up $\frac{1}{4}$ $\frac{1}{3}$ of a cubit. I ask in how many days the pigeon will reach the ground.*[198] Answer: 120 (instead of 113, at the end of the daily progress).

In all these problems, the distance is divided by the daily mean motion; such erroneous computing was common. Some authors, however, had doubts about the result obtained. Thus, an Italian text of the 14th century has the same problem as above, but criticizes the result given by *molti maiestri*. Its author himself computes

$$\frac{10 - \frac{2}{3}}{\frac{2}{3} - (\frac{1}{3} + \frac{1}{4})} = \frac{9 + \frac{1}{3}}{\frac{1}{12}} = 112;$$

that is, he begins by reducing the distance to be covered by the daily forward movement. After 112 days, the pigeon has still $\frac{2}{3}$ to cover, which it will do on the 113rd day.[199]

Towards the end of the 15th century, we find a similar treatment and, again, criticism of the erroneous one, in Chuquet's *Triparty*:[200]

[93] *There is a man who wants to go from Lyons to Paris, distant by 100 leagues. He walks in such manner that he advances 15 leagues in the day and returns 10 leagues in the night. Required the time in which he will have completed his journey, that is, in how many days he will be able to reach Paris without going any further.*

Answer and rule for such problems. Subtract the 15 leagues he advances every day from 100 leagues, it remains 85. You are to divide it by 5, which is how much further he goes than returns; the quotient is 17. Then add to it the 1 day you have subtracted from 100, it makes 18. He will arrive in Paris in 18 days, that is, 17 natural days and 1 artificial day.[201] *The reason is that in 17 days and as many nights he has walked 5 times 17 leagues, which is 85. The day after, he walks 15; with the 85, this makes 100. Therefore he completes his journey in 18 days.*

Some want to proceed as follows. They subtract 10 from 15, which leaves 5. Then they divide 100 by 5, which gives 20. They say that he has made the journey

---

[196] MS. Milan Trivulziana 90, fol. $37^v - 38^r$.

[197] MS. Rouen I 58, fol. 86^v, 89^v (with Roman numerals: *le XVe*, *le XVIe*, and (answer) XXIIII^M days, see *Récréations*, p. 114). On the 'teis', see above, p. 73*n*.

[198] *Algorismus Ratisbonensis*, No. 65.

[199] *Rascioni d'algorismo*, No. 67.

[200] *Appendice*, p. 447; MS., fol. $195^r - 195^v$.

[201] Artificial day: daylight (taken to be half a full day).

in 20 days. But doing that, the walker has gone beyond Paris and returned to it; he has already arrived there, though, on the 18th day.

Chuquet has thus explained the correct reasoning; like his predecessor, he has considered the distance reduced by the forward movement and so needs only to add one day to the result. An earlier Italian text, attributed to Paolo dell'Abbaco, apparently proceeds by trial and error in order to determine the (correct) integral number of days:[202]

[94] *There is a well 30 cubits deep. A snake went into it, and wants to come out. It climbs each day $\frac{2}{3}$ cubit and goes down in the night $\frac{1}{5}$ cubit.*

Computation of the duration of its way out is first expressed as

$$\frac{30}{\frac{2}{3} - \frac{1}{5}} = \frac{30}{\frac{7}{15}} = 64 + \frac{2}{7},$$

and concluded by *it is done*. But the author adds: *Note, however, that one may raise objections to this way of solving by saying the following: By how much does this snake climb in 63 days? You will answer that it climbs 29 $\frac{2}{5}$ cubits. Then you will say: $\frac{3}{5}$ cubit is missing; in how much time will it climb $\frac{3}{5}$ cubit? (...)*[203] *Thus, in $\frac{9}{10}$ of a day he will climb $\frac{3}{5}$ cubit. You may thus say that this snake will climb 30 cubits in 63 days and $\frac{9}{10}$ of (the daylight part of) a day. This is true because on the last day it does not need to descend $\frac{1}{5}$ cubit since it does not reach the night and it has finished climbing during said time, so that its descent does not take place. (...) Note this solving and remark the astuteness required to solve some problems* (e questa ragione osserva, e vedi la malizia che alquante ragioni comettono).[204]

Fig. 92

In the next century, the German Adam Ries(e) reproduces a correct solving which, he says, has been transmitted on to him.

---

[202] Edition, No. 191.

[203] Since $\frac{3}{5} < \frac{2}{3}$, it is clear that it will be out during daylight on the 64th day. The fraction of daylight is found as $\frac{9}{10}$ by a rule of three ($\frac{3}{5} : x = \frac{2}{3} : 1$).

[204] Illustration from the MS. Rome Accad. naz. dei Lincei Cors. 1875, fol. 51^{v}; snake climbing on a tree 40 high, $\frac{3}{4}$ up and $\frac{2}{3}$ down, with the result 480 days.

[95] *A snail is in a well 32 cubits deep. It climbs by crawling every day 4 $\frac{2}{3}$ cubits but falls back every night 3 $\frac{3}{4}$ cubits. In how many days will it be out?*[205]

The text divides the distance reduced, this time, by the (nightly) return:

$$\frac{32 - (3 + \frac{3}{4})}{(4 + \frac{2}{3}) - (3 + \frac{3}{4})} = \frac{32 - \frac{15}{4}}{\frac{14}{3} - \frac{15}{4}} = \frac{\frac{384}{12} - \frac{45}{12}}{\frac{56}{12} - \frac{45}{12}} = \frac{384 - \mathbf{45}}{\mathbf{56} - 45} = \frac{339}{11} = 30 + \frac{9}{11}.$$

The answer is then given as

$$30 + \frac{45 + 9}{56} = 30 + \frac{27}{28}.$$

The fraction in this last computation, which is formed by the numbers in bold above, represents the part of daylight on the 31st day. Ries adds that the doubtful reader may verify the result in two ways. One is by computing numerically. For the other, he will draw a segment of straight line, divide it into 32 parts, each part being then divided into 12; then he will take two compasses, with the distance between the open legs being, respectively, $4 + \frac{8}{12}$ and $3 + \frac{9}{12}$, and will then reproduce the movement accordingly.

The above formula is noteworthy, for it leads to an exact result. But, as said, it does not originate with Ries, for he tells us after obtaining the result: *Such is the time taken by the snail to come out; this has been done correctly, and was first discovered by Hans Conrad, (coinage) inspector in Eisleben.*[206] According to another report of the same Ries, the Nuremberg mathematician (*Rechenmeister*) N. Kolberger was first, before Conrad, to reach an understanding of this problem, despite 'slight errors'.[207]

As a matter of fact, this exact formula already occurs in the mid-12th century, namely in the following problem:[208]

[96] *A ship moves from one place to another at a distance of three hundred miles. It sails daily twenty miles but is driven back five miles daily by the wind. In how many days will it reach this place?*

You will do the following. Add five to twenty; this makes twenty-five. Next, subtract five from three hundred; this leaves two hundred and ninety-five. Divide it by the difference between five, by which it goes back, and twenty, which it sails,

---

[205] *Rechenbuch auff Linien uund Ziphren*, last problem (ed. 1581, fol. 72$^\text{v}$ – 73$^\text{r}$, 'Schnecken gang').

[206] Ries mentions Conrad several times for his help in solving algebraical problems around the year 1515. See Berlet, *Coß von Adam Riese*, pp. 29–30, 56.

[207] 'hat etwas geirrett'; see Berlet, *Coß von Adam Riese*, p. 61 (he used the data $4 + \frac{1}{4}$ and $3 + \frac{1}{3}$).

[208] *Liber mahameleth*, B.370.

that is, by fifteen; this gives nineteen and a remainder of ten. Add this ten to the five miles, thus making fifteen. Denominate it from twenty-five; this gives three fifths of a day. Adding these three fifths of a day to nineteen will make nineteen and three fifths of a day, and in so many days will it reach the intended place.

Let indeed $d = 300$ be the distance to the destination, $a = 20$ the daily progress, $b = 5$ the daily return, and let t be the required number of days. The solution is then computed as follows, in two steps:

$$\frac{d-b}{a-b} = q + \frac{r}{a-b}, \quad \text{thus} \quad \frac{295}{15} = 19 + \frac{10}{15}$$

giving $q = 19$ as the integral quotient and $r = 10$ as the remainder. We form then

$$t = q + \frac{b+r}{a+b}, \quad \text{that is} \quad t = 19 + \frac{5+10}{20+5} = 19 + \frac{3}{5},$$

which is the (correct) answer.

Let us now consider this problem generally. We shall of course suppose that the daily return is less than the daily progress, that the progress precedes the return and that there is no more returning once the destination is reached. We shall further suppose that the two movements take place at the same speed. In this case, the full day will be divided into two unequal parts, having the same ratio as the two movements, namely

$$\tau_1 = \frac{a}{a+b} \quad \text{for the forward movement}$$

$$\tau_2 = \frac{b}{a+b} \quad \text{for the backward movement.}^{209}$$

Since the daily forward movement is a, it is certain that the distance $d - a$ will be covered by the mean daily progress only, namely $a - b$. So let us form

$$\frac{d-a}{a-b} = q + \frac{r}{a-b},$$

with q integral and $r < a - b$ the remainder.

First case: $r = 0$. There remains to be covered the distance a which we had initially subtracted from d. Since this corresponds to the daily forward movement, the moving object will reach its destination at the end of said movement on the day $q + 1$ ('daylight' part). Therefore, the complete movement will last, in days and fraction of a (full) day,

$$t = q + \frac{a}{a+b}.$$

---

[209] We are therefore not to suppose that the first part is always daylight and the second, night.

Second case: $r \neq 0$. There remains to be covered, after q days, the distance $a + r > a$. Since this cannot be done during the day $q + 1$, there will remain for the day $q + 2$ the distance $b + r$ (less than a since $r < a - b$). Therefore the duration of the full movement will be

$$t = q + 1 + \frac{b + r}{a + b}$$

and thus end during the forward movement ('during the daylight part') of the day $q + 2$.

Summarizing, in the case of constant speed, the answer will be computed directly in the following way.

— If $r = 0$, we may write

$$t = \frac{d - a}{a - b} + \frac{a}{a + b} = \frac{d - b}{a - b} + \frac{b - a}{a - b} + \frac{a}{a + b} = \frac{d - b}{a - b} - \frac{b}{a + b},$$

so that the number of days will be expressed by either one of the two equivalent expressions

$$t = \frac{d - a}{a - b} + \frac{a}{a + b} \quad \text{and} \quad t = \frac{d - b}{a - b} - \frac{b}{a + b}, \qquad (*)$$

the first expression giving the integral number of days q increased by the whole 'daylight part' of the day $q + 1$, and the second the integral number of days $q + 1$ reduced by the whole 'nightly part' of this last day.

Remark. We may also simply compute $\frac{d-b}{a-b}$, bearing in mind that the last day is its daylight part.

— But in the case $r \neq 0$, the duration of the journey will be, with the square brackets indicating that the integral part must be taken,

$$t = \left[\frac{d - a}{a - b}\right] + 1 + \frac{b + r}{a + b} = \left[\frac{d - b}{a - b}\right] + \frac{b + r}{a + b},$$

which will be written as

$$t = \left[\frac{d - b}{a - b}\right] + \frac{b + r}{a + b}, \qquad t = \left[\frac{d - b}{a - b}\right] + \frac{b + r}{a}, \qquad (**)$$

according to whether we consider the whole last day or only its daylight part. As to the formula for $r \neq 0$, its fractional part can be obtained by a simple rule of three: if in 1 day the forward movement is a, how much time will be needed to progress by $b + r$?

Remark. The *Liber mahameleth*, though earlier than the *Liber abaci* and all the other (erroneous) examples we have seen, gives the correct formula for the case $r \neq 0$. However, oddly enough, it uses this same formula for the case $r = 0$, and has thus

$$t = \frac{d - b}{a - b} + \frac{b}{a + b}$$

instead of $(*)$, whereby its results are in excess by twice the fraction which should be subtracted.[210]

2. Meeting

[97] *There is here a tower 50 cannes high.*[211] *At the bottom of the tower there is a cat which climbs by day $\frac{3}{4}$ canne and goes down in the night $\frac{1}{2}$ canne. Next, there is at the top of the tower a rat which goes down $\frac{2}{3}$ canne during the day and up $\frac{1}{4}$ canne during the night. I ask in how many days they will meet and what distance each will have covered.*[212]

Fig. 93

The answer to the first question is calculated as expected, namely

$$\frac{50}{\left(\frac{3}{4} - \frac{1}{2}\right) + \left(\frac{2}{3} - \frac{1}{4}\right)} = 75,$$

whereas the meeting will take place during daylight of the previous day.[213]

Sometimes the data give fractions of the whole distance:

[98] *Consider that there is on a high tower a snail which, from sunrise to sunset, goes down the tower by $\frac{1}{3}$ of the whole height of this tower, and by night, from sunset to sunrise, goes back up by $\frac{1}{5}$ of the whole height of this tower. Next, there is at the bottom of this tower another snail which every day climbs up this tower*

---

[210] *Liber mahameleth*, B.371–B.374.

[211] About the *canna*, or *canne*, see p. 96n.

[212] MS. Paris BNF fr. 2050, fol. 86^v; original French text in *Récréations*, p. 121. Illustration (compare with Fig. 91, p. 96) from Muscarello's *Algorismus*, MS. Pennsylvania LJS 27, fol. 74^v (data: $\frac{1}{3}$ & $\frac{1}{4}$ for the cat, $\frac{1}{2}$ & $\frac{1}{3}$ for the rat, whence 240 days for a tower 60 cannes high; ed., II, pp. 188–189).

[213] A Byzantine text (first half of the 15th c.) has a similar problem: two persons, walking $\frac{1}{4}$, $-\frac{1}{5}$ and $\frac{1}{3}$, $-\frac{1}{4}$ respectively, are 40 miles apart. According to the text, the meeting will take place after 300 days (instead of 297). See Deschauer's ed., No. 70.

by $\frac{1}{4}$ of its whole height and goes back down each night by $\frac{1}{6}$ of the whole height of this tower.[214]

Calculating as above, the text finds that the meeting will take place after $4 + \frac{8}{13}$ days. That will in fact take place on the third day, a little before sunset.

The next three examples treat, but each time differently, one and the same problem. First, by Fibonacci.

[99] *A snake is at the bottom of a tower 100 palms high and climbs daily $\frac{1}{3}$ palm and goes down $\frac{1}{4}$. There is at the top of the tower another snake which goes down daily by $\frac{1}{5}$ and goes up by $\frac{1}{6}$. Required in how much time they will meet.*[215]

The computation uses the method of false position, which in fact corresponds to calculating

$$\frac{100}{(\frac{1}{3} - \frac{1}{4}) + (\frac{1}{5} - \frac{1}{6})} = \frac{100}{\frac{1}{12} + \frac{1}{30}} = \frac{100}{\frac{7}{60}} = 857 + \frac{1}{7},$$

whereas the meeting will take place during daylight of the 854th day. Fibonacci proves his result by calculating the point of meeting; since the sum of the distances covered by the *mean* movement corresponds to

$$\left(857 + \frac{1}{7}\right)\frac{1}{12} + \left(857 + \frac{1}{7}\right)\frac{1}{30} = 71 + \frac{3}{7} + 28 + \frac{4}{7} = 100,$$

we indeed find the height of the tower. This may explain the long life of these erroneous reasonings: using the effective daily progress will of course confirm the calculated result.

Remark. In order to determine the integral part we may, as before, consider as a forward movement a the sum of the daily progresses and as a backward movement b the sum of the nightly returns. In the present case, this would give

$$\left[\frac{d - b}{a - b}\right] = \left[\frac{100 - (\frac{1}{4} + \frac{1}{6})}{(\frac{1}{3} + \frac{1}{5}) - (\frac{1}{4} + \frac{1}{6})}\right] = 853.$$

Fig. 94

---

[214] MS. Nantes 456, fol. 68^{r}; original text in *Récréations*, p. 121.

[215] Fibonacci, *Liber abaci*, pp. 177–178. Same problem in F. Calandri's *Opusculum*, from which the illustration (depicting just one 'snake') is taken.

[100] Pacioli has just the same problem with ants.[216] He subtracts from the distance the mean movements and then computes

$$\frac{100 - \left[\left(\frac{1}{3} - \frac{1}{4}\right) + \left(\frac{1}{5} - \frac{1}{6}\right)\right]}{\left(\frac{1}{3} - \frac{1}{4}\right) + \left(\frac{1}{5} - \frac{1}{6}\right)} = \frac{99 + \frac{53}{60}}{\frac{7}{60}} = 761 + \frac{6}{7},$$

to which he adds one day because of the term subtracted. His computation is doubly erroneous: the result of his division is wrong and, what is worse, subtracting the divisor from the dividend makes no sense since a unit is added at the end.

We have already seen Tartaglia attacking Pacioli (p. 73). This was a perfect opportunity to do so again.[217]

[100′] *Friar Luca proposes the following problem in these terms. Two ants are on a flat terrain 100 cubits long, one at one end and the other at the other. The first goes forward daily $\frac{1}{3}$ cubit and back nightly $\frac{1}{4}$ cubit; the other goes forward daily $\frac{1}{5}$ cubit and back nightly $\frac{1}{6}$ cubit. He asks in how many days they will meet.*

Said Friar Luca makes in his solving several computation errors, so that at the end —according to his rule— he concludes that these two ants will meet in $762\frac{6}{7}$ days, although —according to his own rule— they should meet in $857\frac{1}{8}$ days. But both these conclusions are erroneous; the first is erroneous because of mistakes in the computation and the second is erroneous because of the rule given by him for solving such (problems), which is far from being the correct way.[218]

Fig. 95

Let us now see how Tartaglia solves this problem. He continues as follows: *Now, in order to solve correctly this problem and other similar ones, add first the $\frac{1}{3}$ cubit which one goes forward daily and the $\frac{1}{5}$ cubit of the other; this makes $\frac{8}{15}$, and this is what they will do daily between them. This being done, add*

---

[216] *Summa*, fol. 42^{r}, No. 22.

[217] *General trattato*, II, fol. 10^{r} – 10^{v}. Illustration from F. Calandri's *Opusculum* (same problem as Pacioli).

[218] Pacioli's computational error was to take 88 instead of 99 in the dividend. With the latter number, and the addition of one day, he would have found, as in [99], $857 + \frac{1}{7}$. (Tartaglia's $857 + \frac{1}{8}$ must be a typographical error —hopefully, for otherwise it would hardly be for him to criticize Pacioli.) But the error in Pacioli's reasoning, which leads him to the meaningless subtraction in the dividend, is more serious; it would seem that he misunderstood a method, maybe correct, encountered elsewhere.

also the $\frac{1}{4}$ of the nightly regression of one with the $\frac{1}{6}$ of the regression of the other; this makes $\frac{5}{12}$, which is the distance they will go back between them in the night. This being done, subtract the $\frac{5}{12}$ of the nightly regression from the $\frac{8}{15}$ they make during the day. This leaves $\frac{21}{180}$, which, reduced, will be $\frac{7}{60}$, and such will be their progress during day and night. But since the last day, when they will meet, will not be followed by a nightly return, subtract from the 100 cubits the $\frac{5}{12}$ of the regression; this leaves $99\frac{7}{12}$ cubits. This being done, you will say: if $\frac{7}{60}$ requires 1 day, what will require $99\frac{7}{12}$ cubits? Computing, you will find that they require 853 full days and $\frac{4}{7}$ of a day. But do not be concerned with these $\frac{4}{7}$ of a day, for this fraction does not correspond to what is correct: the true fraction for the next day is the following one. Consider by how many cubits these ants will have gone forward in 853 full days at the rate of $\frac{7}{60}$ a day. Computing, you will find that they have gone forward by $99\frac{31}{60}$ cubits. Subtract it from 100 cubits, this leaves $\frac{29}{60}$. See now how much time these two ants will require to cover the fraction $\frac{29}{60}$ of a cubit at the rate of their daily movement —disregarding the way back they would make the next night— which is $\frac{8}{15}$ of a cubit, as found above. Then you will say: if $\frac{8}{15}$ of a cubit requires 1 day, what will require $\frac{29}{60}$ of a cubit? Computing, you will find that it requires $\frac{29}{32}$ of a day. Adding this to the 853 whole days will make altogether $853\frac{29}{32}$, and after these $853\frac{29}{32}$ days the two ants will meet, this time being much less than that which Friar Luca (should have) attained, as obviously seen.

After providing the proof by calculating the sum of the two movements during the time found, Tartaglia observes that the same reasoning error as Pacioli's is found with Pietro Borghi in a problem about a sparrowhawk which, initially at the bottom of a tower 60 cubits high, flies up $\frac{2}{3}$ cubit by day and back $\frac{1}{2}$, while a pigeon flies down from the top of this tower by $\frac{3}{4}$ and up by $\frac{2}{3}$. The answer given is 240, whereas Tartaglia finds, *per la nostra regola disopra*, $235 + \frac{15}{17}$.[219]

Remark. It is clear, both from the computation and from Tartaglia's text, that the fractions $\frac{29}{32}$ and $\frac{15}{17}$ are of the daylight part.

[101] Some twenty years earlier, Sfortunati attempted to determine when two snakes will meet on a tower 80 cubits high, given that the one beginning from the top crawls down during daylight by 5 cubits and up during the night by 2, while the one starting from the bottom crawls up by 3 and down by 1.[220] Some, he writes, compute the meeting time as

---

[219] See indeed the arithmetical work of Pie(t)ro Borg(h)i, fol. 110.
[220] *Nuovo lume*, fol. 88^{r} (ed. 1544), No. 23.

$$\frac{80}{(5-2)+(3-1)} = \frac{80}{5} = 16,$$

and he mentions in particular Pacioli, Borghi and F. Calandri (who are just the three he knows best). He himself computes that in 15 days they have moved altogether 75 cubits, and, since during daylight their common progress is 8 cubits, $15 + \frac{5}{8}$ will be sufficient. The answer of the three other authors would be correct, he adds, only if there were no returning in the night, that is, if one just went forward daily by 3 and the other by 2.

Rudolff has a similar problem:[221]

[102] *A tower is* 99 $\frac{59}{60}$ *cubits high. At the bottom there is a worm which by crawling climbs half a cubit by day and comes down* $\frac{1}{3}$ *in the night; in the morning of the same day, a snail begins to crawl from the top, going down each day* $\frac{1}{4}$ *and up each night* $\frac{1}{5}$.

We have then:

movement	daylight	night	daily
worm	$a_w = \frac{1}{2}$	$b_w = \frac{1}{3}$	$\frac{1}{6}$
snail	$a_s = \frac{1}{4}$	$b_s = \frac{1}{5}$	$\frac{1}{20}$
total	$a = \frac{3}{4}$	$b = \frac{8}{15}$	$\frac{13}{60}$

Fig. 96

Let us form the quotient in our own way, with d now representing the height of the tower; we obtain

$$\frac{d-a}{a-b} = \frac{\frac{5999}{60} - \frac{45}{60}}{\frac{45}{60} - \frac{32}{60}} = \frac{5954}{13} = 458.$$

Now the sum of the mean movements in 458 full days is $458 \cdot \frac{13}{60} = \frac{5954}{60} = 99 + \frac{7}{30}$; there remains, in order to make up the height of the tower, the distance $\frac{59}{60} - \frac{7}{30} = \frac{45}{60}$; since this is precisely a, the meeting will take place on the 459th day, at the end of its daylight part.

This is the result obtained by Rudolff, or rather Stifel, who has somewhat simplified the original problem; for, he says, by putting $d = 100$ Rudolff greatly complicated his computation because of the fraction. As for an earlier problem ([88]), Stifel tells us that he deals reluctantly with such ridiculous problems (*so spötlichen exempeln*). That is probably why he did not mind changing it.[222]

---

[221] *Die Coß*, fol. 258ʳ – 259ʳ, No. 118. Since the fractions proposed have once again consecutive denominators, the mean movements are aliquot fractions.

Fig. 97   Fig. 98   Fig. 99

3. Pursuit

The problems of pursuit with back-and-forth movement are less common than those of meeting, maybe because the situation seems even more absurd or because the only difference is that the addition in the divisor becomes a subtraction. Therefore, the authors who do treat such a problem have generally also treated the corresponding one about meeting.

Thus Fibonacci, who had solved a meeting problem with two snakes (above, [99]), later uses the same data for a pursuit between two ants, the slower one having a given lead:[223]

[103] *Two ants were on a flat ground at a distance of 100 paces, and made their way towards the same place. The first went forward daily $\frac{1}{3}$ of a pace and back $\frac{1}{4}$; the other (which was ahead) went forward $\frac{1}{5}$ and back $\frac{1}{6}$. Required after how many days they will be together.*[224]

Fibonacci's computation, using the false position, corresponds to

$$\frac{100}{\left(\frac{1}{3} - \frac{1}{4}\right) - \left(\frac{1}{5} - \frac{1}{6}\right)} = \frac{100}{\frac{1}{12} - \frac{1}{30}} = \frac{100}{\frac{3}{60}} = 2000,$$

which is in excess by two days.

Right after the corresponding meeting problem, F. Calandri has the one of pursuit. It is just like Fibonacci's except that he adds a destination.[225]

---

[222] Illustrations from Tagliente's *Componimento* (ed. 1525, 1547, 1554), with a tree $26 + \frac{1}{4}$, or 26, cubits high, a cat covering $\frac{1}{6}$ ($= \frac{1}{2} - \frac{1}{3}$) and a squirrel $\frac{1}{20}$ ($= \frac{1}{4} - \frac{1}{5}$), said to meet in $26 : \frac{13}{60} = 120$ days.

[223] *Liber abaci*, p. 182.

[224] The 'pace' (*passus*) —in fact, double step— is the thousandth part of a mile, thus 1.5 m. See p. 72*n*.

[225] *Opusculum*, with the illustration of Fig. 100.

Fig. 100

[104] *Two ants are on a flat ground to reach a heap of wheat, one being 100 paces ahead of the other. That which was closer to the heap went daily $\frac{1}{5}$ pace and returned $\frac{1}{6}$ pace; that which was farther went daily $\frac{1}{3}$ pace and returned $\frac{1}{4}$ pace. The heap of wheat was in such a place that, crawling in this manner, they reached it at the same time. I want to know how many days they needed for that and how far the heap was from the ant (initially) closer to it.*

Calandri calculates as does his predecessor and thus finds 2000 days and (therefore) $66 + \frac{2}{3}$ paces as initial distance between the heap and the first ant.

4. Variable distance

[105] *A rat is at the top of a tree 60 cubits high, and a cat is at the bottom on the ground. The rat goes down during the day $\frac{1}{2}$ cubit and returns nightly $\frac{1}{6}$ cubit. The cat climbs 1 cubit during the day and descends nightly $\frac{1}{4}$ cubit. And, between the rat and the cat, the tree grows during the day by $\frac{1}{4}$ cubit and shrinks during the night $\frac{1}{8}$ cubit. I ask in how many days the cat will catch up with the rat, that is, when they will meet, and how many cubits the tree will have attained as a consequence of this growth, and how many cubits the rat and the cat will have each covered.* [226]

The data of the problem are then as follows (with $+$ and $-$ to indicate by how much the distance between them decreases or increases, respectively; as to the tree, its growing further separates them).

movement	daylight	night	daily
rat	$a_r = +\frac{1}{2}$	$b_r = -\frac{1}{6}$	$+\frac{1}{3}$
cat	$a_c = +1$	$b_c = -\frac{1}{4}$	$+\frac{3}{4}$
tree	$a_t = -\frac{1}{4}$	$b_t = +\frac{1}{8}$	$-\frac{1}{8}$
total	$a = 1 + \frac{1}{4}$	$b = -\frac{7}{24}$	$+\frac{23}{24}$

Fig. 101

(*i*) Solution by Pacioli. Let us consider the part covered by the (mean) daily forward movement, thus $60 - \frac{23}{24} = 59 + \frac{1}{24}$. The corresponding duration of the movement will be

---

[226] Pacioli, *Summa*, fol. 42^r, No. 23.

$$\frac{59 + \frac{1}{24}}{\frac{23}{24}} = 61 + \frac{14}{23}.$$

To this is added one day to cover the part subtracted. The answer is then verified:

— The rat progresses by $\frac{1}{3}$ daily, thus in $62 + \frac{14}{23}$ days it will have covered $20 + \frac{20}{23}$.

— The cat progresses by $\frac{3}{4}$ daily, thus in $62 + \frac{14}{23}$ days it will have covered $46 + \frac{22}{23}$.

— The tree grows by $\frac{1}{8}$ daily, thus in $62 + \frac{14}{23}$ days it will have grown by $7 + \frac{19}{23}$, and will thus measure $67 + \frac{19}{23}$. Now this is indeed the sum of the distances covered by the two animals.

Pacioli's treatment, as in an earlier problem ([100]), just amounts to dividing the 60 cubits by the whole daily mean movement $\frac{23}{24}$. He has thus, here again, continued the movement after the meeting, as Tartaglia reproaches him.[227] As to his proof, it has the same weakness as that of Fibonacci for the meeting of the two snakes (above, [99]).

(*ii*) Solution by Tartaglia. Between day and night, the two animals come closer by $\frac{23}{24}$ cubits, but they move away from one another by $\frac{7}{24}$ during the night. We shall thus consider the quotient of the initial height of the tree reduced by this latter distance, thus $59 + \frac{17}{24}$, divided by the mean movement. This gives

$$\frac{59 + \frac{17}{24}}{\frac{23}{24}} \left(= \frac{d - |b|}{a - |b|} \right) = 62 + \frac{7}{23},$$

of which we shall keep the integral part only. We are left with determining the fraction added to it, which we shall do by examining the situation at the beginning of the 63rd day.[228] At the end of the (whole) 62nd day, the distances covered altogether will be

— for the rat: $62 \cdot \frac{1}{3} = 20 + \frac{2}{3}$ cubits;
— for the cat: $62 \cdot \frac{3}{4} = 46 + \frac{1}{2}$ cubits;

---

[227] *Tutto il suo errore procede per non haver rispetto a l'ultimo giorno, che loro si incontrano, che non ritornaranno più indietro, ne l'alboro non calarà niente di quello sarà cresciuto in quel governo tra il toppo & la gatta (General trattato, II, fol. 10ᵛ – 11ʳ, No. 18).*

[228] *Hor dico che li giorni 62 integri sono giusti, ma il rotto, in queste sorte di ragioni, è sempre falso, perché in quella parte di giorno —perché non camina la notte che seguirà, che li fa ritornar indietro, & scemar l'alboro— bisogna di tal parte di giorno farne il conto a quello che caminano il giorno puro, senza il ritorno della notte & similmente del crescer de l'alboro.*

— for the tree: $60 + 62 \cdot \frac{1}{8} = 67 + \frac{3}{4}$ cubits.

The difference between the increased height of the tree and the sum of the distances covered by the two animals is then

$$67 + \tfrac{3}{4} - \left(20 + \tfrac{2}{3} \ + \ 46 + \tfrac{1}{2}\right) = 67 + \tfrac{3}{4} - \left(67 + \tfrac{1}{6}\right) = \tfrac{28}{48}.$$

Now during one of the twelve daylight hours the forward progresses α_r, α_c, α_t are, respectively, $\frac{1}{24}$, $\frac{1}{12}$, $-\frac{1}{48}$ (again: the growth of the tree further separates them), $\frac{5}{48}$ altogether. The remainder will then be covered in $\frac{28}{5} = 5 + \frac{3}{5}$ hours. The meeting will therefore take place after 62 days, 5 hours, 36 minutes (*in giorni 62 $\frac{7}{15}$ si incontraranno li detti animali*, writes Tartaglia, considering thus that the forward movement indeed lasts 12 hours). His result can be verified:

movement	62 days	5 hours	36 minutes	total
rat	$20 + \frac{2}{3}$	$\frac{5}{24}$	$\frac{3}{5 \cdot 24}$	$20 + \frac{108}{120}$
cat	$46 + \frac{1}{2}$	$\frac{5}{12}$	$\frac{3}{5 \cdot 12}$	$46 + \frac{58}{60}$
tree	$67 + \frac{3}{4}$	$\frac{5}{48}$	$\frac{3}{5 \cdot 48}$	$67 + \frac{208}{240}$

Fig. 102

(*iii*) As a matter of fact, the problem proposed by Pacioli is not original, for it is found in a manuscript written a few decennia earlier.[229] The subject is just the same, only the data (in cubits) differ: $d = 30$, $a_r = \frac{1}{4}$, $b_r = \frac{1}{6}$, $a_c = \frac{1}{2}$, $b_c = \frac{1}{3}$, $a_t = -\frac{1}{6}$, $b_t = \frac{1}{12}$, whence $a = \frac{7}{12}$, $b = \frac{5}{12}$, with the difference $\frac{1}{6}$ (mean movement). The computation made is then

$$\left[\frac{d-a}{a-b}\right] + 1 = \left[\frac{30 - \frac{7}{12}}{\frac{1}{6}}\right] + 1 = \left[176 + \frac{1}{2}\right] + 1 = 177.$$

Now at the end of the 177th day the sum of the mean movements is $177 \cdot \frac{1}{6} = 29 + \frac{1}{2} = 29 + \frac{6}{12}$. Since the purely forward movement is $\frac{7}{12}$, the meeting will take place after $177 + \frac{6}{7}$ days (*e chosì diremo che in 177 dì $\frac{6}{7}$ si schontreranno insieme*, the author concludes). He does not specify that it is during daylight, but this is evident since the movement is only forward. Thus we may say that the reasoning for solving problems with three back-and-forth movements was already correctly dealt with one century before Tartaglia.

---

[229] MS. Florence BNC Palat. 573, fol. 141$^\mathrm{v}$ – 142$^\mathrm{v}$.

Chapter VIII. Large numbers

§ 1. First occurrences

One way of catching the attention of pupils was, apparently, to obtain impressively high results. Another advantage, when the computation was left to the pupil, was that the teacher could have some rest. Thus, the earliest memory C. F. Gauß (1777-1855) had of elementary mathematics was having to sum an arithmetical progression.[230] His teacher, however, was not left in peace for long: the ten-year-old Gauß noted that the sum of terms equidistant from the two extremes was the same, whereby the calculation reduced to multiplying this number by that of such pairs.[231]

What the boy had just discovered was the law of summation of arithmetical progressions: multiply the sum of the two extreme terms by half the number of terms. But this was by far not a new discovery. The early Mesopotamian texts (*c.* 1800 BC) knew that already.[232] In Greek antiquity, it is applied by Heron of Alexandria for calculating the number of seats in a theatre knowing the number for the bottom row ($a_1 = 40$), the number of rows ($n = 50$) and the difference between the numbers of seats in two successive rows ($r = 5$); from these data he could determine the number of seats in the last row ($a_{50} = 285$), and also the number of seats altogether S.[233] Apart from these particular examples, knowing the summation of arithmetical progressions serves to reach arbitrarily large numbers. Actually, it was one of two main ways of reaching them. The other, much more powerful, is the successive multiplication of some given number by itself or by another, fixed one; this is the geometrical progression, the summing of which was also already known in antiquity (see p. 248).

The occurrence of large numbers was, however, not always linked to progressions. It could be that a sequence of numbers arose by successive multiplications of an initial term by various numbers, or by converting units of measurement of different orders. Thus, the late-antique *Anthologia Graeca* informs us about the number of Greek soldiers in the Trojan War: in front of seven hearths were 50 spits, each roasting 50 pieces of meat, before each of which were 900 Acheans.[234] In Anania S̲h̲irakatsi's

---

[230] ß, the mediaeval contraction for two consecutive s, is still in use in Germany.

[231] Sartorius, *Gauss*, pp. 12–13.

[232] Vogel, *Vorgr. Mathematik*, II, p. 36.

[233] In his *Stereometrica*, see *Opera omnia*, V, pp. 48–51; or the *De mensuris*, *ibid.* pp. 180–181. Heron's calculations rely on the formulae $a_n = a_1 + (n-1)r$ and $S = \frac{1}{2}n\left[2a_1 + (n-1)r\right]$.

[234] XIV, No. 147. There will thus be 15 750 000 banqueters.

collection of problems (above, p. 33), we are told that a barn containing 200 kaiths of wheat, thus 82 944 000 grains, was totally ravaged by mice; one of them, being caught, admitted to having eaten 80 grains. From this is inferred that the number of mice was 1 036 800.[235]

But it was handier to repeat the multiplication using the same number, mainly with that number as first term and constant factor of multiplication, for instance seven. Such a sequence is found at various times and in various countries, required being either the last term or the sum of all.[236] There is no need to suppose any connection: the choice of 7 is quite natural, being the number of the celestial bodies known in antiquity and the Middle Ages. Thus we find the following variants:

[106] Ancient Egypt. In one of the two main papyri preserved, written by the scribe Ahmes c. 1800 BC, we find towards the end some kind of inventory, adding up to 19607:[237]

Houses	7
Cats	49
Mice	343
Ears	2401
Bushel	16807.

The interpretation may be as follows: there are in each of seven houses seven cats, each of which has eaten seven mice, each of which has eaten seven ears of wheat comprising seven grains each, altogether a bushel.

[107] At the beginning of the 13th century, Fibonacci has a similar problem:

Septem vetule vadunt Romam	7
Quarum quelibet habet burdones 7	49
Et in quolibet burdone sunt saculi 7	343
Et in quolibet saculo panes 7	2401
Et quilibet panis habet cultellos 7	16807
Et quilibet cultellus habet vaginas 7	117649
Queritur summa omnium predictorum	137256.

Fig. 103

---

[235] Problem 23 (the *kaith* is a unit of capacity). Here we disregard the unattainable large numbers proposed by Archimedes, such as those of the Sun's oxen or of the grains of sand which could fill up the Universe.

[236] These two questions are usual for progressions, in particular the geometrical ones. See pp. 117–118.

[237] So-called *Rhind papyrus*, by the name of its English discoverer. Various editions, initiated by Eisenlohr (*Handbuch*, I, pp. 202–204).

This time, old women are going to Rome with mules, bags, loaves, knives, sheaths; the sum of all is required.[238]

[108] In 17th-century Russia, with peasant women, sticks, knots, shoulder bags, pies, sparrows, gizzards.[239]

Идёт семь баб	7
У всякия бабы по семи посохов	49
На всяком посохе по семи сучков	343
На всяком сучку по семи кошелей	2401
Во всяком кошеле по семи пирогов	16807
Во всяком пироге по семи воробев	117649
Во всяком воробе по семи пупков	823543
И всего	960799.

[109] In an old English riddle: wives, sacks, cats, kittens.[240]

As I was going to St. Yves	
I met a man with seven wives	7
Every wife had seven sacks	49
Every sack had seven cats	343
Every cat had seven kits	2401
Kits, cats, sacks, and wives	
How many were going to St. Yves?	2800.

§ 2. Selected mediaeval examples

1. Successive multiplications

[110] *A tree has 12 branches, each branch bears 4 nests, each nest holds 7 birds. Required, between branches, nests and birds, what their number is. (...) Now put all that together, that is, the 12 branches, the 48 nests, the 336 birds; there will be altogether 396.*[241]

[111] *Here are 12 houses, in each house 12 windows, in each window 12 smaller windows. Required, between houses, large and small windows, their number. (...) Now put all that together, that is, the 12 houses, the 144 large windows, the 1728 small ones; there will be altogether 1884.*[242]

---

[238] *Liber abaci*, pp. 311–312. An Italian version of it in MS. Siena BCI L.IV.21, fol. 225[r] (Fig. 103). Same problem —but with baronies, valleys, towns, houses, dormitories, soldiers— in the *Subtilitates*, No. 31. Similar problem in the *Arte del alguarismo*, p. 202 (*Syete romeros van a Roma*).

[239] Mentioned by Yushkevitsh, *История*, p. 36.

[240] Reported by Ore, p. 118.

[241] MS. Lyons 59 (127), fol. 65[r]–65[v]; original text in *Récréations*, p. 134.

[242] MS. Lyons 59 (127), fol. 65[r]; original text in *Récréations*, p. 134.

Note that the choice of 12 here, as of 7 in the previous examples, is not surprising: because of its division into integral parts, base 12 was often preferred to base 10 in everyday life (even today some objects are still sold by the dozen). Of course base 10 does also appear, as in the next problem.

[112] *A tree has 100 branches, on each branch there are 100 nests, in each nest there are 100 eggs, and in each egg there are 100 birds.*[243] The total sum is therefore 101 010 100.

[113] *A rich man had in his henhouse 12 poles for roosting. On each one were roosting 20 cocks, above each cock 15 hens, and above each hen 15 chicks. Required how many cocks, how many chicks and how many hens there are. Answer. There are 240 cocks, 3600 hens, 54000 chicks.*[244]

[114] *There are 40 monks in a convent. Each monk has each day 5 eggs. Multiply 40 by 5; it gives 200, which is the number of eggs the convent has for one day. Multiply this sum, that is, 200, by 365, which is the number of days in the year; and such is the number of eggs, 73000, which the convent has altogether for one year.*[245]

The absence of any question here is not an omission. The author has taught before, when the numbers of eggs, loaves, or quarts of wine a monk receives daily are known, how to infer from it the annual consumption for a convent with a given number of monks; or also, for state employees, their annual wages given the daily one.[246]

[115] *If one asks how many hares can be on an acre (*arpant*) of land given that each hare occupies one foot in length and half a foot in width.*[247] *Answer. We are to consider how many feet the acre contains in length and width. If the acre contains in length 40 perches, with to each perch 18 feet, the acre will have in length 40 times 18 feet. Thus if you multiply 40 by 18, or 18 by 40, you will have the number of feet in length, namely 720 feet. Next you are to consider how many feet the acre has in width. It has 4 perches, thus it will have 4 times 18 feet; so multiply 4 by 18, and you will have the number of feet in width, namely 72. Thus double the number of feet of the width, and you will have the sum of the half-feet, namely 144. Thus there are as many half-feet in width as there are hares arranged in*

---

[243] Fibonacci, *Liber abaci*, p. 312.

[244] MS. Tours 399, fol. 129^r – 129^v; original text in *Récréations*, p. 135.

[245] MS. Tours 399, fol. 129^v; *Récréations*, p. 135 (text).

[246] See *Récréations*, p. 135.

[247] A foot is about 32 cm. Here the 'perch' is 18 feet, thus slightly less than 6 m, and the acre 40 × 4 square perches, or 720 × 72 square feet. (Elsewhere the value of such units of measurement may vary greatly, depending on time and place.)

width, with their heads in direction of the acre's length. Next multiply the number of half-feet in the width of the acre, thus 144, by the number of whole feet in the length of the acre, namely 720; this will make precisely 103680. Such is the number of hares which can be in an acre of land: 103680. [248]

2. Use of arithmetical progressions

[116] *A clergyman carried a book to a library one day, the second day he carried two, the third day he carried 3, and so on, increasing always by one, up to 15 days. Required then how many books there will be altogether, adding them up. (…) If you have done well, you will find that there are 120.* [249]

[117] *A clock strikes, during a naturel day, 24 times. Required the number of times it has struck altogether, adding up all the strokes of each hour. (…) You will see that in 24 hours the clock has struck 300 times.* [250]

[118] *I saw a ladder that had 100 rungs. On the first was one pigeon, on the second 2 pigeons, on the third 3, on the fourth four, and so on up to 100. Required how many pigeons were on the whole ladder. Answer: 5050.* [251]

The text then explains how to obtain the sum of an arithmetical progression beginning with 1: if it ends with an even number, multiply its half by the next odd number; if it ends with an odd number, multiply it by its greater half.[252]

Similar problems, at least for the data, are found in a Byzantine source:

[119] *I have bought a ship which advances each day by 1 mile more than the day before; how many miles did it cover in 100 days.* [253] The same treatise proposes later on another form of this problem: 100 horsemen (χαβαλλάριοι) find an apple tree on their way and, according to the same progression, strip

---

[248] MS. Tours 399, fol. 130^r–130^v; original text in *Récréations*, pp. 135–136. Such a problem, in Latin version, and with 14 feet to a perch, is found in the 14th-century *Subtilitates*, No. 32. A similar problem, with sheep, occurs in Alcuin's Collection (No. 21).

[249] MS. Lyons 59 (127), fol. 64^v; *Récréations*, p. 136.

[250] MS. Lyons 59 (127), fol. 64^r–64^v; *Récréations*, p. 136. The first mechanical (cathedral) clocks appeared during the 13th century.

[251] MS. Tours 399, fol. 134^v–135^r; *Récréations*, pp. 136–137.

[252] The 'greater half' is half the next (even) number. These two forms of the formula $1+2+\cdots+n = \frac{n(n+1)}{2}$ are commonly found in mediaeval texts, their purpose being just to avoid working with fractions, by dividing the even factor by 2. The same problem is already found in Alcuin's Collection (No. 42); the result is obtained by summing $(1 + 99) + (2 + 98) + \ldots (49 + 51)$ and adding to $49 \cdot 100$ the pigeons on the 50th and the last rungs.

[253] Vogel, *Byz. Rechenbuch*, No. 23.

it of all its apples. In the next two problems, their harvest follows the progressions of the even numbers from 2, then of the odd numbers from 1, giving as the results 10 100 and 10 000 apples, respectively.[254]

3. Use of geometrical progressions

As far as the size was concerned, the answer in Alcuin's ladder problem was not excessive: it followed an arithmetical progression. But Alcuin also has a problem involving a geometrical progression, and this time the result is quite impressive.

[120] *A king ordered his servant to gather from thirty towns an army in such a way that he should take in each town as many men as he had brought. He himself came alone to the first town, with another to the second, and four arrived at the third. Let he who can say it how many men were gathered from these thirty towns.*[255]

Fig. 104

The army gathered by this singular way of recruiting will therefore come to $1 + 1 + 2 + 4 + \cdots + 2^{29} = 2^{30} = 1\,073\,741\,824$ men (half of whom from the last town visited) —incidentally an army larger by far than the world's population at that time, thought to be a few hundred million individuals.

We have already seen two multipliers (and initial terms) chosen for their common use (7 and 12, see [106]–[109], [110]–[111]). The same holds for the limit 30, also familiar to any reader as being the approximate duration of a month. It appears already in a Mesopotamian problem: a stick measuring 30 digits, used for evaluating a distance, when applied decreases each time by one digit.[256]

---

[254] *Ibid.* Nos. 111–113.

[255] *Propositiones*, No. 13. Illustration from F. Calandri's manuscript (MS. Florence Biblioteca Riccardiana 2669, fol. 98^r; text, ed. pp. 144 (illustration) & 196 —other problem).

[256] Vogel, *Vorgr. Mathematik*, II, p. 36.

There are other forms of this successive doubling, where an everyday event gives rise to an unrealistic result:[257]

[121] *A carter fell, who was alone, and could pick up neither his cart nor his horses. He saw nearby 8 houses of good folk. He came to the first and asked for God's sake that one man be lent to him; they were thus 2. He came to the second and asked that as many men were lent as themselves; they were thus 4. At the third house, as many as they were; they were then 8. And so on doubling to the (last of the) 8 houses. Required how many men there were to pick up his cart and his horses. Answer. They were 256.   And this brainteaser (cautèle) is like doubling the chessboard.*

This last sentence brings us to the most known use of the geometrical progression with constant factor 2, and that which historically led to the most impressive results: doubling on the chessboard's squares —that is, doing so to the 64th term. As with similar problems (p. 112), there will be two questions in connection with this one: finding the last term, that is, what is on the last square, or finding what is on the whole chessboard.

§ 3. Doubling on the chessboard

We know today that the game of chess must have been invented in 6th-century India. One or two centuries later it reached Persia, then the Islamic world and, through Spain, Europe in the 12th century.[258]

Fig. 105

Doubling the chessboard's squares (اضعاف بيوت الشطرنج), as it was called, occurs early in Arabic texts. It is often associated with a legendary account about the invention of the game. Kḫāzinī, whom we have already met in connection with the problem of weights, is one of the many sources

---

[257] MS. Tours 399, fol. 126^{v}; text in *Récréations*, p. 138.

[258] Illustration from the MS. Heidelberg Pal. germ. 848, fol. 13^{r} (Manesse Handschrift, first half of the 14th century). Its author cannot have been a chessplayer.

reporting it.[259] Some, he writes, say that after the inventor of the game presented it to his king, the latter was so impressed that he asked him to choose his own reward. Here, there are two versions: according to Khāzinī, the man required a sum of *dirhams* such that 1 would be on the first square, 2 on the second, and so on doubling each time; according to other sources, grains of wheat, not dirhams, were to be used. In any event the outcome was the same: the resources of the kingdom, or even of the whole earth, could not meet the inventor's requirement (or, more probably, challenge), since the sum of $1, 2, 2^2, 2^3, \ldots, 2^{63}$ produces the huge number

$$2^{64} - 1 = 18\,446\,744\,073\,709\,551\,615.$$

Some two centuries before, around 950, the mathematician al-Uqlīdisī tells us, about successive doubling of the unit, the following:[260] *This is a question asked by many, some requiring to double the unit thirty times, others sixty-four times. But they do not all know the answer, neither for the last term nor for the sum of all; therefore, two questions are to be distinguished: one is the quantity reached (at the end) by the duplication, the other is the sum of all terms arising from the duplication.*

The way of doubling up to the 30th term alluded to by Uqlīdisī clearly goes back to antiquity; we have just seen an example of it, in Alcuin's Collection. The limit reached then, arising from the number of days in the month, could later easily be extended to sixty-four terms after the invention of the chess. This extension must have already been common at the beginning of the 9th century, since at the end of his *Algebra* Abū Kāmil (*c.* 890) reports that his predecessor al-Khwārizmī (*c.* 820) had (in a work lost today) explained a shorter way of calculating either the last term or the whole sum.[261] As we know, since the occupants of the 64 squares are $1, 2, 2^2, 2^3, \ldots, 2^{63}$, their sum will be $2^{64} - 1$. The short cut proposed by Khwārizmī is to start with 2, the content of the first cell increased by 1, and to proceed with successive squaring: 2, 4, 16, 256, and so on to the sixth. We thus obtain the sequence $2, 2^2, 2^4, 2^8, 2^{16}, 2^{32}, 2^{64}$. Dividing then 2^{64} by 2, we obtain the last term, thus the number on the last square; subtracting 1 from 2^{64}, we obtain the sum on the chessboard. This answers the two questions mentioned by Uqlīdisī.

As said, this doubling was an extension of the doubling up to 30

---

[259] Arabic text of the *Mīzān al-ḥikma*, p. 74; German translation by Wiedemann, *Beitrag XIV*, 4, pp. 48–49 [= *Aufsätze*, I, pp. 447–448].

[260] *Arithmetic*, translation, p. 337; Arabic text, p. 423.

[261] *Algebra*, fol. 109ᵛ – 110ᵛ (pp. 218–220 of the manuscript reproduction). On Khwārizmī, see p. 267.

terms. But another way was found to reach an even larger number than by simply doubling; the 10th-century mathematician and astronomer al-Qabīṣī alludes to it.[262] In this new mode, each square no longer contains twice the one before, but twice the sum of all the ones before. This arrangement subsequently appeared along with the previous one. Fibonacci knows both, for he writes: *The doubling on the chessboard is, in fact, proposed in two ways. In one, the next square contains the double of the preceding, in the other the next square is said to contain the double of all preceding squares.*[263]

The result of summing up the second way is indeed notably larger than that of the first. Starting from

$$q_k = 2 \cdot \sum_{i=1}^{k-1} q_i, \quad \text{with } q_1 = 1,$$

we find that

$$
\begin{aligned}
q_k &= 2\big[q_1 + q_2 + \ldots + q_{k-1}\big] \\
&= 2\big[q_1 + q_2 + \ldots + q_{k-2}\big] + 2q_{k-1} \\
&= 2\big[q_1 + q_2 + \ldots + q_{k-2}\big] + 2 \cdot 2\big[q_1 + q_2 + \ldots + q_{k-2}\big] \\
&= 2 \cdot 3\big[q_1 + q_2 + \ldots + q_{k-2}\big] \\
&= 2 \cdot 3\big[q_1 + q_2 + \ldots + q_{k-3}\big] + 2 \cdot 3 q_{k-2} \\
&= 2 \cdot 3\big[q_1 + q_2 + \ldots + q_{k-3}\big] + 2 \cdot 3 \cdot 2\big[q_1 + q_2 + \ldots + q_{k-3}\big] \\
&= 2 \cdot 3^2\big[q_1 + q_2 + \ldots + q_{k-3}\big] \\
&= \ldots \\
&= 2 \cdot 3^{k-3}\big[q_1 + q_2\big] \\
&= 2 \cdot 3^{k-2} \quad \text{(with } k \geq 2\text{)}.
\end{aligned}
$$

The sum of all terms will in this case be

$$\sum_{k=1}^{64} q_k = 1 + \sum_{k=2}^{64} q_k = 1 + 2 \cdot \sum_{i=0}^{62} 3^i = 1 + 2 \cdot \frac{3^{63} - 1}{3 - 1} = 3^{63}.$$

The total sum on the chessboard is therefore

$$S_{64} = 3^{63} = 1\,144\,561\,273\,430\,837\,494\,885\,949\,696\,427,$$

---

[262] See our *A treatise*, pp. 486 & 492–493. Although this first appears in Qabīṣī's text, it must be earlier since he does not attribute it to himself.
[263] *Liber abaci*, p. 309.

which is about 10^{12} times larger than the sum on the chessboard according to the first mode; or, as observed by Pacioli, here there are 11 more digits.[264]

Fig. 106   Fig. 107

But this other kind of doubling also occurs with the limit 30. Thus, Cardan calculates the quantity of wheat obtained after 30 days starting with 1 grain and doubling each day the quantity already possessed.[265] To calculate it, Cardan considers the sequence 1, 2, 6, 18, 54, ..., and the sum S_k of the first k consecutive terms as, for instance, $S_4 = 3^3 = 27$. Then, he says, we shall use the property $S_k^2 = S_{2k-1}$ and thus calculate successively

$$S_4^2 = 27^2 = 729 = S_7 = 3^6,$$
$$S_7^2 = 531\,441 = S_{13} = 3^{12},$$
$$S_{13}^2 = 282\,429\,536\,481 = S_{25} = 3^{24}.$$

We are left with computing S_{30}, separated from the previous one by five terms. To do that, we take $S_6 = 243 = 3^5$, and form

$$S_6 \cdot S_{25} = 243 \cdot 282\,429\,536\,481 = S_{30} = 68\,630\,377\,364\,883 = 3^{29}.$$

This is the same principle as seen above (p. 118): reducing the number of computations by a repeated squaring. As we shall see (below, p. 129), Fibonacci had done the same three centuries earlier. With our symbolism, it is banal; without the explicit expression of exponents, it requires some reflection.

But this sum, as well as those arising from the two ways of doubling on the chessboard, as impressive as the number of digits is, will remain abstract if not put in relation with something more familiar. An obvious

---

[264] *La summa de quel modo de tutto el scachieri erano 20 figure e la summa de questo sonno 31* ('trenta e una') *figura* (*Summa*, fol. 43ᵛ). Illustrations from Jehan de Vignay's French version of Jacobus de Cessolis' treatise, MS. Paris Bibliothèque de l'Arsenal 5107 rés., fol. 76ʳ & 79ᵛ. See p. 256.

[265] *Practica arithmetice*, Ch. LXVI, No. 9.

comparison was the size of the earth. We shall examine that after a digression on early estimates of its size.

§ 4. Earth's size, ancient and mediaeval estimates

When the first men wanted to have a better view of their region, they were naturally led to climb to some height; since what they saw was limited all around by the horizon, they considered the earth to be a disk with their region in the centre. Since, later on, by going far enough they reached the sea, they were led to suppose that the terrestrial disk was in the middle of water. Such was the common belief in remote antiquity, particularly in early Greece, before the first philosophers. Thus we know that Thales of Milet (about 600 BC) wanted to explain the existence of earthquakes by supposing that the terrestrial disk was floating.

About 580 BC, Anaximandros added a further dimension to the disk, considering it to be a cylinder, the inhabited part being its upper end. Even more significant is that this cylinder was thought by him to be isolated in space, in a kind of equilibrium at the centre of the Universe. This step was of considerable importance, for it made the earth a celestial body.

A century later, around 480 BC, the idea of a spherical earth appeared, with Parmenides and, surely also, among the Pythagoreans. Arguments in favour of that were of various kinds, from the analogy with the celestial sphere and bodies to the harmony and esthetics of the spherical body as geometrical figure.

The ingenuity of ancient Greeks was to provide the next step, namely proofs of the earth's sphericity. The arguments are anterior to Aristotle (around 330 BC) since he reports some of them and by that time the sphericity of the earth was, in Greece, generally accepted. First, the Greeks knew, from the fifth century BC, that the Moon received its light from the Sun and that lunar eclipses were caused by the interposition of the earth between Sun and Moon. What was seen on the Moon during an eclipse must thus have been, projected onto it, the contour of the earth.[266] Another proof originated in sailors' reports. Whether they sailed east or west, the stellar sky remained familiar to them; it changed, however, with latitude: going north made the southern stars disappear, going south showed new constellations. These two arguments are found in

---

[266] This has nothing to do with the lunar crescent, which depends on the position of the Sun. Astronomy being no longer required in daily life, such elementary knowledge is sometimes lacking today —whence our remark.

Aristotle's treatise 'On the heavens'. But others were surely known, such as the one mentioned in Ptolemy's astronomical work (*Almagest*, around 150 AD): a phenomenon seen everywhere, like a lunar eclipse, is not seen all over the world at the same time, but earlier in the east. Furthermore, when the sea is calm, remote heights, or the mast of a distant ship, are partly hidden, clearly because of the earth's sphericity.

The question of the direction of gravity had still to be settled. For the disk or the cylinder, it was obvious. Maintaining the same direction for a spherical earth would imply the disappearance of ships moving too far from their point of departure —considered to be on the upper part of the earth. Since this was contrary to both reason and experience, the direction of gravity must have been, at whatever place, towards the centre of the earth. An inference of this was to regard the earth as the centre of the Universe.

Admitting thus the sphericity of the earth, the Greeks wanted to measure the terrestrial globe. We know of four results and two methods. The oldest result is reported by Aristotle (who does not mention its origin): 400 000 stadia for its circumference.[267] In his 'Sand-reckoner', Archimedes writes that his predecessors had estimated its length to be 300 000 stadia. His contemporary Eratosthenes found it to be 252 000 stadia. Finally, Poseidonios (100 BC) puts it at 180 000 stadia. These measurements, or at least the last two, are not as divergent as it would seem, for the stadion may vary in value depending on where it is used. As to how these last two measurements were obtained, we know the methods, which in fact are based on just the same principle.

Eratosthenes' estimation is based on four observations or mensurations:

(1) First, it was common knowledge that in Syene (today Aswān[268]) on the day of the summer solstice, at noon, the obelisks did not throw any shadow, and also that the Sun was reflected at the bottom of a well; in other words, the Sun must then have been at the zenith.

(2) Next, the Sun culminates at the same time in Syene and in Alexandria; in other words, these two towns must be on the same meridian, and the plane passing through them and the Sun at noon must also pass through the earth's centre.

(3) Furthermore, according to Eratosthenes' measurement, the shadow projected in Alexandria by a vertical stick at noon of the summer solstice made an angle equal to one fiftieth of the whole circumference.

---

[267] The 'stadion' (στάδιον) is a Greek measure of length (see below).

[268] From the Arabic article *al-* (here *as-*) and the Greek *Suene* (Συήνη).

(4) Finally, the distance between Syene and Alexandria is 5000 stadia.

Therefore (Fig. 108), with the parallelism of the Sun's rays and assuming perfect terrestrial sphericity, the angle of the shadow measured at Alexandria must equal the angle $\widehat{ACS}$ at the centre of the earth, formed by the radii passing through the two towns. Since its arc $\widehat{AS}$ measures 5000 stadia and is a fiftieth of the circumference, the great circle of centre C passing through A and S must equal $50 \cdot 5000 = 250\,000$ stadia. Now, for practical reasons, it is convenient to know the value of the meridian degree, that is, the distance between two places on the same meridian 1° apart; so Eratosthenes rounded up his value to $252\,000$ stadia, thus obtaining, as the value of the meridian degree, 700 stadia instead of the less convenient $694 + \frac{4}{9}$.

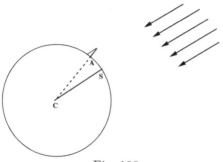

Fig. 108

This changing of the measurement may seem at first sight rather casual; but Eratosthenes must himself have been aware that the values he used were approximations. Indeed, as we now know,

(1) Syene is not exactly on the Tropic of Cancer, and so the angle of the shadow thrown at noon on the summer solstice cannot really be zero.

(2) The two towns are not exactly on the same meridian (there is a longitudinal difference of 3°).

(3) The angle of the shadow is difficult to measure precisely, even by a careful observer.

(4) Assessing the distance between two towns depended on the reliability of the *bematists*, those professional walkers employed since the time of Alexander the Great to determine terrestrial distances by counting their footsteps (βήματα, *bemata*).

Apart from these (ultimately, minor) discrepancies, the quality of the measurement depends on the value of the stadion used by Eratosthenes; which, since it may vary almost from simple to double, is more relevant than the above approximations. Eratosthenes being in Alexandria, he must have used the Alexandrian stadion, which corresponds to 157.5 metres. The 252 000 stadia would then correspond to 39 690 km, thus very close to the modern value (40 077 for the equatorial circumference, 40 009 for the meridian). Note that, omitting the rounding up to 252 000, the value obtained, 39 375 km, would still be quite acceptable.

Poseidonios' figure also relies on four observations or mensurations.

(1) Rhodes and Alexandria are situated on the same meridian.

(2) The star Canopus, not visible in Greece, culminates just above the horizon at Rhodes.

(3) Canopus culminates at $7°30'$ in Alexandria.

(4) Rhodes and Alexandria are 3750 stadia apart.

Fig. 109

Thus (Fig. 109), the angle at which Canopus is seen from these two towns will be found in the centre of the earth as $\overset{\frown}{ACR}$. Since $360 : 7.5 = 48$, the earth's circumference will equal $48 \cdot 3750 = 180\,000$ stadia, and the meridian degree will have the convenient value of 500 stadia.

Here too, approximations are necessarily involved.

(1) Rhodes and Alexandria are not exactly on the same meridian.

(2) The presence of Canopus just on the horizon is difficult to verify, and subject to error anyway since refraction there makes the stars appear higher up (as must have been noticed in antiquity[269]).

(3) For the same reasons, its culmination in Alexandria can only be measured approximately.

(4) The evaluation of maritime distances between two towns is even less reliable than that of terrestrial ones.

The principle of this second method is much the same as that of the first and the origin of the inaccuracies, no different. As to the result, it seems very different. But the stadion used by Poseidonios is not the

---

[269] See Toomer's translation of the *Almagest*, p. 421, n. 8.

Alexandrian one; converted to it, the circumference would become 240 000 stadia, or 37 800 km, a result in any event less than Eratosthenes'.

The adoption of this second measurement by the astronomer Ptolemy means that it is what is mainly found in late antiquity. In Roman times, the 180 000 stadia were converted to Egyptian Ptolemaic miles of 7.5 stadia each, and the earth's circumference became equal to 24 000 miles, with the meridian degree thus to $66 + \frac{2}{3}$ stadia. In mediaeval Europe, this measurement was kept; the miles used in Italy, however, reduced the circumference to 36 000 km. In the East too, the measurement was kept. But since the oriental mile, used since the Roman Empire, was about 2 km, the circumference attained some 48 000 km.

Despite the great respect Islamic people had for ancient knowledge, it became obvious that this measurement was inappropriate. For that reason, Caliph al-Ma'mūn (reigned 813-833) ordered it to be recalculated. The idea was to move along a meridian between two places displaying a change of 1° between the culminations of a certain celestial body; for then the distance between these two places would be equal to the length of one meridian degree. The region chosen was the plain of Sinjār, in the north-west of today's Iraq, and the result found, $56 + \frac{2}{3}$ miles. Reports on how that was obtained differ, however: the culmination observed was either that of the North Star, at night, or that of the Sun, at noon; the move along the meridian was made either by a single group heading north, or the mean was taken between two measurements made by two groups, one going north and the other south; to keep in a straight line, they used either three sticks, which, driven into the ground one after the other, had to remain lined up, or two ropes of the same length, of which the first half of the second was, in turn, lined up on the second half of the one ahead. Whatever the case, the terrestrial circumference thus found, 20 400 miles, was indeed better than the ancient measurement taken with Eastern miles.

As said, the reason for re-measuring the meridian degree in the time of Ma'mūn was the inaccuracy of the values received, this inaccuracy being itself due to an inappropriate conversion of units. Exactly the same happened with the Arabic value. Said measurement of 20 400 miles was transmitted to Europe, and confidently adopted.[270] Since, however, Italian miles were in the ratio 3 : 4 to Eastern miles, the terrestrial circumference was reduced by $\frac{1}{4}$, just as it had increased by $\frac{1}{3}$ when passing to the East a few centuries earlier.

---

[270] We have indeed met the value 20 400 in one of Pacioli's problems (p. 88).

Let us now return to a mediaeval problem which computes, with the usual approximation $\pi \cong 3 + \frac{1}{7}$, the number of turns made by a cartwheel over a given distance.[271]

[122] *If one asks how many times the wheel of a cart turns over a league.*

One can know, from (the diameter of) the wheel, how many feet its circumference has, as follows: if the wheel has 7 feet in its diameter, it has 22 feet in its circumference. Then look how many miles make a league, and how many paces make a mile, and how many feet make a pace. You are to know that 2 miles are a league, a thousand paces make a mile, 5 feet make a pace. Therefore a thousand paces make 5000 feet. Therefore the league has 10000 feet. And dividing these 10000 feet of the league by the feet of the wheel, namely 22, you will find, as the so-called quotient,[272] 454, which is how many times the wheel turns over one league, and in addition 12 feet, which is one foot more than half the wheel since the wheel has 22 feet.[273] Therefore the wheel will turn over one league 454 times and half, plus one foot, no more and no less.

We mention this, apparently unrelated problem here as an appropriate introduction to another way of measuring the earth's circumference, that made by the physician Jean Fernel (c. 1497-1558). He measured the length of one meridian degree by counting the number of turns made by a wheel of his vehicle when following the straight, and meridian, road between Amiens and Paris. According to his report, the wheel made *about 17024 rotations, heights and valleys uniformly reduced according to our feeling*. The wheel being in diameter, he says, hardly more than six feet six inches ($= 6\frac{1}{2}$ ft), its circumference was 20 feet, thus 4 paces. Multiplying therefore the number of turns by 4, he found 68 096 paces, which are 68 Italian miles and 96 paces. This seemed to him to be better than the then known values of the meridian degree (remember: altered because of the change in units), namely, he says, $87 + \frac{1}{2}$ (Eratosthenes' figure), $62 + \frac{1}{2}$ (Poseidonios'), $56 + \frac{2}{3}$ (Ma'mūn's).[274]

The ellipsoidal form of the earth was proved during the 18th century.

---

[271] MS. Tours 399, fol. 131^{v} (original text in *Récréations*, pp. 146–147). In his *Measurement of a circle*, Archimedes has demonstrated that the ratio of a circumference to its diameter, thus π, is between $3 + \frac{10}{71}$ and $3 + \frac{10}{70}$; the upper limit became the usual approximation in the Middle Ages. Note here the length units: 1 league = 2 miles, 1 mile = 1000 paces, 1 pace = 5 feet (see p. 72n).

[272] The word may have been unknown to the reader.

[273] The quotient is indeed $454 + \frac{12}{22} = 454 + \frac{1}{2} + \frac{1}{22}$.

[274] *Cosmotheoria*, fol. 3^{v} and 2^{r}, respectively. The measurements by Eratosthenes and by Poseidonios (that adopted by Ptolemy) become, using the conversion factor 1 : 8, 31 500 and 22 500 miles, giving as the meridian degree 87.5 and 62.5 miles. These two values of the earth circumference are also mentioned by Cardan (below, p. 130).

In the century before, there was hesitation between an ellipsoid stretching out towards the poles (hypothesis of Descartes) and towards the equator (hypothesis of Newton). Since the meridian degree should be longer where the earth is flatter (for the radius of curvature is greater there), a French expedition was organized in the years 1735-1737 to measure it in Peru and Lapland. This decided in favour of Newton's hypothesis.

The difference between the longer axis $(2 \cdot 6\,378.388 \text{ km})$ and the shorter one $(2 \cdot 6\,356.912 \text{ km})$ is small, and so therefore is the difference between the meridian degree at the pole (111.700 km) and at the equator (110.576 km). It was therefore undetectable with ancient means. All the more troubling is a passage of Bishop Jacobus of Edessa in his *Hexameron*, written towards the end of the 7th century, who observes that the shape of the earth is not quite spherical: *In order to provide a picture helping you to understand, I shall give one which will serve to better grasp what I say. Let any wise, clever and learned man among you take a piece of dough, as much as his two hands can contain; let him work on it and make out of it a perfect sphere. Next, with his hand, he will press this sphere of dough in order to give it another shape, and make it pass from the perfect sphere to a sphere a little bit oblong and a little bit narrower. With his fingers, he will form in it, here and there, holes, undulations and heights. Such is more or less the shape of the earth which men must imagine and represent in thought.* Elsewhere he makes the surprising assertion that *people say that facing Spain and Hercules' columns* (Gibraltar), *up to the country of the Chinese, which is eastwards of India, there is an unknown and uninhabited land.*[275]

§5. Some illustrations of the result of doubling

Ed eran tante, che 'l numero loro
Più che 'l doppiar de li scacchi s'inmilla.[276]

We have seen that the reward for the discoverer of the chess game was, in theory, to be paid either in coins or in wheat grains. These same two objects are found again to illustrate the sum arising from doubling on the chessboard.

1. Covering the globe with coins

Qabīṣī, the mid-10th century Arabic author already encountered (p. 119), imagines the terrestrial globe (considered as a solid sphere) covered by *dirhams* and compares the result with the two sums on the chessboard:

---

[275] Martin, *L'Hexaméron*, pp. 454, 456.

[276] Dante, *Divina Commedia, Paradiso* xxviii, 92-93 (*inmillare* = grow by thousands).

[123]^{277} *It was established during the observations of Ma'mūn, according to the testimony of the members of his entourage (of scholars), (...) that they found the terrestrial length corresponding to one degree on the celestial sphere to be* $56\frac{2}{3}$ *miles. (...) Therefore, by virtue of this, the periphery of the earth will be, since its circumference is 360 degrees, 20 400 (miles). Now since the circumference is* $3\frac{1}{7}$ *times the diameter —according to what Archimedes has shown— the terrestrial diameter will be 6491 miles.*^{278} *As it appears also in the teachings of Archimedes that the product of the diameter of the sphere and the circumference of a great circle on it gives the area of its entire surface,*^{279} *the area of the entire surface will be, by virtue of this, 132 416 400 square miles. Since the square mile is 4000 cubits by 4000 cubits, the (square) mile is 16 000 000 (square) cubits; therefore the area of its whole surface, (including) land and sea, plains and mountains, populated and uninhabited areas, will be 2 118 662 400 000 000 (square) cubits. And since the measurement in silver coins of a cubit by a cubit is —according to my trials— 500 dirhams,*^{280} *the total measurement of (the surface of) the earth, (including) land and sea, plains and mountains, will be, in coins, 1 059 331 200 000 000 000 dirhams. Now (the total of) doubling the chess squares equals about* $17\frac{1}{5}$ *times this figure.*^{281} *This (concerns) the doubling of chess squares which is (...) the smaller; as for (the other), (...) it is many times this quantity.*

2. Filling chests with coins

After calculating the sum on the chessboard according to the first mode, Fibonacci remarks that *the number (arising) from doubling on the chessboard, because of its size, cannot be conceived; we shall give the way to conceive it in a more accessible manner.*^{282} He too uses coins, namely bezants (*bizantii*, see p. 36n).

[124] At the end of the first two horizontal rows of the chessboard, thus after 16 squares, we shall have $2^{16} - 1 = 65\,535$ bezants; taking one bezant elsewhere, we shall have 65 536 of them, and this, he says, enables us to fill a chest (*arca*). Thus, the first two rows together (rounded up to 2^{16}) fill a chest, and the first square of the next row by itself contains 2^{16} bezants, thus one chest. Continuing the doubling, but this time of chests, we shall have, together with the chest resulting from the first two rows, 65 536 chests at the end of the fourth row, whereby we may fill a house (*domus*).

---

^{277} See his treatise, pp. 489–490 of our translation (text, ed. Anbouba, pp. 191–190); repeated by a 12th-century author, see our *Une compilation*, pp. 159–160 & 177–178.

^{278} Exactly: $6490 + \frac{10}{11}$.

^{279} Corollary to Proposition I.34 of his *On the sphere and cylinder*.

^{280} The dirham (see p. 41n) was then a silver coin about 2 cm in diameter.

^{281} Rather: $17 + \frac{2}{5}$ times.

^{282} *Liber abaci*, p. 310.

Therefore we shall also have a house on the first square of the fifth row, and, at the end of the sixth row, 65 536 houses, that is, a town (*civitas*). Likewise, we shall have at the end of the eighth row, 65 536 towns, each full of bezants —just with one bezant more, which will be removed from one of the chests. All these bezants thus result from doubling on the chessboard.

3. Transporting wheat by sea

[125] Another of Fibonacci's illustrations is the number of ships required to carry the same quantity of wheat grains, this time estimated with units of capacity. According to Fibonacci, 4 wheat grains correspond to one carob, 6 carobs to one denarius of càntera, 25 denarii of càntera to one ounce, 12 ounces to a pound, 140 pounds to a sextarius, 24 sextarii to a Pisan bushel, while a ship can carry 500 Pisan bushels. Accordingly, a ship can carry 12 096 000 000 wheat grains. In order to transport a number of wheat grains equal to the whole result of doubling (according to the first mode), we shall need 1 525 028 446 ships, all completely filled but for the last one. *It thus appears clearly enough how this number of ships is (so to say) infinite and innumerable.*[283]

Fig. 110   Fig. 111

Fibonacci does not illustrate the second mode of doubling. He just calculates its sum and gives a shorter way of obtaining it. This is analogous to K͟hwārizmī's for the first mode: starting from $q_1 + q_2 + q_3 = 9 = 3^2$, we are to calculate the results of successive squaring, namely 3^4, 3^8, 3^{16}, 3^{32}, 3^{64}. Dividing the last result by 3 gives the required sum. We have seen that, some three centuries later, Cardan proceeded in the same manner (p. 120).

§ 6. Estimating the earth's weight

[126] An eastern illustration involving the earth's size originates with the 12th-century scholar K͟hāzinī, already encountered twice (pp. 25, 117).

---

[283] *Qui navium numerus quam infinitus et innumerabilis sit, satis liquido hic depre-henditur* (*Liber abaci*, p. 310). Illustrations from Tagliente's *Componimento*, ed. 1547 and 1554.

After setting out a list of the number of *mi̱ṯh̲qāls* (1 mi̱ṯh̲qāl ≈ 4.5 gr.) contained in a cubic cubit of various metals, he computes the weight in mi̱ṯh̲qāls that the earth would weigh if it were in pure gold.[284] At the origin of his research there is, he writes, a Koran surah, according to which even this quantity of gold would be insufficient to pay for the sins of those who lived and died as infidels.[285] With the estimated terrestrial circumference $c = 20\,400$ miles, K̲h̲āzinī can calculate its diameter d (= $\frac{c}{\pi}$), area S (= $d \cdot c$) and volume V (= $\frac{1}{6} d \cdot S$); converting it to cubic cubits, and multiplying the result by the quantity of mi̱ṯh̲qāls contained in a cubic cubit of pure gold, he obtains for the earth's weight in gold approximately $5 \cdot 10^{27}$ mi̱ṯh̲qāls.

As to Cardan, he wishes to estimate the weight of the earth considered as a mixture of earth and water.[286]

[127] *Supposing the size of the earth* (= its circumference) *to be, following Ptolemy, 180 000 stadia, thus 22 500 miles —for this opinion is more true than that attributing 31 500 miles, as we have shown in the explanation of Ptolemy's Geography—*[287] *I wish to know how much the whole earth would weigh, with the water, if it were on a scales. We are here to lay as foundations three things: first, (determination of) the space contained by a (cubic) mile;*[288] *second, (determination of) the weight of a (cubic) cubit of earth mixed with water;*[289] *third, (taking for granted) uniformity of the (terrestrial) body. This being set, reduce the terrestrial sphere to cubic miles* (miliaria corporea).

Let (in our terms) d be the diameter of the earth, c its circumference, S its surface, V its volume. Cardan then calculates, successively,

$$d = \left(\frac{c}{\pi} \cong\right) \frac{22\,500}{3 + \frac{1}{7}} \cong 7159 \ \text{(miles)};$$

$$S = d \cdot c \cong 7159 \cdot 22\,500 = 161\,077\,500 \ \text{(square miles)};$$

$$V = \tfrac{1}{6} d \cdot S \cong 1193 \cdot 161\,077\,500 = 192\,165\,457\,500 \ \text{(cubic miles)}.$$

Since 1 mile = 1000 paces, and thus 1 cubic mile = 1 000 000 000 cubic paces,

$$V = 192\,165\,457\,500\,000\,000\,000 \ \text{cubic paces}.$$

---

[284] See his *Mīzān al-ḥikma*, p. 73. Translation in Wiedemann's *Beitrag XIV*, 4, pp. 46–47 [= *Aufsätze*, I, pp. 445–446], or in Khanikoff, p. 79.

[285] See Koran, 3.85 —though the Koran (71.19) considers the earth to be flat.

[286] *Practica arithmetice*, Ch. LXVI, No. 25.

[287] See above, p. 126*n*.

[288] Rather: estimating the volume of the earth in cubic miles and paces; this will be done below.

[289] No such estimate given.

Having then the weight (in pounds) of a (cubic) pace[290] *of earth, supposed to be more or less uniform, and multiplying it by the above number, you will obtain the number of pounds that weighs the sea with the earth, surely not quite exactly since there are mountains and other, but such that there will not occur an error larger than one ten-thousandth.*[291]

§ 7. Perfect numbers

1. Perfect numbers in antiquity

Mediaeval mathematicians were also led to finding large numbers of quite a specific kind. Their characteristics, established during Greek antiquity, became known in the Christian Middle Ages quite early on, being part of the little that had survived of ancient mathematics before the Latin translations from the Arabic (and the Greek) were made in the 12th century. Indeed, in the first years of the 6th century, Boëtius had adapted into Latin the *Introduction to arithmetic* written by Nicomachos of Gerasa (*c.* 100). As a matter of fact, Nicomachos' book was a second-rate work, containing correct information (when taken from good authors) but also some errors (when inferred by the author from numerical examples). This work, however, is important as being the sole ancient mathematical work known in Latin in the early Middle Ages. As the title indicates,[292] it contained assertions on the nature and properties of various types of integers, including perfect numbers: Nicomachos reported their definition and the way to form them, illustrated by examples.

Let N be a natural number. Consider, in our symbolism, the sum of its divisors $\sigma(N)$, as well as $s(N) = \sigma(N) - N$, thus the sum of the divisors of N excluding itself. For the ancient Pythagoreans (between 550 and 350 BC) there were three categories of natural numbers: those having the sum of their divisors less than the number itself, thus with $s(N) < N$, such as 8 and 10, which they called 'deficient' (ἐλλιπεῖς); the numbers with $s(N) > N$, such as 12, 18, 24 —thus those with a large number of divisors— which they called 'abundant' (ὑπερτέλειοι); while

---

[290] Here 'pace' (*passus*), previously 'cubit' (*cubitus*).

[291] *Habito igitur pondere unius passus terre, ex mediocri uniformitate supposita, multiplicando per suprascriptum numerum habebis quot libras ponderat mare cum terra, non quidem omnino precise cum adsint montes & reliqua, sed ita quod non accidet error ex 10000 partibus in una.* Newton had hypothesized that the earth should weigh between five and six times its volume in water. Experimental estimations made between the second half of the 18th century and the second half of the 19th came progressively closer to this interval.

[292] The ancient Greek word 'arithmetic' referred to number theory, not to computational arithmetic.

the others, being neither deficient nor abundant, were called 'perfect' (τέλειοι). Accordingly, in the last definition of Book VII of the *Elements*, Euclid (*c.* 300 BC) defines them as follows: *A perfect number is that which is equal to (the sum of) its divisors* (Τέλειος ἀριθμός ἐστιν ὁ τοῖς ἑαυτοῦ μέρεσιν ἴσος ὤν).

Euclid would not have defined them had he not later a theorem about them. It is the proposition concluding Book IX, which is the last of the three 'books' (or: chapters) dealing with number-theoretical properties of the natural numbers. This remarkable theorem is expressed by him as follows: *If, starting with the unit, any quantity of numbers in double ratio is added up and the sum is a prime, the product of the sum by the last will be a perfect number* (Ἐὰν ἀπὸ μονάδος ὁποσοιοῦν ἀριθμοὶ ἑξῆς ἐκτεθῶσιν ἐν τῇ διπλασίονι ἀναλογίᾳ, ἕως οὗ ὁ σύμπας συντεθεὶς πρῶτος γένηται, καὶ ὁ σύμπας ἐπὶ τὸν ἔσχατον πολλαπλασιασθεὶς ποιῇ τινα, ὁ γενόμενος τέλειος ἔσται). In our terms this means that, if the sum of the sequence of numbers $1, 2, 2^2, \ldots, 2^{n-1}$ (thus $2^n - 1$) is a prime, then $2^{n-1}(2^n - 1)$ will be a perfect number.

As we infer from Nicomachos' *Introduction* (I.XVI.3), the Greeks knew at least the first four such numbers:

$$2\,(2^2 - 1) = 6 \,(= 1 + 2 + 3)$$
$$2^2\,(2^3 - 1) = 28 \,(= 1 + 2 + 4 + 7 + 14)$$
$$2^4\,(2^5 - 1) = 496$$
$$2^6\,(2^7 - 1) = 8128.$$

2. Perfect numbers in the 17th and 18th centuries

Evidently, the search for perfect numbers having the form as defined by Euclid will mean finding prime numbers of the form $2^n - 1$. Since this search depends itself on the form of the exponent n, it may at once be restricted.

— First, n cannot be an even number, thus of the form $2k$ $(k > 1)$; since $2^{2k} - 1 = (2^k + 1)(2^k - 1)$, then $2^n - 1$ will be divisible by at least these two factors.

— Next, n could not be an odd composite number either. If $n = p \cdot q$ $(p, q \neq 1)$, applying the identity $a^t - 1 = (a - 1)(a^{t-1} + a^{t-2} + \cdots + a + 1)$ and taking $t = q$ and $a = 2^p$, or $t = p$ and $a = 2^q$, shows that $2^n - 1$ will certainly be divided by $2^p - 1$ and $2^q - 1$.

Therefore, n must be prime. But this condition is far from being sufficient; indeed, primes of the form $2^n - 1$ are rare, and become increasingly so as we gradually advance in the sequence of natural numbers. Nowadays, we know some fifty of them, but still do not know whether their quantity is infinite or not.

Primes of the form $2^n - 1$ are called 'Mersenne primes', and are therefore designated today as M_n. Marin Mersenne (1588-1648), theologian and mathematician, played an essential part in the correspondence between mathematicians, informing them about the state of current research and passing on contemporary unsolved problems (just as Euler's friend Christian Goldbach was to do a century later). The *Grand Dictionnaire universel du XIXe siècle*, vol. X, tells us that Mersenne died due to the ignorance of a physician who bled him by confusing vein and artery.

As to said primes, Mersenne had set out a list of admissible exponents n, namely:[293]

$$2, 3, 5, 7, 13, 17, 19, 31, 67, 127, 257.$$

Vbi fuerit operæpretium aduertere XXVIII. numeros à Petro Bungo pro perfectis exhibitos, capite XXVIII. libri de Numeris, non esse omnes Perfectos, quippe 20 sunt imperfecti, adeout solos octo perfectos habeat videlicet 6. 28. 496. 8128. 13550336. 8589869056. 137-438691328, & 2305843008139952128; qui sunt è regione tabulæ ←
Bungi, 1, 2, 3, 4, 8, 10, 12, & 29 : quique soli perfecti sunt, vt qui Bungum habuerint, errori medicinam faciant.

Porrò numeri perfecti adeo rari sunt, vt vndecim dumtaxat potuerint hactenus inueniri: hoc est, alii tres à Bongianis differentes : neque enim vllus est alius perfectus ab illis octo, nisi superes exponentem numerùm 62, progressionis duplæ ab 1 incipientis. Nonus enim perfectus est potestas exponentis 68 minus 1. Decimus, potestas exponentis 128, minus 1. Vndecimus'denique, potestas 258, minus 1, hoc est ←
potestas 257, vnitate decurtata, multiplicata per potestatem 256.

Fig. 112

At first, his list does not seem to be very impressive. For the first four numbers were known to the Ancients and Mersenne's list turns out to be both incomplete (61, 89, 107) and erroneous (67, 257). However, Mersenne's merit lies in the vast number of computations made by him. Indeed, his omissions and errors were not detected before the end of the 19th and beginning of the 20th century. Furthermore, the number corresponding to the exponent 127, $2^{127} - 1$, written with 77 digits, was to remain the largest known prime till 1952. Mersenne had, moreover, determined that, from a list of twenty-eight numbers given as perfect by Pietro Bongo, only eight actually were.[294] These eight numbers, which

---

[293] *Cogitata physico-mathematica*, general preface, No. XIX (Fig. 112).

[294] As a matter of fact, Bongo's list comprised only 24 numbers said to be perfect: he had calculated the numbers of the form $2^{n-1}(2^n - 1)$ for $n - 1 = 1, 2, \ldots, 28$ —but did not include the values $5, 11, 17, 23$ (*Numerorum mysteria*, XXVIII, p. 468).

correspond to the first eight numbers of Mersenne's list of exponents, are (Fig. 112, top) the four known to the Greeks and

33 550 336 (2 in Fig. 112 is a printing error)

8 589 869 056

137 438 691 328

2 305 843 008 139 952 128 (3 at the end is a printing error).

(Nos. 1, 2, 3, 4, 8, 10, 12, 19 on Bongo's list, see Fig. 113). It is after these eight numbers that Mersenne adds three others, or, rather, the corresponding exponents (Fig. 112, bottom).

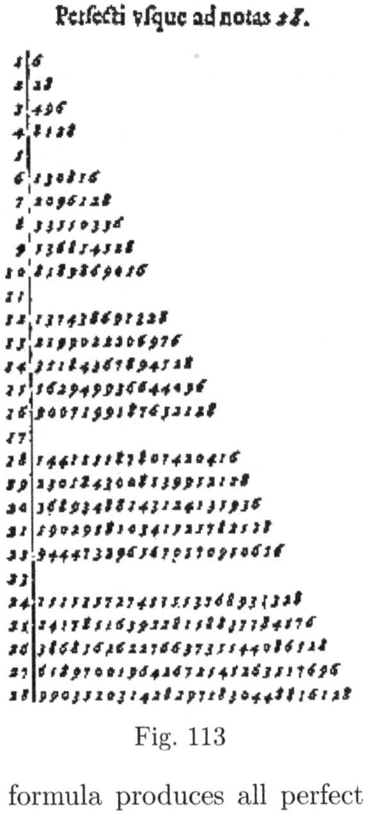

Fig. 113

Whether Euclid's formula produces all perfect numbers is still unknown today. Obviously, all such numbers must be even. And Euler has demonstrated that all even perfect numbers must have the form found by Euclid. His reasoning was the following. Suppose N to be an even perfect number. It will then have the form $N = 2^h \cdot a$, with a odd and $h \neq 0$. Since these two numbers cannot have any common divisor other than 1, the sum of the divisors of N, $\sigma(N)$, will be the product of the sum of the divisors of 2^h by the sum of the divisors of a; that is, $\sigma(N) = \sigma(2^h) \cdot \sigma(a)$. Now

$$\sigma(2^h) = 1 + 2 + 2^2 + \cdots + 2^h = 2^{h+1} - 1,$$

$$\sigma(a) = 1 + a + d_1 + d_2 + \cdots + d_j,$$

where $d_1 + d_2 + \cdots + d_j = D$ is the sum of the divisors of a other than the banal ones 1 and a (thus $D = 0$ if a is prime).

Since, by definition, $s(N) = \sigma(N) - N$ while, by hypothesis, $s(N) = N$, we must have $\sigma(N) = 2N$, and therefore

$$(2^{h+1} - 1)(1 + a + D) = 2^{h+1} \cdot a,$$

from which is inferred that

$$a = (2^{h+1} - 1)(1 + D).$$

Now if we suppose that a is not prime, thus that $D \neq 0$, a must be divisible by these two integral factors —both certainly different from 1 since $h \neq 0$ and $D \neq 0$. Therefore, they should be among the divisors of which D is the sum. But

$$(2^{h+1} - 1) + (1 + D) = 2^{h+1} + D > D.$$

This means that the supposition that a is not prime must be rejected. But then, if a is prime, we must have $D = 0$. Accordingly, a takes the form $a = 2^{h+1} - 1$, and the even perfect number considered initially must be of the form $N = 2^h(2^{h+1} - 1)$, which is precisely that found by Euclid.

With this demonstration, the question of the generality of Euclid's formula reduces to that of the existence of odd perfect numbers. Now, none has been found, though it has not be demonstrated that they could not exist. Which is exactly what Euler asserted in the conclusion of his demonstration.[295]

3. Perfect numbers in the Middle Ages

As said above, the existence of perfect numbers was already known in the early Middle Ages through the work of Nicomachos-Boëtius. They are therefore also mentioned by later authors, some of whom we have already encountered. Fibonacci gives the rule for forming them and computes the first three; he writes that, proceeding likewise, we can find others

---

[295] *Quamobrem alii numeri perfecti pares reperiri non possunt, nisi qui contineantur in formula prius inventa $a = 2^n (2^{n+1} - 1)$, existente $2^{n+1} - 1$ numero primo; haecque est ipsa regula ab Euclide praescripta. Utrum autem praeter hos dentur numeri perfecti impares nec ne, difficillima est quaestio: neque quisquam adhuc talem numerum invenit, neque nullum omnino dari demonstravit (Opera postuma, I, p. 88).*

to infinity.[296] What he meant exactly we do not know. In any case, the illustration of Fig. 114, taken from a late 14th-century manuscript, is revealing:[297] some authors considered that, in order to find perfect numbers larger than the first four, we were simply to calculate the products $2^{n-1}(2^n - 1)$: here $128 \cdot 255 = 32640$ and $256 \cdot 511 = 130816$ are followed by the instruction *e così fa tute* —after all, Bongo was still doing the same two centuries later.

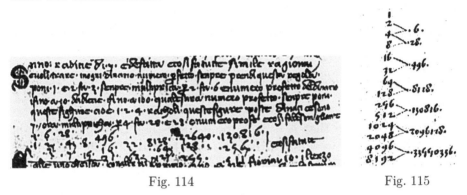

Fig. 114   Fig. 115

Other authors, however, made more constructive attempts. But the rarity of perfect numbers led to dealing with fairly large numbers, and, since the numerical verification is difficult, it is hardly surprising that some numbers claimed to be perfect were not.

[128] Chuquet reports correctly the definition and the rule for forming perfect numbers.[298] For those following 6, Chuquet computes successively: $2^2(2^3 - 1) = 28$; $2^3(2^4 - 1)$ not perfect since $2^4 - 1 = 15$ is not prime; $2^4(2^5 - 1) = 496$; $2^5(2^6 - 1) = 8128$, *and in this way innumerable perfect numbers may be found*, as was already said by Fibonacci; but Chuquet then observes that, relative to 'imperfect' numbers, there are very few of them. Next, referring to a figure of his (Fig. 115[299]), he observes that perfect numbers end alternately in 6 and 8.

This needs commenting. First, in his table, only the first four and the last are perfect; it would thus seem that Chuquet considered 511 and 2047 to be prime numbers. As to the alternation of 6 and 8, already stated by Nicomachos (below, p. 138*n*), who apparently inferred it from

[296] *Et sic semper faciendo poteris in infinitum perfectos numeros reperire* (*Liber abaci*, p. 283).

[297] MS. Florence BNC Conv. soppr. G 7 1137, fol. 109^v.

[298] *Triparty*, p. 620; MS., fol. 20$^{bis\,v}$–21^r. It appears from this place in the MS. that Chuquet takes his information directly from Boëtius, for he writes: *Et tout ce* (= ceci) *dit boece en son arismetique* (fol. 20$^{bis\,r}$), and later *selon ce que dit Boece* (fol. 21^r).

[299] *Ibid.*, fol. 20$^{bis\,v}$.

the first four perfect numbers, it is simply wrong. It is true, though, that only 6 and 8 can be the last digits of perfect numbers: since the powers of 2 end with 2, 4, 8, 6 and the subsequent power less 1, with 3, 7, 5, 1, and 5 is excluded, N may have only, as its last digit, that of the products $2 \cdot 3$, $4 \cdot 7$, $6 \cdot 1$, thus 6 and 8.

[129] Pacioli, after explaining in detail the theory, wished to verify that $9\,007\,199\,187\,632\,128$ is indeed perfect.[300] He thus carried out successive divisions by 2, until he ended with an odd number, namely $134\,217\,727$, which he declared to be *indivisibilis*. Now a simple division by 7 would have proved the contrary. But we have already noted some weaknesses in Pacioli's computations (p. 104n).

1				
2	3	primus	6	perfectus
4	7	primus	28	perfectus
8	15	compositus		
16	31	primus	496	perfectus
32	63	compositus		
64	127	primus	8128	perfectus
128	255	compositus		
256	511	compositus		
512	1023	compositus		
1024	2047	compositus		
2048	4095	compositus		
4096	8191	primus	33550336	perfectus
8192	16383	compositus		
16384	32767	compositus		
32768	65535	compositus		
65536	131071			

Fig. 116

[130] Before him, however, the fifth and perhaps another perfect number occurred in a table.[301] This table (Fig. 116) gives the two factors of Euclid's formula, then indicates whether the second is prime or composite, and, in the first case, the corresponding perfect number. Note that this indication is missing in the last line. Now $131\,071$ is indeed prime, and

---

[300] *Summa*, fol. 8^r.

[301] Curtze, *Studienreise*, pp. 288–289 & 305 (MS. Vienna Palat. 5203, see below p. 257).

thus 8 589 869 056 is the sixth perfect number. One possible reason for this omission is that this perfect number is the first to break the alternation of 6 and 8 conjectured by Nicomachos, and believed later, on the basis of his assertion, to hold generally.[302]

Let us conclude this chapter on large numbers by mentioning a strange marginal note in a 14th-century manuscript.[303]

[131] *If you wish to know how many angels fell from Paradise into Hell:*[304] *Suppose an empire in which there are a thousand thousand kingdoms, in each kingdom a thousand thousand provinces, in each province a thousand thousand towns, in each town a thousand thousand castles, in each castle a thousand thousand towers, in each tower a thousand thousand refectories, in each refectory a thousand thousand diners. That many fell. Let he who can do it calculate their number, and let him then drink if he is thirsty.*[305]

Remark. Since the word 'million' did not exist in Latin, it was customary to use 'a thousand thousand' for 10^6. This being repeated seven times (again number 7, see p. 112), the total number of angels in disgrace is 10^{42}.

---

[302] *Introduction*, I.XVI.3 (ed., p. 40): καὶ παρέπεται αὐτοῖς μίαν παρὰ μίαν εἰς ἑξάδα ἢ ὀγδοάδα καταλήγειν καὶ πάντως εἶναι ἐν ἀρτίοις; in Boëtius version (see p. 131 above): *Et semper hi numeri duobus paribus terminantur, VI et VIII, et semper alternatim in hos numeros summarum fine provenient* (I.20; ed., p. 42).

[303] MS. Reims G.559, fol. 88. See the Catalogue of this library, II.1, p. 46.

[304] Fallen angels, thus angels expelled from heaven, according to the Book of Enoch (altogether 72 there).

[305] *Si vis scire quot angeli ceciderunt de paradiso in inferno, suppone unum imperium in quo sunt mille milia regna, in quolibet regno mille milia provincie, in qualibet provincia mille milia civitates, in qualibet civitate mille milia castra, in quolibet castro mille milia turres, in qualibet turre mille milia tinella, in quolibet tinello mille milia comestores. Tot ceciderunt. Et computet numerum qui computare poterit, et postmodo bibet si sitit.*

Chapter IX. Arrangements

§ 1. Musical chairs

The problem of permuting places being quite clearly expounded by the 16th-century mathematician Buteo, we shall just reproduce his inductive reasoning.[306]

[132] *Twelve guests were sitting on a bench to eat; it is asked in how many different ways they may change the order of their places.*

You will set out the search for this change as follows. One alone may sit in just one way. Two in two, namely each occupying in turn the first place. Three may vary their places in six manners; for, the first remaining at his place, the two others will change their seats twice, and each of the others taking likewise the first place this will produce twice a different arrangement; therefore, there will be for three a placing which can be modified six times. Consequently, the placing for four can be varied in twenty-four manners; for if the first is placed on the first seat, the other three will change among them six times their place; likewise then: the three others occupying in turn the first place, there will be three times six manners for the change.

The rule for any quantity proceeds from all that: if a number of ways of sitting is multiplied by the immediately following number of persons sitting, this will give the number of ways of sitting for this latter number.[307] For instance, in the present case, as follows. Multiplying one way of sitting by two persons gives two ways of sitting. Multiplying likewise two ways of sitting by three persons sitting gives six ways of sitting for three persons. Likewise, multiplying 6 by 4 gives 24 ways of sitting. Multiplying then 24 by 5 gives 120. Next, multiplying 120 by 6 gives 720. Multiplying then 720 by 7 gives 5040. Proceeding always likewise the multiplication will reach 12, giving 479001600 different ways of sitting, which can be realized with twelve guests sitting on a bench. That is what was required.

Pacioli also has a problem of guests, in which he computes each of the cases up to eleven persons.[308]

§ 2. The broken eggs

A characteristical example is the following.

[133] *A girl was carrying eggs to sell in the market, and met a young man who wanted to play with her, and broke all the eggs and did not want to pay for them. She had him summoned before the judge, who condemned him to pay for the eggs.*

---

[306] *Logistica*, Ch. IV, No. 22, pp. 219–221.

[307] In our terms: $P_n = n \cdot P_{n-1}$, with $P_1 = 1$, and therefore $P_n = n!$.

[308] *Summa*, fol. 43$^\mathrm{v}$–44$^\mathrm{r}$, No. 29.

But the judge did not know how many eggs there were, and asked the girl. She said that she did not know, for she was young and could not count; but she had arranged them 2 by 2, and one egg remained; then 3 by 3, and one egg remained; then 4 by 4, and one egg remained; then 5 by 5, and one remained; then 6 by 6, and one remained; then 7 by 7, and nothing remained. I ask how many eggs she had.[309]

To find the answer, we are told to multiply the numbers of the arrangements leaving one egg and to add to the result this remaining egg, thus $2 \cdot 3 \cdot 4 \cdot 5 \cdot 6 + 1 = 721$. Such was the payment ordered by the judge.

Fig. 117   Fig. 118

Contemporary French texts have a similar problem, with the same data and (of course) the same material.[310]   At most the context may differ. Thus a woman may be knocked accidentally by someone, or a person on a horse, and that person asks her the number of eggs to be repaid. Although the data are the same as before, the result found in these two texts is 301.

Now since the girl, or woman, when arranging her eggs by 2, 3, 4, 5, 6 has one egg left, their number must be some multiple of the common multiple of 2, 3, 4, 5, 6, plus 1; thus it must have the form $m \cdot 2^2 \cdot 3 \cdot 5 + 1 = 60m + 1$, with m natural. But since the required number must be divisible by 7, we shall put $60m + 1 = 7n$, with n natural as well. Solving the equation $7n - 60m = 1$ using Euclid's algorithm or, more simply, by trial and error, one finds as the smallest solution 301 ($m = 5$). Generally, since 420 is the lowest common multiple of 2, 3, 4, 5, 6, 7, the next solutions will be $301 + 420 \cdot k$, with k natural, thus 721 for $k = 1$.

Trial and error means here varying m or n in the equation $7n - 60m = 1$. For instance, one may take the successive multiples of 60 and look for the first which, divided by 7, will leave 6 as the remainder; adding 1 will

---

[309] *Livre de chiffres et de getz*, No. 20 (text in *Récréations*, p. 160). The illustrations are from Tagliente's *Componimento*, ed. 1547 and 1554.

[310] MSS. Rouen I 58, fol. 85ᵛ & 89ʳ (*ce sont III C et un oeufs*), and Paris BNF fr. 1339, fol. 78ʳ–78ᵛ.

give the required number. Such is the rule given by Fibonacci.[311] One may also, as Chuquet does, take the odd multiples of 7 (since the result of the subtraction must be odd) and consider those which, when divided by the integers lower than 7, leave 1. Like Fibonacci, Chuquet thus reaches the smallest solution 301. But he also suggests, as a computing rule, calculating (in our terms) $6! + 1 = 721$, as in our first example. Indeed, as he points out, such problems may have many different answers.[312] Now adopting his solving generally would imply that $m! + 1$ is *always* divisible by $m + 1$; which is true in the case $m = 6$ and always if $m + 1$ is prime, as was to be asserted three centuries later (Wilson's theorem).

There was another way of finding the solution, used for instance by Tagliente in the same problem of broken eggs (illustrations above). He reaches the correct result 301 by calculating $(6 \cdot 7 + 1) \, 7$.[313] This infuriated Tartaglia, who wrote that it is ridiculous (*cosa ridiculosa*) to give a rule which is only applicable in one case instead of trying to find a general one (*regola generale*). This is what he did, and he reached the result, he writes, while his treatise was being printed, on the 14th of June 1554.[314] Oddly enough, his rule is merely that already proposed by Fibonacci, seen above: finding the common multiple of 2, 3, 4, 5, 6, thus 60, and choosing from among its multiples those, or one, which will leave the remainder 6 when divided by 7; increasing it by 1 will give the required number.

[134] In order to convince the reader of the validity of this rule, Tartaglia adds two problems. The first, of a gentleman and a peasant women, with a quantity of eggs leaving the remainder 1 with 2, 3, 4 and 0 with 5. Of the multiples of 12 he takes 84, with the smallest suitable number then being 85 (he does not consider 25; because it would not fill a basket?). The other problem is about a gentleman with a horse and a peasant with a mare carrying eggs. Now the horse 'put himself behind the mare' (*si misse dietro a questa cavalla*), with the inevitable happening to the eggs. The compensation was calculated according to the peasant's claim: that there was always one egg left except for the rows of 11. As Tartaglia calculates, 2520 is the common multiple, 25 200 divided by 11 leaves 10, so 25 201 is the required number.

[135] His next example is slightly different. A gentleman's treasurer must count his master's gold ducats. To do so, he puts them in rows of 2 to

---

[311] *Liber abaci*, p. 282.

[312] *Appendice*, pp. 452–453; MS., fol. 202^v–203^r.

[313] *Farai così: moltiplica 6 fia 7, fa 42; dapoi aggionge 1 sopra 42, fanno 43, & questo 43 moltiplica per 7, fa 301. & 301* ('trecento e uno') *ovo era nel cesto.*

[314] *General trattato*, I, fol. 257^v.

9 and notes that one is always missing to complete the last row (thus the remainder is always one less than the divisor). This time, 1 is to be subtracted from the common multiple. Tartaglia thus finds 7559 as the smallest number (but 2519 is in fact the smallest).

[136] The case in which the successive remainders are 1, 2, 3 and so on to the penultimate counting, but with zero as the last remainder, is also considered. With 7 as the last divisor, Fibonacci finds the answer $59 + k \cdot 60$, with 119 as the first admissible answer.[315] Whereas Fibonacci has the problem with an abstract number, Benedetto da Firenze considers a basket of eggs. He gives as the possible answers $119 + t \cdot 420$; for the actual answers, he adds, the basket will have to be taken into account.[316]

§3. Unwanted passengers

This one is often called the Josephus problem. Emperor-to-be Vespasian, at that time one of Nero's generals, went to Judea in 67 to quash a revolt. Josephus, then governor of Galilee, went to Jotapata (today: Yodfat) to defend it. He did so for some fifty days, but then Jotapata fell. With forty of his companions, he took refuge in a pit, from which a passage led to a cavern where they could hide. A woman having given away their hiding place, Vespasian ordered them to surrender, and let Josephus know that he would be safe. His companions, however, were ready to commit collective suicide rather than surrender to the Romans, and began to threaten him. Now, as Josephus himself tells us, God had ordered him to remain alive.[317] The Romans indeed treated him well, since he received Roman citizenship and could stay at one of the imperial palaces of Titus, then emperor, in Rome.

How Josephus managed to survive is the subject of this problem, explained in detail by Bachet de Méziriac in the introduction to his work on recreational problems. His source is, he tells us, Hegesippus.

[137] *Josephus, who left us an account of the same story, is known for both his literary writings and his military ability. Being then governor of the very important town of Jotapata, he had to resist the assault of the Roman army led by Vespasian; in spite of the many proofs of his competence and valour displayed during the siege,*

---

[315] *Liber abaci*, p. 282.

[316] *Et però al giudichare si vuole avere riguardo al paniere.* See Arrighi, *Scritti scelti*, pp. 347–348 (from MS. Florence BNC Magl. XI 76).

[317] Περὶ τοῦ Ἰουδαϊκοῦ πολέμου ('The Jewish War'), III, 351–354, 361. Our two sources for the whole story are Josephus' own account in that work, III, 340–398, and the *Historiae* of Hegesippus, a Christian writer of the second half of the 2nd century, III.15, 4 & 18, 1–2.

he could not prevent the town being conquered after several assaults. Not knowing what to do in the face of an evident danger, and followed by a group of soldiers who wanted to share his fate, he withdrew into a cavern, or an underground cave, which the author, in the Hebrew manner, calls a lake.[318] Having been there a few days, and forced by necessity, he suggests to his companions that it would be more opportune to surrender to the Romans and to be at the mercy of the victors instead of ending their days miserably there, waiting for starvation, which might make them eat one another like wild animals.

But these soldiers, having conceived in their weak minds the hope to gain, by a generous act, immortal fame, and preferring, as it seemed to them, an honourable death to a shameful life, refused altogether. They were resolved to die by their own hand, killing one another courageously, rather than give in, defeated by starvation, or surrender out of cowardice to an idolatrous people, enemy of God's people, and bear all the indignities that an insolent victor might inflict on those defeated. Josephus then attempted to dissuade them from doing so, explaining that such a plan showed more cowardice and lack of courage than generosity. They stood by their plan and even began threatening, should he not agree, to force him, and make him the first victim of their tragic fate. Josephus, not seeing any way to escape, thought up the following ruse. He pretended to agree, but convinced them, in order to avoid the disorder and confusion which might arise while killing one another without co-ordination, to stand in a row and, beginning to count from one end, kill always a given one —but the author does not say which given one.[319] Proceeding thus, in the end he alone with one other remained alive; since he was eloquent and uncertain,[320] it was easy for him to impose his will upon his companion, by choice or by force.

Now Josephus no doubt escaped an inevitable death by the device of my twentieth problem.[321] For he had foreseen that, organizing the killing as he had ordered, two only were to remain, and he made sure of being among his soldiers, in all innocence, in one of those two places.

This is quite a remarkable story, and it teaches us that one should not hold in contempt those little subtleties that sharpen the mind and enable men to do higher things, and may even be put to an unforeseen use. Bachet has thus, in the

---

[318] At first sight, this does not seem to make sense. In fact, Josephus' text has λάκκος, which means cavity but also reservoir; Hegesippus' text has *lacus*, which has the same meaning but, in addition, that of 'lake'.

[319] Indeed, it is only said that the one drawn by lot will have his throat cut by the next one: *ut is qui sorte exierit ab eo qui sequitur interimatur* according to Hegesippus (III.18, 2), ὁ λαχὼν δ' ὑπὸ τοῦ μετ' αὐτὸν πιπτέτω according to Josephus (III, 388).

[320] He did not want to kill the other and feared to be killed, the sources report.

[321] Where, however, Bachet specifies that each ninth will be killed.

introduction to his work, justified the study of recreational problems, with this example where it was a matter of life or death.

The context, in the mediaeval version of this problem, is different, although it always represents a threatening situation. The most common is the following. A ship is in the middle of a storm and near to sinking. The captain considers that the only possibility is to throw overboard half the passengers. Now half are Christians and the other half foreigners, the nature of which depends on the epoch. In times of famine and scarcity, when usury becomes extreme, they are Jews; in times of piracy, they are Moors or Saracens; from the 16th century on, when Ottomans were attacking Europe, they are Turks —the rôles being of course inverted in the opposite party, with Moslems or Jews wanting to drown Christians.[322] Moreover, the choice of disposable passengers may be more or less restricted: they may be just pagans, or else monks of a different order.[323] Anyway, this problem was older than these time-dependent enmities, and the name of Josephus was only later associated with a problem no doubt long known; ancient solutions survived in the 10th and 11th centuries, even with other choices than every ninth.[324] As to the problem with two survivors, it occurs later (below, pp. 146–147).

Fig. 119

Chuquet explains quite clearly how to *find* the appropriate arrangement. We may therefore just reproduce his reasoning.[325]

---

[322] For this latter case (examples of the first are to follow), see Murray, *Hist. of chess*, p. 280*n*; Moshe ben Maimon (Maimonides, d. 1204), *Hilkhoth Akum* X, 1 (*non licet misereri eorum*, transl. Pranaitis).

[323] Arrighi, *Scritti scelti*, p. 360, according to the treatise of Benedetto da Firenze (*su una nave erano 15 cristiani et 15 paghani*, from MS. Florence BNC Magl. XI 76); or (different monks), illustration from F. Calandri's manuscript, MS. Florence Biblioteca Riccardiana 2669, fol. 103ᵛ (text, ed. p. 205: *15 frati di sancto francesco, 15 monaci di Camaldoli*, thus Franciscans and Camaldolites).

[324] Curtze, *Zur Geschichte des Josephspiels*; Tropfke, pp. 652–653.

[325] *Appendice*, pp. 453–454; MS., fol. 203ᵛ – 204ʳ. For a mathematical solution of the

[138] *There is a captain who has as passengers 15 Christians und 15 Jews. Because of the storm he is obliged to throw off from his boat half the people in order to save the other half; otherwise, all would die and the boat would sink into the sea. He would like to find a way for only the Jews to be thrown into the sea and the Christians to remain safe, but he fears that the Jews will rebel and not accept his will. In order, however, not to be seen as having a preference for certain persons to the detriment of the others, he had them sit side by side in a circle with the Christians mixed in with the Jews in a certain order. He then said: 'It is necessary that half of you be thrown into the sea in order to save the other half. You see in what extreme danger we are. I order that, counting continuously from 1 to 9, then again from 1 to 9, and so on always, the ninth be thrown into the water.' They all agreed with this way of proceeding. He then began counting 1, 2, 3, 4, and so on to the ninth, who was thrown into the sea. And from the tenth he resumed, 1, 2, 3, and so on to the ninth, who likewise was thrown into the sea. And this counting was repeated until 15 persons were thrown into the sea, and this was done so skilfully that the 15 Christians were kept and saved whereas the 15 Jews were drowned in the sea. Required the arrangement adopted.*

Solution to perform such computing. One may use 30 ○ instead of 30 persons and put them in a circle or in a line, then say where the beginning is to be set and count as seen above, always crossing out the ninth ○, like this: ⊘. One is to continue likewise till 15 ○ are crossed out. Then consider the ⊘, crossed out, for the Jews, and the ○, not crossed out, for the Christians. For as their order is in the figure, in such order were placed the Christians and the Jews by the captain. The beginning in the figure is marked by a cross (Fig. 120; a line in our transcription, Fig. 121), *and the subsequent 4 ○ towards the right, not crossed out, stand for 4 Christians, and the five ⊘, crossed out, stand for 5 Jews, and so on for the others to the 30th, which is next to the cross, which is ⊘, crossed out.*

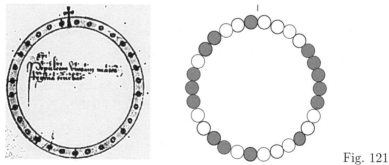

Fig. 120   Fig. 121

general problem of n persons of whom each pth is successively removed, see Schubert, *Zwölf Geduldspiele*, pp. 124–132.

The way for the 15 Christians and the 15 Jews may likewise be applied to 18, and 24, and all the other numbers; moreover, one could take the 10th, or the 11th, or whichever we wish, as was done with the 9th.

Such is the way Chuquet teaches for establishing the above figure: starting from the line (his cross), we shall count the circles and cross out each ninth, then continue disregarding the circles crossed out.

Now once the figure is constructed one should be able to remember its arrangement. To that end, a Latin verse may be memorized, the number of each vowel (from 1 to 5) indicating the sequence of circles in the same state. In our case, said verse is

Populeam virgam mater regina tenebat

giving the sequence 4, 5, 2, 1, 3, 1, 1, 2, 2, 3, 1, 2, 2, 1, summing up to 30. See Fig. 120, inside the circle.

The same arrangement is found in several other texts, the only differences being the kind of people and the mnemonic verse, as in the following three examples.

— In MS. Rouen I 58, fol. 88^v, the unwanted passengers are Saracens and the verse finishes with *ferebat*.

— In P. M. Calandri's *Tractato d'abbacho*, the mnemonic verse is

Noue la pinta dà e certi mantenea.

Here half are *cristiani*, as expected, while the other passengers are *giudei* at the beginning of the problem but *mori* at the end.[326] Nice example of uncritical borrowing from different sources.

— In the *Livre de chiffres et de getz*, involving Saracens, the circle becomes a straight line (as suggested by Chuquet) and the mnemonic is a set of Latin verses; indeed, the successive quantities of same figures are expressed in words:

Post quattuor quinque da, post duos unum collega.
Tres numerabis, postea unum collocabis.
Unum dic pariter, et duo consequenter.
Duos post apponas, et tres simul hic apponas.
Semel dic ante bis, post duos unum terminabis.
Primi christiani sunt, sarracenique secundi.

At the end of his text, Chuquet alluded to different numbers of persons n and different numberings p. But it may also happen that the final remainder is to differ. Thus Pacioli has the arrangement for thirty Jews

---

[326] See his treatise, ed. pp. 142–143.

and two Christians sitting one next to the other. He tells us to begin from the fifth following them and count, but going towards them, every ninth. Indeed, in the end, only the two Christians will remain of the initial 32 persons.[327]

In the middle of the 15th century, with 30 persons, one finds different numberings p:[328]

— $p = 6$: *Larga dei pietas bene manes omnia papam* (below, Fig. 125)

— $p = 8$: *Arte parare mea veniant adistere sorte* (below, Fig. 127)

— $p = 9$: *Non dum pena minas a te declina degeas* (above, Fig. 121)

— $p = 10$: *Rex angli cum veste bona dat signa serena* (below, Fig. 128)

— $p = 12$: *Ibant per montes q(u)erebant desidiosa* (below, Fig. 130).

Tartaglia has a larger choice of both numbering and mnemonics.[329] In his case, we are to either discard black counters or throw off Turks. The answers are given for $p = 3$ to $p = 12$, once again to be memorized with obscure verses, Latin or Italian, three for each arrangement (four for the common case $n = 9$).[330] Omitting the figure for the case $n = 9$ seen above, they are the following. (We begin with the white circles, except for the cases $p = 5, 7, 12$.)

— $p = 3$: 2, 2, 1, 1, 1, 2, 2, 2, 1, 1, 1, 2, 2, 2, 1, 1, 2, 1, 2, 1 (Fig. 122)

— $p = 4$: 1, 1, 1, 2, 1, 2, 2, 3, 2, 1, 1, 1, 1, 1, 2, 2, 3, 2, 1 (Fig. 123)

— $p = 5$: 2, 2, 2, 1, 1, 1, 1, 1, 1, 2, 2, 1, 1, 1, 1, 2, 3, 4, 1 (Fig. 124)

Fig. 122   Fig. 123   Fig. 124

— $p = 6$: 1, 1, 2, 3, 3, 2, 1, 2, 2, 1, 2, 4, 3, 1, 1, 1 (Fig. 125)

— $p = 7$: 1, 3, 1, 1 3, 2, 3, 3, 2, 1, 2, 4, 2, 1, 1 (Fig. 126)

[327] *De viribus quantitatis*, I, LVI (Agostini, p. 187).

[328] *Algorismus Ratisbonensis*, No. 80.

[329] *General trattato*, I, fol. 264^v – 265^r.

[330] See *General trattato*, ad loc., or *Récréations*, pp. 170–171.

— $p = 8$: 1, 2, 1, 1, 2, 2, 1, 2, 3, 1, 1, 3, 2, 2, 4, 2   (Fig. 127)

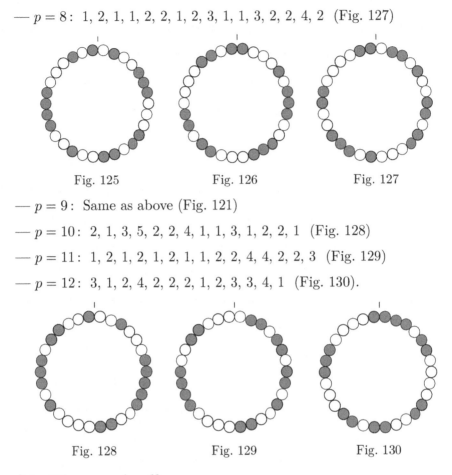

Fig. 125   Fig. 126   Fig. 127

— $p = 9$: Same as above (Fig. 121)

— $p = 10$: 2, 1, 3, 5, 2, 2, 4, 1, 1, 3, 1, 2, 2, 1   (Fig. 128)

— $p = 11$: 1, 2, 1, 2, 1, 2, 1, 1, 2, 2, 4, 4, 2, 2, 3   (Fig. 129)

— $p = 12$: 3, 1, 2, 4, 2, 2, 2, 1, 2, 3, 3, 4, 1   (Fig. 130).

Fig. 128   Fig. 129   Fig. 130

§ 4. The nuns' cells

J. Ozanam's *Mathematical and physical recreations* (see p. 262) begins with the following problem.

[139] *A blind abbess, visiting her nuns, who are equally divided up among eight cells placed at the four corners of a square and in the middle of each side, finds everywhere an equal number of persons in each row of three cells; visiting them another time, she finds in each row the same number of persons, although four men have entered; visiting them a third time, she finds once again in each row the same number of persons, though the four men have left, each with one nun. Required how this can and must be.*

The same problem also opens a new edition, revised by Grandin, who completes the solution. However, the later reviser Montucla puts it in the tenth chapter (No. 20): he blames the previous authors for starting with a somewhat indecent problem just to catch the reader's attention. His own version asks how, when a square is divided into nine cells, counters must be arranged in its outer ones in order to have nine in each border row with the total number varying from 20 to 32.

Ozanam's original solution is as follows. In the beginning, there are three nuns in each cell (Fig. 131; the abbess is in the central one). Then one nun in each corner cell lets a man come in while the two others move into the next median cell (Fig. 132). After the couples have all left, four nuns leave the median cells and go into the empty corner cells (Fig. 133). There are thus each time nine persons in each side, even though the total number was successively 24, 28, 20.

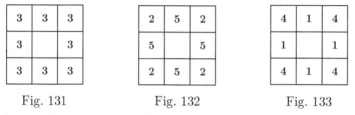

Fig. 131   Fig. 132   Fig. 133

The later editors continue the story. Taking the situation of Fig. 133, Grandin has each departed nun now return with two men; after they have each entered a median cell, they are joined by three nuns from the corners (Fig. 134); there are now 32 people altogether. In this situation (but with counters) Montucla has four more men entering, with all, nuns and visitors, being grouped together in the median lateral cells (Fig. 135), thus now 24 nuns and 12 men altogether; then all the men leave with six nuns, the remaining eighteen nuns dividing themselves up as appropriate (Fig. 136). Montucla concludes: *It is certainly superfluous to show how the illusion of the abbess arises. Indeed, the numbers in the corner cells are counted twice since these cells are common to two rows. Thus, the more crowded the corner cells are when the middle ones are emptied, the more double-counting there is; this is why there seems to be always the same number although it is reduced. The contrary occurs by filling the median cells when the corner cells are emptied, whereby one has to add units so as to have 9 in each row.*

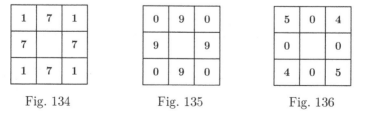

Fig. 134   Fig. 135   Fig. 136

Indeed, let us suppose that there is c in each corner cell and m in each median one. We must have $2c + m = 9$ in each row. Let us then put, with

t integral, $c = 3 + t$ and thus $m = 3 - 2t$. This last relation excludes the values $t > 1$ and the first, the values $t < -3$. We may then choose the values $t = 0$, $t = -1$, $t = 1$, $t = -2$, $t = -3$; they correspond to the above figures 131–135 (in the last, c and m no longer have the same values on each side).

Chapter X. Crossing a river

§1. Goat, wolf and cabbages

As we have seen above (p. 2), Firri, trying to fool Tirri, had proposed the question of a river being crossed by a man ferrying over a goat, a wolf and a cabbage. The answer to this riddle was, as Tirri pointed out, well known at that time. It is thus no surprise to find this problem in Alcuin's Collection:[331]

Fig. 137

[140] *A man had to take across a river a wolf, a goat and a load of cabbages. He could only find a boat allowing two of them to pass. Now he had to carry over all three without any damage. Let the one who can do it tell how he could have them cross safely.*

Solution. In the same manner,[332] *I have the goat cross first, leaving behind the wolf and the cabbages. I return and take the wolf across. Taking it out of the boat, I put the goat in the boat again and take it to the other side. Letting it out, I take the cabbages over. Rowing again, I take the goat and bring it to the other side. That way, the crossing has been performed safely, avoiding any loss.*

Although being more a question of logic than of mathematics, such problems are frequently encountered in mediaeval mathematical works. As Bachet wrote after treating the problems of decanting and of crossing a river (below, [142]), although these problems are 'somewhat ridiculous' (*soyent comme ridicules*), they require subtle reasoning and are therefore not out of place in mathematical textbooks.

But the most frequent version in the Middle Ages is that of the jealous husbands; the conditions are different, but the problem can thus be extended. Consider then that a river must be crossed by two groups of m persons by means of a boat with p places ($p \geq 2$), and under the condition that *the members of one of these two groups may never be in greater number than the members of the other in the same place.* Thus, in the

---

[331] *Propositiones*, No. 18. We find in the 15th-century MSS. Tours 399 and Paris BNF fr. 1339 a French version of it; see *Récréations*, p. 175. Illustration, for the same problem, from the MS. New York Columbia X511.AL3 (ed. Vogel, *Ital. Rechenbuch*, p. 130).

[332] The previous problem is also about crossing; see below, [141].

problem of husbands and wives, or of brothers and sisters, the women may never outnumber the men, either on the banks or on the boat. Thus the rôles of the two groups are not interchangeable, and therein lies the difficulty.

§ 2. Two couples and a boat with two places

There are two possibilities, namely:

AaBb				AaBb		
Bb	Aa →			AB	ab →	
Bb		Aa		AB		ab
Bb	← A	a		AB	← a	b
ABb		a		AaB		b
b	AB →	a		a	AB →	b
b		AaB		a		ABb
b	← a	AB		a	← A	Bb
ab		AB		Aa		Bb
	ab →	AB			Aa →	Bb
		AaBb				AaBb

Fig. 138

By the way, there is no difference between the two solutions since one merely inverses the steps of the other.

§ 3. Three couples and a boat with two places

As already noted (p. 151n), Alcuin has another crossing problem, namely the present one. Fig. 140 represents the various possibilities, which differ only at the beginning (I and I′) and at the end (V and V′, VI and VI′); indeed, steps III & IV are common while steps II-II′ are not fundamentally different. Alcuin's solution, which follows, describes the steps I, II, III, IV, V, VI.[333]

Fig. 139

[141] *There were three (unrelated) brothers, each with his sister, who had to cross a river. Each felt desire for the other's sister. Having attained the river, they found a boat enabling only two of them to pass. May the one who can do it tell how they crossed the river without any of the sisters being offended.*

---

[333] *Propositiones*, No. 17. * * indicates where we have emended the extant Latin text. Illustration from the MS. New York Columbia X511.AL3 (ed. Vogel, p. 132).

I *I* (A) *and my sister* (a) *entered the boat first and crossed; I let my sister out and rowed the boat back to the other bank.* **II** *Then the sisters of the two men remained on the bank crossed; when they came out of the boat, my sister, who had passed first, boarded the boat and brought it to me.* **III** *When she got out, the two brothers* (BC) *boarded and crossed the river; one of them boarded the boat with his sister* (Bb) *and came back to us.* **IV** *I myself and he who had crossed* (AB), *leaving* *our sisters* *behind, crossed; upon arrival on the bank,* *the woman stayed there* (c) *came back with the boat.* **V** *She took my sister with her* (ac) *and they both came to us, and the one whose sister had remained* (B) *got on the boat (and crossed).* **VI** *He took her with him. The crossing was thus made without any unseemly contact.*

I	AaBbCc			**I′**	AaBbCc		
	BbCc	Aa →			ABCc	ab →	
	BbCc		Aa		ABCc		ab
	BbCc	← A	a		ABCc	← a	b
II	ABbCc		a	**II′**	AaBCc		b
	ABC	bc →	a		ABC	ac →	b
	ABC		abc		ABC		abc
	ABC	← a	bc		ABC	← a	bc
III			AaBC		bc		
			Aa	BC →	bc		
			Aa		BbCc		
			Aa	← Bb	Cc		
IV			AaBb		Cc		
			ab	AB →	Cc		
			ab		ABCc		
			ab	← c	ABC		
V	abc		ABC	**V′**	abc		ABC
	b	ac →	ABC		c	ab →	ABC
	b		AaBCc		c		AaBbC
	b	← B	AaCc		c	← a	ABbC
VI	Bb		AaCc	**VI′**	ac		
		Bb →	AaCc			ac →	
			AaBbCc				AaBbCc

Fig. 140

The same combination I, II, III, IV, V, VI, with jugglers instead of brothers and wives instead of sisters, occurs in a 15th-century manuscript.[334]

The combination with step I′ occurs in the *Annales Stadenses*, then in the 14th century, each time with the same mnemonic:[335]

> *Bine, sola, due, mulier, duo, vir mulierque*
> *Bini, sola, due, solus, vir cum muliere.*

These same steps (thus with I′) are found later, in Chuquet's *Triparty* and in Tartaglia.[336] In his *De viribus quantitatis*, Pacioli has steps I′ and V′, VI′.[337] Finally, Bachet has steps I′, V′ and VI′, but with the same woman (our *c*) taking at the end successively the two other women. He is the first to explain, after giving the answer, the reasoning step by step, as follows.[338]

[142] *It would seem that this problem too* (like the preceding one, on decanting) *does not rely on any reasoning. The set condition, however, that no woman should be with a man other than her husband if he is not there, can guide us to the solving by an infallible argument. Indeed it is obvious that for crossing two by two there must be together two men, or two women, or a man with his wife.*

Now, on the first crossing, one cannot have two men passing; for then one man would stay alone with three women, against the condition. Thus it is necessary that either two women cross (our I′) *or that one man crosses with his wife* (I); *but these two ways come about to the same: if two women cross, one has to return with the boat, leaving one on the other bank, and the same will happen if a man crosses with his wife since he must bring back the boat, for if the woman were to do it she would find herself without her husband with the two other men.*

On the second crossing two men cannot pass, for one of them would leave his wife with another. Another with his wife cannot cross either for, having passed, he would be alone with two women. It is therefore necessary that two women cross. Then, after three women have crossed, one of them must bring back the boat.

This being done, there remain to pass on the third crossing (III, →) *the three men and one woman. Two women cannot cross since there is just one. A man with his wife cannot pass either since, having crossed, he would find himself alone with three women. Thus two men must cross and join their wives, leaving behind one*

---

[334] MS. Paris BNF fr. 1339, fol. 77ʳ – 77ᵛ (text in *Récréations*, pp. 178–179); similar text in MS. Tours 399, fol. 140ᵛ – 141ᵛ.

[335] *Annales*, p. 334; *Subtilitates*, No. 26.

[336] *Appendice*, pp. 459–460, MS., fol. 209ʳ; *General trattato*, I, fol. 257ʳ.

[337] MS. Bologna BU it. 250, fol. 104ᵛ – 105ʳ.

[338] *Problemes plaisans*, pp. 140–142 (ed. 1612), 212–215 (ed. 1624). The reasoning is further commented by Labosne in his reedition, pp. 148–150, then by Lucas in his *Récréations mathém.*, I, pp. 5–18.

man with his wife. Now who is to bring back the boat (III, ←)? *A man cannot do it since he would leave his wife with another man. A woman cannot either for she would join another man, leaving her husband. If two men were to bring it this would mean doing nothing since they would return to where they came from. Thus the only way is that a man and his wife bring the boat back.*

For the fourth crossing there remain to pass two men with their wives. It is certain that a man with his wife must not pass, for that would mean doing nothing. The two wives cannot pass either, for then the three women would be alone with one man. Therefore the two men must cross.

Next, to bring back the boat (IV, ←), *two men cannot be used for that would mean returning to where they came from. A man alone cannot either, for doing that he would be alone with two women. Thus it must be the woman who fetches one at a time the two remaining women, and such are the fifth and the sixth crossings.*

Therefore they will have all crossed in six times without infringing the condition.

§4. Four couples and a boat with two places

Having treated the case of three couples, Tartaglia proceeds with that of four couples (Fig. 141–142).

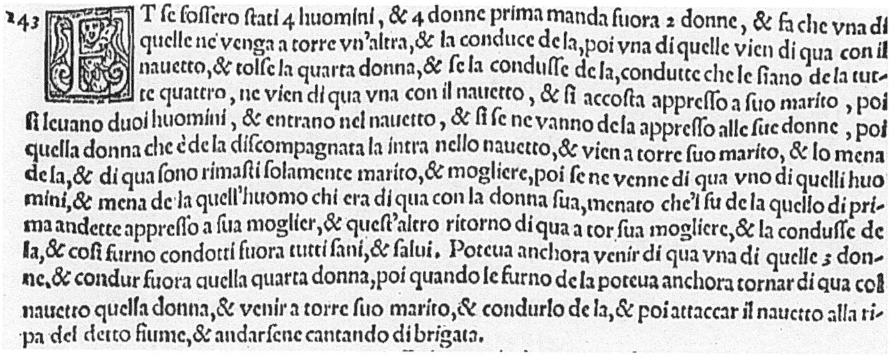

143 T se foſſero ſtati 4 huomini, & 4 donne prima manda fuora 2 donne, & fa che vna di quelle ne venga a torre vn'altra,& la conduce dela,poi vna di quelle vien di qua con il nauetto,& tolſe la quarta donna,& ſe la conduſſe dela,condutte che le ſiano dela tutte quattro, ne vien di qua vna con il nauetto, & ſi accoſta appreſſo a ſuo marito, poi ſi leuano duoi huomini, & entrano nel nauetto, & ſi ſe ne vanno dela appreſſo alle ſue donne, poi quella donna che è de la diſcompagnata la intra nello nauetto,& vien a torre ſuo marito, & lo mena dela,& di qua ſono rimaſti ſolamente marito,& mogliere,poi ſe ne venne di qua vno di quelli huomini,& mena de la quell'huomo chi era di qua con la donna ſua,menato che'l ſu de la quello di prima andette appreſſo a ſua moglier,& queſt'altro ritorno di qua a tor ſua mogliere,& la conduſſe de la,& coſi furno condotti fuora tutti ſani,& ſalui. Poteua anchora venir di qua vna di quelle 3 donne,& condur fuora quella quarta donna,poi quando le furno de la poteua anchora tornar di qua coſi nauetto quella donna,& venir a torre ſuo marito,& condurlo de la,& poi attaccar il nauetto alla tipa del detto fiume,& andarſene cantando di brigata.

Fig. 141

[143] *Supposing 4 men and 4 women, let first two women cross over* (ab). *Have one return, fetch another and take her to the other side* (bc), *then one of them return with the boat, fetch the fourth woman and take her to the other side* (cd). *All of them being on the other side, one* (d) *returns with the boat and lands by her husband. Two men stand up* (AB), *board the boat and go to their wives. Then the unaccompanied woman there* (c) *boards the boat, comes to fetch her husband and takes him to the other side. On the first bank have remained only husband and wife* (Dd). *Then one of these men* (A) *returns and takes to the other side the man left with his wife* (AD); *and the first one after returning goes to his wife while the*

other (D) *returns to fetch his wife and take her to the other side. Therefore all of them crossed safely. One of the three women* (a) *could also return*[339] *and fetch the fourth* (ad); *then, when they were on the other side, this woman* (d) *could return with the boat to fetch her husband* (Dd), *take him to the other side and tie up the boat at the riverbank. Then the whole group goes off singing.*

AaBbCcDd					
ABCcDd	ab →				
ABCcDd		ab			
ABCcDd	← b	a			
ABbCcDd		a			
ABCDd	bc →	a			
ABCDd		abc			
ABCDd	← c	ab			
ABCcDd		ab			
ABCD	cd →	ab			
ABCD		abcd			
ABCD	← d	abc			
ABCDd		abc			
CDd	AB →	abc			
CDd		<u>AaBbc</u>			
CDd	← c	AaBb			
CcDd		AaBb			
Dd	Cc →	AaBb			
Dd		AaBbCc	Dd		AaBbCc
Dd	← A	<u>aBbCc</u>	Dd	← Aa	BbCc
ADd		aBbCc	AaDd		BbCc
d	AD →	aBbCc	ad	AD →	BbCc
d		AaBbCcD	ad		ABBbCcD
d	← D	AaBbCc	ad	← b	ABCcD
Dd		AaBbCc	abd		ABCcD
	Dd →	AaBbCc	a	bd →	ABCcD
		AaBbCcDd	a		ABBbCcDd
			a	← A	BbCcDd
			Aa		BbCcDd
				Aa →	BbCcDd
					AaBbCcDd

Fig. 142

As seen in Fig. 142, left, there are two errors (also with Tartaglia's other way). The second, though, when woman 'a' is on the second bank with BC and their wives, could be avoided, as may be seen from the sequence of crossings we have added on the right.

---

[339] Thus instead of A at the end, when Dd are alone on the first bank.

As to the first erroneous situation, it is unavoidable.[340] In order to verify the impossibility, consider the case of five persons having crossed over to the second bank, who will necessarily be of both sexes. It is certain that the case of five persons being on the second bank will occur since, after each crossing, the number of persons on that bank will either remain the same (if two return) or increase by one unit (if only one returns). Now, the possibilities when five persons are on the second bank, and therefore three on the first, are those given in Fig. 143.

	first bank			second bank		
case	women	men	possibility	women	men	possibility
I	0	3	*possible*	4	1	*excluded*
II	1	2	*possible*	3	2	*excluded*
III	2	1	*excluded*	2	3	*possible*
IV	3	0	*possible*	1	4	*possible*

Fig. 143

Of the situations of Fig. 143, the first three are excluded since there are on one bank more women than men. The only remaining case is the last. But it must also be excluded on account of the preceding situation:

— if one woman and one man have crossed, said man would have been with four women on the first bank;

— if two men have crossed, they must have been with three women on the first bank.

There is therefore no way of reaching the number of five persons on the second bank if we are to keep the conditions.

§5. Five or six couples and a boat with two places

The same situation will occur if there are five couples and six persons have crossed, or if there are six couples and seven persons have crossed. In the first case, the situation will be as above (Fig. 144).

Here, only III and V are possible; but either the subsequent situation or the previous one leads to an impossibility. Indeed, after III, the boat has to return either with a woman, and there would then be one woman more on the first bank, or with a man, and there would be one woman more on the second bank. As to V, it is excluded, as before, by the preceding situation on the first bank.

---

[340] The argument that woman 'c' boards the boat immediately upon its arrival, is not valid; see Bachet's 1624 ed., pp. 214–216. The following explanation is Labosne's, pp. 150–151 in his reedition of Bachet's work.

	first bank			second bank		
case	women	men	possibility	women	men	possibility
I	0	4	*possible*	5	1	*excluded*
II	1	3	*possible*	4	2	*excluded*
III	2	2	*possible*	3	3	*possible*
IV	3	1	*excluded*	2	4	*possible*
V	4	0	*possible*	1	5	*possible*

Fig. 144

In the other case, only VI is possible (Fig. 145); but there would previously have been, on the first bank, one man and six women or two men and five women.

	first bank			second bank		
case	women	men	possibility	women	men	possibility
I	0	5	*possible*	6	1	*excluded*
II	1	4	*possible*	5	2	*excluded*
III	2	3	*possible*	4	3	*excluded*
IV	3	2	*excluded*	3	4	*possible*
V	4	1	*excluded*	2	5	*possible*
VI	5	0	*possible*	1	6	*possible*

Fig. 145

As a matter of fact, it is no longer possible to make the crossing under the prescribed conditions for $m \geq 4$ and a single boat with two places. One place has then to be added.

§6. Four couples and a boat with three places

Five crossings ($\rightarrow$) are necessary.[341] As before, the three places on the boat will not always be occupied during the crossing (Fig. 146).

§7. Five couples and a boat with three places

Six crossings are needed (Fig. 147).

The cases of crossing in a boat with three places stop at five couples. With six couples, more room would be needed, namely a boat with four places; but then each time one couple could ferry another across and there would be no limit to the number of couples.

---

[341] Labosne, p. 151.

```
AaBbCcDd
ABCDd      abc →
ABCDd                  abc
ABCDd      ← a         bc
AaBCDd                 bc
ABCD       ad →        bc
ABCD                   abcd
ABCD       ← a         bcd
AaBCD                  bcd
Aa         BCD →       bcd
Aa                     BbCcDd
Aa         ← Bb        CcDd
AaBb                   CcDd
a          ABb →       CcDd
a                      ABbCcDd
a          ←A          BbCcDd
Aa                     BbCcDd
           Aa →        BbCcDd
                       AaBbCcDd
```

Fig. 146

```
AaBbCcDdEe
ABCDdEe    abc →
ABCDdEe                abc
ABCDdEe    ← a         bc
AaBCDdEe               bc
ABCDE      ade →       bc
ABCDE                  abcde
ABCDE      ← ab        cde
AaBbCDE                cde
AaBb       CDE →       cde
AaBb                   CcDdEe
AaBb       ← Cc        DdEe
AaBbCc                 DdEe
abc        ABC →       DdEe
abc                    ABCDdEe
abc        ← d         ABCDEe
abcd                   ABCDEe
d          abc →       ABCDEe
d                      AaBbCcDEe
d          ← D         AaBbCcEe
Dd                     AaBbCcEe
           Dd →        AaBbCcEe
                       AaBbCcDdEe
```

Fig. 147

In Pacioli's *De viribus quantitatis*, we find, after treatment of the case $p = 2$ with $m = 2$ then $m = 3$ couples, the following observation:

[144] *You might also imagine a case involving 4 husbands and 4 wives, or else 5 husbands and 5 wives, and likewise any number. But for the 4 you need the boat to be able to carry 3, for 5 to carry 4 or 3. Otherwise, your efforts will be in vain. It was nice, however, to ask for 4 with a boat for 2, which is impossible, so also for 5* (Fig. 148).[342]

Thus, two generations before Tartaglia, Pacioli already knew that with a boat with two places at most three couples can cross, and that a boat with three places is required for four or five couples.

*Porrai anchora date per tuo ingegno negotiado disponere di 4. mariti et 4. mogli et cosi di s. mariti et s. mogli et sic de quotlibet 2.
Ma per li 4. bisogna la barca possi portar. 3. et per. s. ne porti. 4. o 3. aliter ad impossibile laboratur. et pero fu bello a propore de. 4. cō la barca de z. g, non potest et ancho de. s.*

Fig. 148

---

[342] MS. Bologna BU it. 250, fol. 105ᵛ; Agostini, p. 188.

Chapter XI. Miscellanea

§ 1. Sharing with a traveller

Fibonacci reports how a problem frequently encountered in mediaeval mathematics was treated erroneously.[343]

Fig. 149

Fig. 150

[145] *Of two men one had three loaves, worth one bezant each, and the other had two of them.[344] Whilst walking they reached a spring and sat down to eat. They invited a soldier passing by to share their meal. After they had eaten all the loaves, the soldier took leave of them and left 5 bezants for his share. The first, that of the three loaves, received 3 of them and the second took the two remaining bezants for his two loaves. Required whether the sharing was fair or not.*

*Some inexperienced people (*quidam imperiti*) hold that the sharing was fair since each obtains one bezant for each loaf. But this is wrong. Indeed, between all three they ate five loaves, so that each had $1\frac{2}{3}$; therefore, the soldier ate $1\frac{1}{3}$ loaf, thus $\frac{4}{3}$, of the loaves of the one who had three whereas he ate only $\frac{1}{3}$ of those of the other. That is why to the first man are due 4 bezants and to the other, 1.*

Indeed, it is not a sale, but a sharing. Same problem in the *Liber mahameleth*, thus in the middle of the 12th century;[345] but there it is about two men having respectively $n_1 = 60$ and $n_2 = 40$ sheep, who admit a third partner for an equal share of the animals; since the last one gives $S = 60$ *nummi*, required are the two parts of this sum.[346] Since they gave, of their original parts, respectively

$$n_1' = n_1 - \frac{1}{3}\,(n_1 + n_2) = 26 + \frac{2}{3}, \quad n_2' = n_2 - \frac{1}{3}\,(n_1 + n_2) = 6 + \frac{2}{3},$$

they are to receive, respectively,

---

[343] *Liber abaci*, p. 283. The illustrations are taken from Tagliente's *Componimento*, ed. 1547 and 1554 (the latter with the table suspended in the air).

[344] It is clear from this example that 'bezant' cannot always designate a gold coin.

[345] No. B.116.

[346] The *nummus* is a Roman coin, considered here as a monetary unit, just as the bezant is by Fibonacci.

$$s_1 = S \, \frac{n_1'}{n_1' + n_2'} = 48, \quad s_2 = S \, \frac{n_2'}{n_1' + n_2'} = 12.$$

Same problem as the previous one in the Castilian *Arte del alguarismo*, with three men having 3, 4, 5 loaves, each worth one *dinero*, and a passer-by leaving 12 *dineros*. Therefore the first will receive nothing and the second, half of the third's part.[347]

Fig. 151

§2. Multiplication of rabbits

The origin of the Fibonacci sequence is a mathematical recreation, for his suppositions on the proliferation of rabbits are hardly realistic. The question is about the number of pairs of rabbits found after one year in an enclosure containing initially one pair, given the following conditions:

(*i*) the time between one generation and the next, the duration of gestation, the time to maturity are uniformly one month;

(*ii*) a birth is immediately followed by further pregnancy;

(*iii*) the rabbits born are each time a mixed pair, one male and one female;

(*iv*) no rabbits die during the whole year.

Here are Fibonacci's problem enunciation and solution.[348]

[**146**] *How many pairs of rabbits are produced in one year by a single pair.*

Someone put a pair of rabbits in a place enclosed on every side in order to know how many descendants this single pair would produce in a year. Now it is in their nature to have one pair a month (iii, ii, i), which have descendants two months after their birth (i).

Since the aforesaid pair has descendants during the first month, you double it, and there will be two pairs at the end of the first month.[349] Of these, the first will

---

[347] Edition, pp. 196–197. The illustration is from the MS. Florence BNC Magl. XI 86 (Paolo dell'Abbaco's treatise, ed. p. 82).

[348] *Liber abaci*, pp. 283–284.

[349] The counting thus begins two months after the birth of the first pair (or one month after conception).

have descendants during the second month, and there will be 3 pairs at the end of the second month. Of these, two will attain maturity during this same month, so two pairs of rabbits are brought into the world in the third month, and there will be 5 pairs at the end of this month. Of these, 3 will attain maturity, and there will be 8 pairs at the end of the fourth month. Of these last, 5 will produce five pairs which, added to the 8 pairs, will give 13 pairs for the fifth month. Of these pairs, 5, brought into the world during the same month, will not procreate during said month, but the other 8 pairs will; therefore, there will be 21 pairs at the end of the sixth month. Adding to it the 13 pairs procreating during the seventh, there will then be at the end of this month 34 pairs. Adding to it the 21 pairs procreating during the eighth month, there will then be at the end of it 55 pairs. Adding to it the 34 pairs procreating during the ninth month, there will then be at the end of it 89 pairs. Adding to it likewise the 55 pairs procreating during the tenth month, there will then be at the end of it 144 pairs. Adding to it likewise the 89 pairs procreating during the eleventh month, there will then be at the end of it 233 pairs. Adding to it also the 144 pairs procreating during the last month, there will then be 377 pairs. Such is the number of pairs produced by the aforesaid pair put in said place after one year.

You may see in this margin our procedure.[350] We have added the first number to the second, thus 1 to 2, the second to the third, the third to the fourth, the fourth to the fifth, and so on to the addition of the tenth to the eleventh, thus 144 to 233, and we have obtained the sum of these rabbits, namely 377 (pairs). You could proceed like that for any number of months.

	1	2	3	4	5	6	7	8	9	10	11	12
1	2	3	5	8	13	21	34	55	89	144	233	377

Fig. 152

Such is the first appearance of the Fibonacci sequence, characterized by the recurrent relation

$$u_n = u_{n-1} + u_{n-2},$$

which may be inferred from Fibonacci's account from the seventh month: the number of pairs at the end of a given month equals the sum of the numbers of pairs in the two preceding months.

As said, Fibonacci begins counting the months from the time of procreation by the first pair. Today we put as initial terms $u_1 = 1$, $u_2 = 1$, and we thus obtain the sequence

$$1, 1, 2, 3, 5, 8, 13, 21, 34, 55, 89, 144, 233, 377, \dots ,$$

whereby Fibonacci's answer corresponds to u_{14}.

---

[350] The margin contains in two columns the sequence of months and numbers of pairs (aligned horizontally in our figure).

Note that there is no mention by Fibonacci of the characteristic links between the terms of this sequence.[351] There is moreover no allusion to its link with the golden ratio, itself known from antiquity; indeed, this link is seen by taking the limit, and that was beyond the mathematical tools existing at the time.[352]

§ 3. The dishonest servant

This problem is presented as follows by Cardan (we shall see other versions).

[147] *Someone was carrying wine in a container for his master. Since he was thirsty, he drank on the first day three pitchers of wine, and put back as much water. The second day, he drank as much and filled with water. The third day, he drank once again as much and completed with water. The same was done on the fourth day. But when he brought the container to his master, the latter understood that the wine was adulterated. He had it tested and it appeared that there was as much water as wine. Required the number of pitchers in the container.*[353]

If a pitchers is the initial quantity of wine found in the container when full, there remained at the end of the first day $a - 3$ pitchers of pure wine. At the end of the second, third and fourth days the remaining pure wine was, respectively,

$$a\left(\frac{a-3}{a}\right)^2, \qquad a\left(\frac{a-3}{a}\right)^3, \qquad a\left(\frac{a-3}{a}\right)^4,$$

for it diminished each time in the same proportion. Now the last quantity above must equal half the content of the whole container. So we shall have

$$a\left(\frac{a-3}{a}\right)^4 = \frac{a}{2}, \qquad (a-3)^4 = \frac{a^4}{2}, \qquad \frac{a}{a-3} = \sqrt[4]{2},$$

whence $a(\sqrt[4]{2} - 1) = 3\sqrt[4]{2}$, and thus

---

[351] Such as

$$\sum_1^n u_k = u_{n+2} - 1, \quad \sum_1^n u_{2k-1} = u_{2n}, \quad \sum_1^n u_{2k} = u_{2n+1} - 1, \quad \sum_1^n u_k^2 = u_n \cdot u_{n+1}.$$

[352] According to the definition,

$$\frac{u_n}{u_{n-1}} = 1 + \frac{u_{n-2}}{u_{n-1}}; \text{ supposing now } \frac{u_n}{u_{n-1}} \to \Phi \text{ for } n \to \infty,$$

we shall have

$$\Phi = 1 + \frac{1}{\Phi}, \ \Phi^2 = \Phi + 1, \text{ and thus } \Phi = \frac{1 + \sqrt{5}}{2},$$

which is the golden ratio (or number), known by the Greeks to be equal to the ratio of the radius of a given circle to the side of the regular decagon inscribed, or also to the ratio of the side of the pentagram to the side of the regular pentagon (the pentagram being obtained by joining every other apex of the pentagon).

[353] *Practica arithmetice*, Ch. LXVI, No. 36.

$$a = \frac{3\sqrt[4]{2}}{\sqrt[4]{2}-1} = \frac{3\sqrt[4]{2}\left(\sqrt[4]{2}+1\right)}{\sqrt{2}-1} = 3\sqrt[4]{2}\left(\sqrt[4]{2}+1\right)\left(\sqrt{2}+1\right)$$

$$= 3\left(2+\sqrt[4]{8}+\sqrt{2}+\sqrt[4]{2}\right) = 6+\sqrt{18}+3\sqrt[4]{2}\left(\sqrt{2}+1\right).$$

The content of the vessel is then, according to Cardan's approximate computation (*secundum propinquitatem*), $18 + \frac{427}{500}$ pitchers.

Remark. Cardan solves this problem by a false position. He considers the successive contents, at the beginning and at the end of each day, to be (writing ours), α_0, α_1, α_2, α_3, α_4, with $\alpha_0 = 2 \cdot \alpha_4$, say $\alpha_0 = 2$ and $\alpha_4 = 1$. Since $\alpha_0 : \alpha_2 = \alpha_2 : \alpha_4$, $\alpha_0 : \alpha_1 = \alpha_1 : \alpha_2$, $\alpha_2 : \alpha_3 = \alpha_3 : \alpha_4$, we infer that $\alpha_2 = \sqrt{2}$, $\alpha_1 = \sqrt[4]{8}$, $\alpha_3 = \sqrt[4]{2}$. Therefore, with $\alpha_0 = 2$, the difference at the end of the first day will be $2 - \sqrt[4]{8}$, whereas, with the true initial content a, it should be 3; then

$$\frac{2-\sqrt[4]{8}}{2} = \frac{3}{a} \quad \text{and so} \quad a = \frac{6}{2-\sqrt[4]{8}},$$

whence, eliminating the roots in the divisor,

$$a = \frac{6\left(2+\sqrt[4]{8}\right)}{4-\sqrt{8}} = \frac{6\left(2+\sqrt[4]{8}\right)\left(4+\sqrt{8}\right)}{8},$$

which brings Cardan to the expression seen above.

Fig. 153   Fig. 154

Similar problem by his contemporary Buteo:

[148] *A drunkard servant took secretly each day, from an eight-congii amphora of wine, one congius.*[354] *To make it up, he poured back as much water. Proof (of the fraud) having been found after five days, required the quantity of (pure) wine remaining in the amphora.*[355]

---

[354] The *congius* was a liquid measure in ancient Rome, about three litres —thus indeed a heavy drinker.

[355] *Logistica*, Ch. IV, No. 85, pp. 296–298. The illustrations are taken from F. Calandri's manuscript, MS. Florence Riccardiana 2669, fol. 84^{v} & 96^{v} (text, ed. pp. 169 & 193).

Buteo finds

$$8\left(\frac{7}{8}\right)^5 = \frac{16807}{4096} = 4 + \frac{423}{4096}.$$

We again find, in the work on mathematical recreations by J. Ozanam —as revised by Montucla— such a problem; the barrel contains initially 100 pints, the daily withdrawal is one pint, and the theft lasts a month. Since Montucla's reasoning is quite clear, we shall reproduce it in its entirety.[356]

[149] *A dishonest butler, every time he went into his master's cellar, stole a pint from a particular cask, which contained 100 pints, and supplied its place by an equal quantity of water. At the end of 30 days, the theft being discovered, the butler was discharged. Of what quantity of wine did he rob his master, and how much remained in the cask?*

It may be readily seen that the quantity of (pure) wine which the butler stole did not amount to 30 pints; for the second time that he drew a pint from the cask, taking the hundredth part of what it contained, it had already in it a pint of water, and as he each day substituted for the liquor he stole a pint of water, he every day took less than a pint of wine. To resolve, therefore, the problem, nothing is necessary but to determine in what (geometrical) progression the (quantity of pure) wine which he every day stole decreased.

For this purpose, we must first observe that after the first pint of wine was drawn, there remained in the cask no more than 99 pints (of pure wine) and the pint of water which had been added. When a pint therefore was drawn from the mixture, it was only $\frac{99}{100}$ of a pint of (pure) wine; but (since) before the pint was drawn the cask contained 99 pints of (pure) wine, consequently, after it was drawn, there remained 99 pints less $\frac{99}{100}$ (of pure wine), that is to say $\frac{9801}{100}$, or 98 pints plus $\frac{1}{100}$. When the third pint was drawn, the (pure) wine contained in it would be only $\frac{98}{100} + \frac{1}{10000}$, which being taken from the quantity of (pure) wine in the cask, viz 98 $\frac{1}{100}$[357] pints, would leave $\frac{970299}{10000}$, or 97 pints & $\frac{299}{10000}$.

It must here be remarked that $\frac{9801}{100}$ is the square of 99 divided by 100; and that $\frac{970299}{10000}$ is the cube of 99 divided by the square of 100, and so on; consequently, when the second pint is drawn, the (pure) wine remaining will be the square of 99 divided by the first power of 100; after the third, it will be the cube of 99 divided

---

[356] *Récréations*, I, Ch. XI, No. 21 (here from the 1803 English translation). The pint is in this case slightly less than one litre.

[357] Other way of writing $98 + \frac{1}{100}$; sometimes also, the two terms are separated by a comma, as found below.

by the square of 100, etc. Whence it follows that after the 30th pint is drawn, the quantity of (pure) wine remaining will be the 30th power of 99 divided by the 29th power of 100. But it may be found, by logarithms, that this quantity is $73,\frac{97}{100}$*; consequently the quantity of (pure) wine stolen is* $26,\frac{3}{100}$ *(a).*[358]

Fig. 155

We have already alluded to the various forms this problem might take. When it is about a dishonest servant, the theft is concealed by adding water; the quantity of liquid thus remains the same, it is only its degree of purity which is altered. But this problem also occurs, and more naturally, with the same fraction being taken each time and without the quantity taken being replaced; the initial quantity will therefore not remain fixed, as before, but will diminish according to the fraction withdrawn, as did before the degree of purity.

[150] *A man having 100 bezants went through 12 towns, in each of which he had to give up a tenth of the bezants he had. Required how much was left to him after leaving the 12 towns.*[359]

According to Fibonacci, he will retain from what he initially possesses the fraction

$$\frac{9}{10}\frac{9}{10}\frac{9}{10}\frac{9}{10}\frac{9}{10}\frac{9}{10}\frac{9}{10}\frac{9}{10}\frac{9}{10}\frac{9}{10}\frac{9}{10}\frac{9}{10} = \frac{282429536481}{1000000000000}.$$

He adds: *A similar problem is that of the cask containing 100 jars of wine from which is taken each month a tenth of the remainder, where it is asked how many*

---

[358] The '(a)' refers to a footnote noting the advantage of using logarithms: *If the usual method of calculation were employed, it would be necessary to find the 30th power of 99, which would contain not less than 59 figures; and to divide it by unity followed by 58 ciphers; whereas, if logarithms be used, nothing is necessary but to multiply the logarithm of 99 by 30, which will give 598690560, and to subtract the product of the logarithm of 100 multiplied by 29, which is 580000000. The remainder 18690560 is the logarithm of the required quantity; which, in the tables, will be found to be nearly* $73,\frac{97}{100}$. The illustration (not a cellar) is taken from Tagliente's *Componimento*, ed. 1525.

[359] *Liber abaci*, p. 313.

jars will remain at the end of the year, thus after 12 months.[360] He has also observed the similarity with the orchard problem (above, p. 33), for the problem immediately following is that of the toll at the exit of a town with ten gateways, at each of which he must leave a fraction of what he possesses plus the same fraction of a bezant.

Same problem, in another form, by Ghaligai.[361] Someone, in order to speak to *un Signore*, must pass ten guards who, 'as usual' (*per consuetudine*), take each a tenth of what he possesses, that is, of what remains of it each time, as a tip. (This practice has apparently survived: it is customary, when visiting monuments in Italy, to find at almost every door another wicket selling another ticket.)

Chuquet has the problem of a vat losing each day a tenth of its contents, and about which it is asked when it will have lost half.[362]

[151] *A vessel has an opening such that, when it is open, a tenth part runs out each day. Required in how many days half of this vessel will have run out.*

Answer and rule for such computations. Let us put for the contents of this vessel a large number of measures of which we may easily remove the tenth part several times, say 10000000. For each time (a tenth runs out) we shall count one day, until we arrive at the two consecutive days of which one (corresponds to an integral quantity) more than half of 10000000 and the other, (a quantity) less. Then the fraction of day must be sought by the rule of three. So doing we shall find that half of said vessel will have run out in 6 days (and) $\frac{314410}{531441}$ of a day.

Or, otherwise.[363] *Take the $\frac{9}{10}$ of 1, which are $\frac{9}{10}$, for the first day. Take again $\frac{9}{10}$ of $\frac{9}{10}$, you will have $\frac{81}{100}$ for the second day. Take again the $\frac{9}{10}$ of $\frac{81}{100}$, you will have $\frac{729}{1000}$ for the third day. Continue this way until you find two successive fractions one of which is more than $\frac{1}{2}$ and the other less. Pursuing in this way to the sixth day, one will find $\frac{531441}{1000000}$ (or: $\frac{5314410}{10000000}$), which is more than $\frac{1}{2}$. For the seventh day, one will find $\frac{4782969}{10000000}$, which is less than $\frac{1}{2}$. Now, put as the numerator what is more than $\frac{1}{2}$, thus 31441 (rather: 314410), and put as the denominator $\frac{1}{2}$ with the excess, thus 531441. You will have $\frac{31441}{531441}$ ($\frac{314410}{531441}$) of a day. Thus half of this vessel will have run out in 6 days and $\frac{31441}{531441}$ ($\frac{314410}{531441}$).*

---

[360] *Ibid.*, p. 316.

[361] *Pratica d'arithmetica*, Ch. IX, No. 29 (ed. 1552, fol. 66^v).

[362] *Appendice*, p. 439; MS., fol. 181^r–181^v.

[363] The difference between these two ways is merely due to the absence of decimal notation for the fractions at the time (see below, p. 247): the first solving uses integers starting from 10^7 (Chuquet already knew the answer) and the other starts from 1 and thus computes with fractions.

Thus Chuquet obtains, for the whole number of days and the fraction of the seventh (linear interpolation), the result

$$6 + \frac{\frac{5314410}{10000000} - \frac{1}{2}}{\frac{5314410}{10000000} - \frac{4782969}{10000000}} = 6 + \frac{\frac{314410}{10000000}}{\frac{531441}{10000000}} = 6 + \frac{314410}{531441} \quad (\approx 6.5916178).$$

He adds, however: *Many are satisfied with this way of doing. However, it seems to be closer to the truth that we are to seek a certain proportional number between 6 and 7 days, which at present is unknown to us.*

Indeed, the computation made today would lead to

$$S\left(\frac{9}{10}\right)^n = \frac{1}{2} S, \quad \text{and then} \quad n \cong 6.57205.$$

The most common form of such a problem, and also the most obvious, is that of a capital placed at compound interest. A 15th-century example is the following touching foresight of a father for his newborn daughter.[364]

[152] *A daughter was born to a man and, immediately after her birth, he went to a banker and gave him money at 10 per 100, and gave him such a quantity that it would give after 15 years, between capital and interest, 1000 livres. I ask how much money he gave initially.*

Answer. Since he said 'at 10 per 100', you may say that 10 gives 11. And since there are 15 years, you are to put (it) 15 times, as here ('*come ycy*', Fig. 156):

$$\frac{11 \cdot 11 \cdot 11 \cdot 11 \cdot 11 \cdot 11 \cdot 11 \cdot 11 \cdot 11 \cdot 11 \cdot 11 \cdot 11 \cdot 11 \cdot 11 \cdot 11}{10 \cdot 10 \cdot 10 \cdot 10 \cdot 10 \cdot 10 \cdot 10 \cdot 10 \cdot 10 \cdot 10 \cdot 10 \cdot 10 \cdot 10 \cdot 10 \cdot 10} \cdot {}^{[365]}$$

Then you are to multiply the (pairs of) digits which are above, it gives 4177224816 415652.[366] Then multiply those below one by the other, it gives 1000000000000000. Then you are to say, by the rule of three: If 4177224816415652 arises from 100000 0000000000, from what will arise 1000? You will find that they arise from 239 livres, 7 sous, 10 denarii, (and) $\frac{1725397655129992}{4177224816415652}$ *of one denarius.*

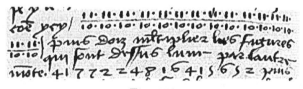

Fig. 156

---

[364] MS. Paris BNF fr. 2050, fol. 93^{v}. The *livre* was worth 20 *sous* or 240 *deniers* (1 *sou* = 12 *deniers*).

[365] The · between the numbers is merely a sign of separation, not a multiplication sign (to appear two centuries later, see p. 252).

[366] 4177248169415651, actually. This distorts the subsequent computation.

§4. The future heirs

This problem is of Roman origin, for it appears in legal texts at the beginning of the Roman Empire.[367] In mediaeval times, both Latin and Arabic texts have such examples. It starts with Alcuin, where a dying father orders that if his pregnant wife gives birth to a son, he will receive thrice as much as his mother, but if it is a daughter, she will receive seven fifths of what the mother is to receive; now the mother gives birth to a boy and a girl.[368]   In later texts, the situation is similar, as in the following example by Chuquet.[369]

| Fig. 157 | Fig. 158 |

[153] *A man makes his will whilst his wife is pregnant. He orders that, of his 100 crowns* (écus, silver coins), *if his wife gives birth to a girl, the mother will take twice as much as the girl; if she bears a boy, he orders that he will have twice as much as the mother. It happens that the mother has a boy and a girl. Required how much, in accordance with the will , the mother, the boy and the girl will have of these 100 crowns.*

The remark about respecting the will means that the same proportions are to be maintained. Thus, since the mother is to receive twice as much as the daughter and the son twice as much as the mother, the boy will have four times more than the daughter and the mother the double of the daughter (proportion 4 : 2 : 1); therefore, the 100 crowns will be divided into seven parts, and the respective shares will be $57 + \frac{1}{7}$, $28 + \frac{4}{7}$, $14 + \frac{2}{7}$.

The same way of solving is seen in contemporary, similar problems:[370] thus, with an estate of 3579 florins, the parts will be $2045 + \frac{1}{7}$, $1022 + \frac{4}{7}$, $511 + \frac{2}{7}$; with a capital of 1200 crowns they will be $685 + \frac{5}{7}$, $342 + \frac{6}{7}$, $171 + \frac{3}{7}$.

---

[367] Tropfke, *Geschichte*, p. 655.

[368] *Propositiones*, No. 35.

[369] *Appendice*, p. 421, & MS., fol. 154ᵛ; different text, but same values, in MS. Nantes 456, fol. 69ᵛ. The illustrations are taken from Tagliente's *Componimento*, ed. 1525 then 1547 (dying man better off).

[370] MS. Paris BNF fr. 2050, fol. 91ᵛ; *Livre de chiffres et de getz*, No. 6. Original texts in *Récréations*, p. 196.

Obviously, such a problem offers little scope for variants. In order to remove the more or less legal conditions, we can for instance pass from the human to the animal kingdom, where the mother may be worth twice the son and three times the daughter.[371]

[154] *Maevius sold to Titius a pregnant cow under the condition that if it were to bear a heifer, Titius would pay forty, if a bull-calf, forty-five. About the price of the cow they agreed that it would be thrice that of the heifer, and twice that of the bull-calf. In due course, the cow bore twins, male and female. Required how much Titius owed to Maevius.* He owes him 55, namely 30 for the cow, 10 for the heifer, 15 for the bull-calf.

Another variant, taking once again the situation of the last will, increases the number of beneficiaries, supposing for instance the offspring to be, not twins, but triplets, for instance two boys and one girl; from seven in problem [153], the number of parts then becomes eleven. Thus for an estate of 900 silver pounds (*libre*), the parts will be $327 + \frac{3}{11}$ (twice), $163 + \frac{7}{11}$, $81 + \frac{9}{11}$.[372]

Fig. 159

Clearly, it is not an advantage for a mother to have numerous offspring, unless the largest part, that of the boy, can be reduced; for example, if the boy himself is reduced:

[155] *A merchant, reaching the end of this brief, sad life, leaves 1200 florins to his pregnant wife, on the condition that if God makes her the mother of a daughter, said daughter would have 300 fl. and the mother the remainder. But if she bears a boy, then the mother will have the 300 fl. and the boy, the remainder. It happened, after the death of the testator, that the widow bore a son, a daughter and a hermaphrodite or androgyne. Required how to satisfy the will of the testator.* The author gives the answer directly: 768 for the son, 192 for the androgyne, 192 also for the mother, 48 for the daughter.[373]

---

[371] Buteo, *Logistica*, Ch. V, No. 12, pp. 341–342.

[372] Anonymous *Libro d'abaco*, pp. 136–137. Illustration from Muscarello's *Algorismus*, MS. Pennsylvania LJS 27, fol. 76ʳ (*uno padre che venne ad morte e lassa la sua donna gravida*; ed., II, pp. 189–191).

[373] Mellema, *L'arithmétique*, II, p. 56. The answer corresponds to the proportion son : mother : daughter = 8 : 4 : 1 (and not 9 : 3 : 1), with the androgyne receiving four times less than the boy but four times more than the daughter.

§5. Sharing camels

A frequently reported anecdote is that of an Arab who wished to share his herd of seventeen camels among three people, of whom one was to receive half, the second a third and the last a ninth. The solution was to borrow a camel from a neighbour, share out the eighteen as required, and return the extra camel to its owner; thus the first person received nine camels, the second six and the third two, that is, each one's due part of the (increased) herd.

Now the foundation of this is not just anecdotal. Roman law, and then later Moslem law, attributed to the heirs (and not only to direct descendants as in the previous problems) specific parts of the inheritance, according to their relationship to the deceased. But the sum of the fractions rarely added up to the unit. Here, in the case of the camels, that sum equals $\frac{17}{18}$, whence the subterfuge of the extra animal.

Consider, generally, that we are to share out a capital S according to given fractions, but such that their sum does not equal the unit:

$$\sum \frac{k_i}{l_i} = \frac{K}{L} \neq 1, \quad \text{then} \quad \sum s_i = \sum \frac{k_i}{l_i} S = \frac{K}{L} S.$$

Since the will of the testator is that the parts conform to the given proportions, we shall put, instead of the theoretical parts s_i of S, the same parts of a fictitious capital S', with

$$\sum s_i = \sum \frac{k_i}{l_i} S' = \frac{K}{L} S' = S,$$

and we shall therefore put $S' = \frac{L}{K} S$. The sum of the parts of this new capital will indeed give the available capital S. In the above case of the camels, $S = 17$, $\sum \frac{k_i}{l_i} = \frac{17}{18} = \frac{K}{L}$, whence $S' = \frac{18}{17} S = 18$.

Mediaeval mathematical treatises have examples of such problems; the next two, with $S' > S$ then $S' < S$, are taken from the *Liber mahameleth*.[374]

[156] *You want to divide ten nummi among two men, for one of them a half and for the other a third.*[375]   The sum of the two fractions being $\frac{5}{6}$, we shall put $S' = 12$, with the parts 6 and 4.

[157] *You want to divide thirty nummi among three men, to one of whom is due two thirds of the amount, to the second as much as the whole, that is, thirty, and, to the third, one and a half times as much.*   Since

---

[374] B.118 & B.119.

[375] About the *nummus*, see p. 161n.

$$\sum \frac{k_i}{l_i} = \frac{2}{3} + 1 + \frac{3}{2} = \frac{19}{6}, \quad \text{we have} \quad S' = \frac{6}{19} \cdot 30 = 9 + \frac{9}{19},$$

with the parts

$$s_1 = \frac{2}{3} S' = 6 + \frac{6}{19}, \quad s_2 = S' = 9 + \frac{9}{19}, \quad s_3 = \frac{3}{2} S' = 14 + \frac{4}{19},$$

which add up to $S = 30$.

Sometimes the given fractions are just not supposed to make up the initial capital or merchandise, as in the following example by Fibonacci.[376]

[158] *Four men buy a pig for 60 shillings (solidi); the first wanted to have a third of it, the second a fourth, the third a fifth, the fourth a sixth. The first paid 20 shillings for a third of the pig, the second 15 shillings for a fourth, the third 12 shillings for a fifth, and the fourth 10 shillings for the sixth. But all added together make 57 shillings, and inexperienced people wonder why there remain 3 shillings to be paid from the 60 and ask who is to pay them. In fact, they fail to realize that these four men did not buy the whole pig.*

§ 6. Striking the hour

A recurrent mediaeval problem, though consisting of a simple computation with fractions, is nevertheless to be considered as recreational because of its oddity.[377] Here follows an example by Ghaligai.[378]

[159] *Two meet in the street. One of them asks the other about the hour, and the other answers that $\frac{1}{3}\frac{1}{4}$ of what the bell has struck is as much as $\frac{1}{5}\frac{1}{6}$ of what is to be struck.* (It would be interesting to give such an answer to someone asking for the time today.) The answer given is as absurd as the question: $9 + \frac{5}{19}$ have been struck (*tante n'era sonate*) and (thus) $14 + \frac{14}{19}$ remain (*tante havevano a sonare*).

In another problem by Piero della Francesca the fractions are $\frac{1}{3} + \frac{1}{4}$ and $\frac{1}{4} + \frac{1}{5}$, and the answer is $10 + \frac{14}{31}$ and $13 + \frac{17}{31}$, respectively.[379]

In yet another text, a chaplain asks his sacristan to ring for matins, but the latter refuses to get up, saying that it is too early: the $\frac{1}{3} + \frac{1}{4} + \frac{1}{5}$ of the time elapsed are the $\frac{1}{5} + \frac{1}{6} + \frac{1}{7}$ of that to come.[380]

---

[376] *Liber abaci*, pp. 142–143.

[377] Such a type of problem already existed in antiquity; see the *Anthologia Graeca*, XIV, No. 6.

[378] *Pratica d'arithmetica*, Ch. IX, No. 33 (ed. 1552, fol. 67^r); the same problem, with the same data, is also proposed in the *Trattato d'abaco* by Piero della Francesca (ed., p. 66).

[379] Edition, p. 102.

[380] MS. Rome Accad. naz. dei Lincei Cors. 1875, fol. 50^v (with illustration, Fig. 160).

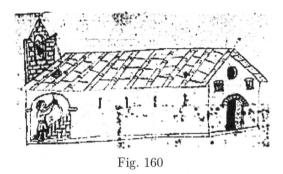

Fig. 160

§ 7. Two workers, same work

In the following example, Chuquet once again resorts to his algebraical way of computing.[381]

[160] *There are two carpenters who want each to build a house of equal size and quality. One of them tells his companion that if he will help him with his house for 8 days, he will finish it in 20 days. The second tells him that if he will help him for 10 days, he will finish his in 15 days. Required in how many days each will build his house by himself.*

In order to solve this problem and similar ones, we are to know that one day of the second compared to one day of the first is such that when multiplied by 8 and added to 20 it makes the same as if it were multiplied by 15 and added to 10. So let us put 1^1; when multiplied by 8 and added to 20, it gives 8^1 plus 20. Next, let us multiply 1^1 by 15 and add 10 to it; we shall have 15^1 plus 10, equal to 8^1 plus 20.

In our terms, designating by a_1 and a_2 their daily achievement, we must have, because of the identity of the two houses, $20a_1 + 8a_2 = 15a_2 + 10a_1$, that is, $20 + 8\frac{a_2}{a_1} = 15\frac{a_2}{a_1} + 10$; this $\frac{a_2}{a_1}$ is therefore Chuquet's unknown. Having thus found that the daily work of the second is worth $1 + \frac{3}{7}$ of the first's daily work, Chuquet computes that the first by himself would finish his house in $8 \cdot \left(1 + \frac{3}{7}\right) + 20 = 31 + \frac{3}{7}$ days, and the second by himself his in $31 + \frac{3}{7} : \left(1 + \frac{3}{7}\right) = 22$ days. This is what the next part of the text explains:

Reduce your (two) sides, you will have on one side 10 and on the other 7^1. Divide 10 by 7, you will have $1\frac{3}{7}$, and such is the day of the second compared to the day of the first. Now multiply 8 by $1\frac{3}{7}$, and then add with 20, you will have $31\frac{3}{7}$ days. In such a number of days will the first finish his house. Next, divide $31\frac{3}{7}$ by $1\frac{3}{7}$, you will have 22 days, and this is the time taken by the second to finish

---

[381] *Appendice*, p. 429; MS., fol. 162^r. We have already seen instances of Chuquet's algebra (p. 81; see also p. 252).

his house without the first one's help.[382]

His next problem is of the same type. There are two bricklayers building two identical houses; if the second helps the first for 5 days, the first will finish his work in 17 days, whereas the second when helped by the first for 6 days will finish the house in 24 days. Required the time taken by each individually. Since $17a_1 + 5a_2 = 6a_1 + 24a_2$, and thus $\frac{a_2}{a_1} = \frac{11}{19}$, Chuquet finds $19 + \frac{17}{19}$ days for the first one's house and $34 + \frac{4}{11}$ days for the second one's.

Fig. 161

§ 8. Same earnings from same items in different quantities

[**161**] *There are three women carrying apples to the market to sell them. One carries fifty, the other thirty, the last ten. They sell them in such a way that each obtains as much as the other by giving as many as the other for one denarius. Required how they sold them.*[383]

Fig. 162   Fig. 163

For this to be possible, they are to sell at two prices, namely seven apples for one denarius (price p_1) and one apple for three denarii (price p_2). Then the first earns ten denarii by selling 49 at p_1 and one at p_2, the second 28 then 2, the third 7 then 3.

---

[382] Illustration from Muscarello's *Algorismus*, MS. Pennsylvania LJS 27, fol. 74^r: a master builder (*maystro*) shows the palace he has built (text: ed., II, p. 188).

[383] *Arithmétique de Pamiers*, problem C.112 (also: Chuquet, *Appendice*, p. 453; *Livre de chiffres et de getz*, No. 13). The illustrations are taken from Tagliente's *Componimento*, ed. 1547 and 1554.

The answer is given directly. But how this was found is clear. We are to solve the three pairs of equations

$$\begin{cases} x_1^{(1)} + x_2^{(1)} = s^{(1)} = 50 \\ p_1 x_1^{(1)} + p_2 x_2^{(1)} = 10 \end{cases} \quad \begin{cases} x_1^{(2)} + x_2^{(2)} = s^{(2)} = 30 \\ p_1 x_1^{(2)} + p_2 x_2^{(2)} = 10 \end{cases} \quad \begin{cases} x_1^{(3)} + x_2^{(3)} = s^{(3)} = 10 \\ p_1 x_1^{(3)} + p_2 x_2^{(3)} = 10. \end{cases}$$

Taking for instance $\frac{1}{7}$ for the first price (thus 7 apples to the denarius), we shall take, for convenience, all $x_1^{(i)}$ divisible by 7 and less than $s^{(i)}$, say then $x_1^{(1)} = 49$ (worth 7 denarii), $x_1^{(2)} = 28$ (worth 4 denarii), $x_1^{(3)} = 7$ (worth one denarius); since the respective remainders $x_2^{(i)}$ are 1, 2, 3 apples whilst 3, 6, 9 denarii are missing, we shall choose $p_2 = 3$ (thus a price 21 times as much as the first!).

However, when the author attempts to extend this problem to three prices, he is obliged to introduce sales of fractions of apples.

Chapter XII. Family relationships

Problems of family relationships originally dealt with a very serious matter and would not, normally, appear among mathematical recreations: the purpose was to ensure the fair and legally based repartition of a legacy among the deceased's heirs. (The Arabic examples we shall report are all taken from treatises on such successions.) But for some this was a unique occasion to divert the reader with curious situations, proposed as riddles or even exercises in law —or simply in reasoning. That is why Alcuin has three such problems among his propositions 'to sharpen the minds of young people', according to the title (see p. 254). One might object that this does not involve mathematics. Once again (see pp. 143–144), we shall have recourse to former authorities: Chuquet, who presents such a problem (see below, [165]), writes about it that, although not solved by means of numbers, it has nevertheless been included since to many it will sound original and amusing.

In any event, the oddness of the various situations means taking some liberties with the vocabulary; thus half-brother and half-sister will just be brother and sister. Conversely, our 'uncle' and 'aunt', which apply to both the paternal and maternal side, are distinguished in the sources, for both Latin and Arabic have specific terms to designate them. Note further that in all texts there is never consanguinity: blood ties between spouses are wholly excluded.

In the examples below, we shall see each of two unrelated men marry a close relative of the other: sister, mother, daughter, then one the mother and the other the daughter. Then father and son will marry a mother and her daughter, or inversely. Finally, the spouses are separated by two generations.

§ 1. Two men marry each other's sister

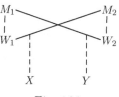

Fig. 164

[162] *If two men marry the sister of the other, tell me, please, what the relationship between their children will be?*[384]

---

[384] Alcuin, *Propositiones*, No. 11. Same problem in the MS. Paris BNF ar. 4441 (15th century), fol. 46ᵛ, *ll.* 9–10.

Each is the child of the other's maternal uncle. For (Fig. 164) M_1, father of Y, is (by his sister W_1) X's maternal uncle.

§2. Two men marry each other's mother

[163] *If, likewise, two men marry the mother of the other, how will their children be related?*[385]

Each of the two boys is the paternal uncle of the other. Indeed (Fig. 165), X is (by his mother W_2) the brother of m_2.

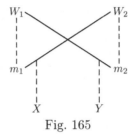

Fig. 165

A more fanciful version appears in the *Annales Stadenses*.[386]

[164] *A servant went from Cologne to Dusseldorf; another encountered him, going from Dusseldorf to Cologne. They greeted each other. After this greeting, each let the other visit the home of his mother and made sure that he was received there hospitably. One's name was Berthold, the other's Stephen.[387] Berthold entered the home of Stephen's mother, and took her as wife. She was called Bertha. Stephen entered the home of Berthold's mother, and took her likewise as wife. She was called Oxane. Each gave his wife a boy. How are they related?*

After the diagram of filiation there are two (Latin) verses which each mother (say W_2) may address to the child of the other (Y): *Sit down, dear child, whose father (m_2) was born of me, and call me 'grandmother', you whose brother (m_1) I call 'husband'.*[388]

[165] Chuquet has the very same problem.[389] It is formulated thus: *There are two women holding each a beautiful son in her arms, of whom it was asked: 'Whose are these beautiful sons you hold?'. They then answered: 'In truth, they are sons of our sons and brothers of our husbands, all by legal marriage'.*

---

[385] Alcuin, *Propositiones*, No. 11′ and MS. Paris BNF ar. 4441, fol. 46ᵛ, *ll.* 4–6.

[386] For the three examples given here ([164], [167], [173]), see Pertz's edition, p. 335.

[387] This may seem irrelevant but in fact makes the situation clearer: there is a diagram of the same kind as ours but with the names.

[388] *Care puer sedeas, tibi qui pater est mihi natus*
    Meque vocas aviam, tibi fratrem dico maritum.

[389] *Appendice*, p. 460; MS., fol. 209ᵛ – 210ʳ; see (text) *Récréations*, p. 205.

§ 3. Two men marry each other's daughter

[166] *Two unrelated men have each married the other's daughter and gave each of them a child; how will the children be related?*[390]

Each of the two is the maternal uncle of the other; for X is the brother of w_2 (Fig. 166).

Same problem in the *Annales Stadenses*:

[167] *Two soldiers having each a daughter marry each the other's daughter. Given that the soldiers are unrelated, and so are their daughters, we are to find out how these two sons might be related.*

§ 4. Of two men, one marries the other's mother and the second the other's daughter

[168] *Two unrelated men have married: one, the other's mother and the other, the daughter of his comrade, and their unions were each blessed by the arrival of a child; what is the relationship between the two children?*

Answer. The son of the one who married the mother is the paternal uncle of the daughter's son and his maternal uncle, and the child of the one who married the daughter is the son of the sister of the child of the one who married the mother and the son of his brother through his mother.[391]

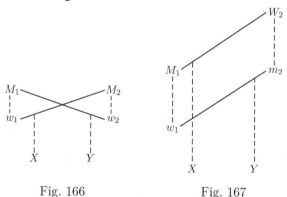

Fig. 166   Fig. 167

From Fig. 167 it indeed appears that since X is the brother of w_1 and m_2, he will be both the maternal and paternal uncle of Y; in other words (and this explains the second part of the answer) Y is the son of X's sister and X's brother. But since now the situations are no longer symmetrical, there is no longer reciprocity.

---

[390] Arabic manuscript mentioned above, fol. 46ᵛ, 7–8.

[391] *Ibid.*, fol. 46ᵛ, 11–15. For the Arabic text of this and the next problem, see *Récréations*, p. 206.

The same manuscript has the following variant of this problem (Fig. 168, equivalent to Fig. 167); in the enunciation Y is addressing X whereas in the solution it is X who explains the situation:

[169] *A person says to another: 'Oh my paternal uncle, oh my maternal uncle'. Solution. My brother* (m_2) *through my mother has married my sister* (w_1) *through my father. He has given her a son. Therefore I am his paternal uncle and his maternal uncle.* [392]

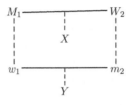

Fig. 168

§5. Father and son marry mother and daughter

[170] *A man having a son marries a woman having a daughter, and then the son marries the daughter.* [393]

Same answer as in the previous situation: X being the brother of m_1 and of w_2 (Fig. 169) will be the paternal and maternal uncle of Y.

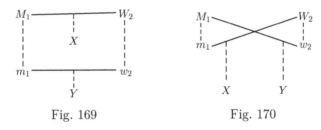

Fig. 169   Fig. 170

§6. Father and son marry daughter and mother

[171] *If a father and his son marry an abandoned woman, or a widow, and her daughter, the son taking the mother and the father, the daughter, tell me, pray, what the relationship between their sons will be?* [394]

Each child will be at the same time uncle and nephew of the other. For X (Fig. 170), being the brother of w_2, is the maternal uncle of Y, but also, since Y is the brother of m_1, his (paternal) nephew. Analogous relationships for Y.

---

[392] *Ibid.*, fol. 42$^\mathrm{v}$, 22–43$^\mathrm{r}$, 1.

[393] *Ibid.*, fol. 46$^\mathrm{v}$, 15–16.

[394] Alcuin, *Propositiones*, No. 11″; also in the Arabic manuscript mentioned above, fol. 46$^\mathrm{v}$, 19–20.

§7. Three generations

[172] *Riddle of the Imām Shāfiʿī, may God be pleased with him, about the question expressed thus:* [395]

I have a paternal aunt of whom I am also paternal uncle. I have a maternal aunt of whom I am also maternal uncle. As to the one of whom I am paternal uncle, my father is his brother through her mother. What jurisconsult knowing the science of successions will show us the right answer and thus free us from this ignorance?

Then, the text continues, Imām Shāfiʿī is said to have given, in a poem, the following explanation of the first two assertions, thus clarifying the situation. Concerning the first ('I have a paternal aunt of whom I am also paternal uncle'): my (Fig. 171, X speaking) brother (m_1) through my mother (w_1) has married the mother (W_2) of my father (m_2), and they have had a daughter (y'); as sister of my father, she will be my paternal aunt, and, as daughter of my brother, I shall be her paternal uncle. Concerning the second ('I have a maternal aunt of whom I am also maternal uncle'): the father (M_1) of my mother (w_1) has married my sister (w_2) through my father (m_2), and they have had a daughter (y''); as sister of my mother, she will be my maternal aunt, and, as daughter of my sister, I shall be her maternal uncle.

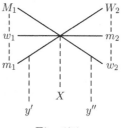

Fig. 171

The last of our riddles about filiation is from the *Annales Stadenses*, again one of the questions two young people ask each other (above, p. 1).

[173] *My master never wanted to take his meal without first hearing some odd news. He sent someone outside. Upon returning, the latter told him what follows: 'Whilst walking, I came across twelve old men. After greeting them, I went on.*

---

[395] In verse in the text (*ibid.*, fol. 43^r):

ولي عمّة وانا عمّها   ولي خالة وانا خالها

فامّا التى انا عمّ لها   فان ابي امّه اخـوهـا

فاين الفقيه الذى عنده   فنّ الفرائض مع علمها

بين لنا نسبًا صايحًا   يكشف للنفس عن غمّها

The Imām Shāfiʿī (d. 820) gave his name to a law school based on his teaching.

Again, I came across twelve middle-aged men. A little while later, for the third time, I came across twelve young people, who preceded an already decrepit man walking together with a young woman. Turning towards me, the decrepit man greeted me. Do you see, he said, this young lady? She is my wife, and the twelve young people are the children of her and me; the twelve middle-aged men you met are my sons, and the maternal uncles of that young lady; the twelve old men are also my sons, and are all paternal uncles of this young lady'.

The underlying scheme is seen in Fig. 172, with the thirty-six children resulting from three marriages of M_1, while w herself is born from children of previous marriages of W_1 and W_2.

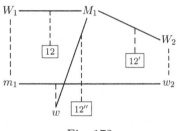

Fig. 172

Chapter XIII. The knight's tour

§ 1. Euler's study

The problem of the knight's tour on the chessboard consists in going once through each of its squares with the knight's move (thus going through two squares vertically and then one horizontally, or two squares horizontally and then one vertically). The solution is still associated with the name of Euler (1707-1783). After hearing about the problem at an evening party, he studied it and wrote the article which describes his method for finding a path.[396] Some twenty leaves in his notebooks preserved in the Saint Petersburg Academy of Science show that his article is the conclusion of a long sequence of trials, often empirical.[397] Indeed, proceeding merely by trial and error makes it possible to cover without much difficulty between fifty or sixty squares; but there will still remain inaccessible ones, mostly on the sides. The most inconvenient situation is that of the corners since they are accessible from two squares only.[398]

Fig. 173

Euler's article describes how, starting from some incomplete tour, to use parts of that and, by rearranging them and including one by one the squares left empty, to arrive at a complete tour. By the same method, Euler could find other complete tours, in particular with the initial and final squares given, as in the first example of his article where they are next to one another (Fig. 174). He could also, as in his second example, obtain 'closed tours', where the initial square is one knight's move away from the last one (Fig. 175). Whereas in the others, 'open tours', there were only two possibilities for the point of departure, namely the first and last squares, in a closed tour any square of the chessboard can be chosen as starting point and the sequence of squares be followed since 1 and 64 can be associated.

---

[396] See his *Solution d'une question curieuse*.

[397] The whole edited and analyzed in our *Euler et le parcours du cavalier* (reproducing Euler's manuscript as well as his printed article).

[398] Chessboard, illustration from the MS. Paris Bibliothèque de l'Arsenal fr. 5107 rés., fol. 73^{v} (de Vignay's translation of de Cessolis' *Liber de moribus*, see p. 256).

42	59	44	9	40	21	46	7
61	10	41	58	45	8	39	20
12	43	60	55	22	57	6	47
53	62	11	30	25	28	19	38
32	13	54	27	56	23	48	5
63	52	31	24	29	26	37	18
14	33	2	51	16	35	4	49
1	64	15	34	3	50	17	36

42	57	44	9	40	21	46	7
55	10	41	58	45	8	39	20
12	43	56	61	22	59	6	47
63	54	11	30	25	28	19	38
32	13	62	27	60	23	48	5
53	64	31	24	29	26	37	18
14	33	2	51	16	35	4	49
1	52	15	34	3	50	17	36

Fig. 174   Fig. 175

Euler then extended his research to tours with supplementary conditions, such as symmetrical tours, where diametrically opposite cells display the uniform difference 32 (Fig. 176).[399] He also considered boards of other dimensions, square or rectangular, and also non-rectangular ones. His printed article gives a few examples, but many more appear in his manuscript notes.[400] A nice result is that of Fig. 177 (his *perfectū*, thus *perfectum*, means that the tour on the 60 squares of this figure is closed).

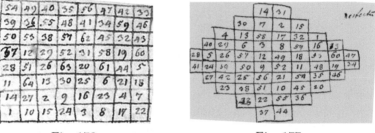

Fig. 176   Fig. 177

In his printed article, after telling us where he heard about this problem for the first time, Euler also reports that the participant mentioning it presented a solution of it. Euler did not recall his tour but remembered that he claimed to be able to perform a complete tour starting from any given cell. This problem therefore did not originate with Euler. As a matter of fact, it had been known long before: its first appearance followed shortly after arrival of the game of chess in Persia, one or two centuries after its invention in India (*c.* 6th century). It would seem that the aim originally was to remove the thirty-two chess pieces from the half-board. Only later was the problem extended to the whole 8 × 8 chessboard, with the aim then being to find a complete tour.

---

[399] See our *Euler et le parcours du cavalier*, pp. 49 & 267.
[400] *Ibid.*, pp. 63–150.

§ 2. Mediaeval tours

Instances of tours, on either half or the whole board, have been preserved from India and the Moslem and Christian worlds. Some closed tours were also known. They were evidently held in high esteem, for there existed Arabic poems to memorize them: imagining a chessboard with a known sequence of sixteen letters along two sides (Fig. 178), the first two letters of each of the sixty-four verses gave the coordinates of the path to follow.[401] There were even four different poems for this same example. This would of course have been disproportionate if some reasoned method for first reaching a complete tour and then transforming it to a closed one had existed. This explains why Euler's article was so fundamental.

35	40	47	44	61	8	15	12	ح (h)
46	43	36	41	14	11	62	9	ز (z)
39	34	45	48	7	60	13	16	و (w)
50	55	42	37	22	17	10	63	ه (e)
33	38	49	54	59	6	23	18	د (d)
56	51	28	31	26	21	<u>64</u>	3	ج (j)
29	32	53	58	5	2	19	24	ب (b)
52	57	30	27	20	25	4	<u>1</u>	ا (a)
ت	ش	ر	ق	م	ل	ك	ى	
(t)	(sh)	(r)	(q)	(m)	(l)	(k)	(i)	

Fig. 178

These Eastern examples of tours are often found in manuscripts teaching chess, and the same happened in mediaeval Europe. One of the earliest example is in an Anglo-Norman text written towards the end of the 13th century in verse —thus enabling the reader to better memorize the fifty-five games described.[402] Now the first of these games is an open knight's tour. We are told that the knight must start in one corner and, after passing through all the squares, end in the next corner (Fig. 179 and (MS.) 180). The author's commentary (also in verse) just says what the game is about and that the squares have been provided with numbers in

---

[401] *Euler et le parcours du cavalier*, p. 159. In our case, the sequence of pairs of letters would be *ia* (or *ai*, the order is irrelevant), *lb*, *ij*, *ka*, ...

[402] MS. London BM Reg. 13A.XVIII, fol. 161ʳ–173ʳ. See van der Linde, *Quellenstudien*, pp. 197 & 205–206, or Murray, pp. 588–600.

order to show the successive moves. He says nothing about any method, nor about the two halves being filled separately —and differently since the point of departure is different.

23	26	11	4	49	52	45	40
10	3	22	25	46	41	48	51
27	24	5	12	53	50	39	44
2	9	28	21	42	47	54	59
29	20	13	6	61	58	43	38
8	1	16	19	32	35	60	55
17	30	7	14	57	62	37	34
•	15	18	_31_	36	33	56	_63_

Fig. 179

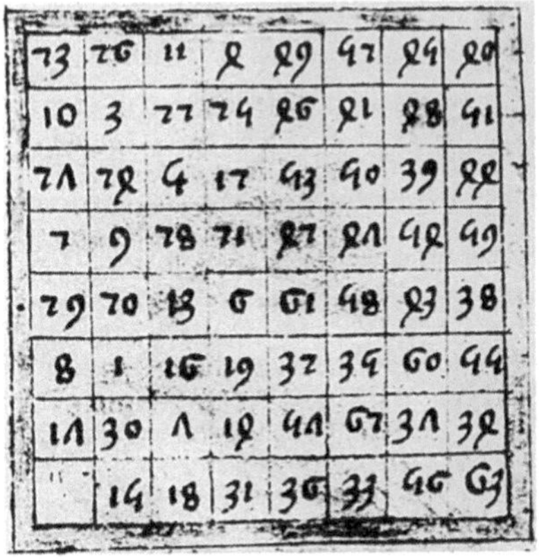

Fig. 180

Rey	roc	poun	poun				
poun	poun	Rey	roc				
chr	ferce	poun	poun				
poun	poun	roc	ferce				
rok	chr	poun	poun				
poun	poun	poun	alfin				
chr	alfin	poun	poun				
chr	poun	alfin	alfin				

Fig. 181

Fig. 182

32	25	12	5
11	4	31	26
24	29	6	13
3	10	27	30
28	23	14	7
9	2	17	20
22	19	8	15
_1_	16	21	18

Fig. 183

We have thus seen two different ways of filling the half-board, or even four if we consider inverting the path. Still another way is given in the same place (Fig. 181-183). As we are told, all men on the half-board will be successively taken by the knight. The way to achieve that is (approximately) described: the knight in the corner (*chivaler*, square 1) first eliminates, by successive moves around the board, both white and black pawns (*poun(s)*, squares 2–17), then the bishops (*alfin(s)*, 18–21), next the remaining knights (22–24), the rooks (*rocs* or *roks*, 25–28), the queens (*ferce(s)* or *reyne(s)*, 29 & 30), and finally the kings (*rey(s)*, 31 & 32).

All these ways of filling the half-board have a common feature. In Fig. 184, with the first number put in the lower left-hand corner, the horizontal rows of the half-board have been divided into pairs of adjacent squares of the same colour. Now we may observe that in the three above examples the white squares are filled with the first sixteen numbers (• to 15, 32 to 47, 1 to 16) whereas the black ones contain the sixteen following numbers. Now this separation is a necessary condition in order to obtain a complete tour on the half-board, for each of these two groups of sixteen squares (white or black) must be run through separately. Note too that if, for each group, the initial square is in one median column, the final one will be in a lateral one, and conversely. It follows that both extremities of a complete tour on the half board must be in lateral columns, whence the impossibility of a closed tour on the half-board.[403]

Q.	K	O.	B	K.	H	F.	Chr
N.	C	I.	I	G.	A	C.	P
L	P.	E	L.	N	A.	G	E.
D	M.	M	H.	F	D.	O	B.

32	11	30	3	26	9	22	1
29	4	25	10	23	2	19	16
12	31	6	27	14	17	8	21
5	28	13	24	7	20	15	18

Fig. 184　　　　Fig. 185　　　　　　　Fig. 186

The ways of filling the half-board are thus subject to more limitations than for the whole board. They are also less difficult to attain: keeping Fig. 184 in mind will greatly facilitate their filling.

About 1530 an anonymous booklet was published in Paris (*S'ensuit jeux partis*), the purpose of which was, in only ten leaves, to teach how to play chess.[404] On the last page there is a table illustrating how a knight, starting from the upper right-hand corner, may reach the other upper corner. The path, indicated in the text by the sequence of letters A, B,..., P, A., ..., Q., as in our Fig. 185 (but with the second sequence in red ink), corresponds to that of our Fig. 186. Note that it differs only slightly from the mediaeval one seen in Fig. 183 (namely with 24, ..., 30); but, maintaining the same points of departure and arrival and in view of the restrictive conditions seen above, one should not be too hasty in assuming that there is a connection between the two.

There may be some link, though, with northern Italy, since the very same path occurs in a text by Paolo Guarini (1464-1520), where the problem is about a soldier's journey.[405]

---

[403] All the above was established by Flye Sainte-Marie in 1877 (but already surmised by de Jaenisch in his *Traité*, II, p. 46); he also inferred that there must be 7772 different tours on the half-board.

[404] One copy at the Nationalbibliothek in Vienna. See also the study by von Heydebrand.

[405] See Franz's study.

1	22	5	28	9	32	15	18
6	25	8	21	4	17	12	31
23	2	27	10	29	14	19	16
26	7	24	3	20	11	30	13

Fig. 187

Another late mediaeval Italian manuscript has the path of Fig. 187.[406]
There is no comment, although this disposition is particularly interesting
since the given path may easily be extended to a complete tour on the
chessboard, namely by taking this same half-board, adding 32 to its numbers, turning it by 180° and putting it on the other half. Then 33 (former
1) will be a knight's move away from 32. The resulting tour will even be
closed since 64 (former 32) will be at a knight's move from 1. This was
indeed one of Euler's ways of obtaining a closed tour: filling a half board
so that 1 and 32 appear on the upper row, with one at a distance δ from
the side and the other at a distance $\delta + 2$ from the other side.[407]

1	26	9	20	3	32	11	22
16	19	2	25	10	21	4	31
27	8	17	14	29	6	23	12
18	15	28	7	24	13	30	5

Fig. 188   Fig. 189

Another such example, once again without any allusion to its possible
extension, and once again concluding a book on chess (this time, more
seriously, with 55 leaves), is found at the end of the 16th century: see
Fig. 188. Fig. 189 is that of the original text (with the numbers added
below).[408]

---

[406] Van der Linde, *Quellenstudien*, p. 196.

[407] See his printed study, pp. 328–331, or also the edition of his manuscript, pp. 64–74.

[408] Della Mantia, *Della maniera di giuocar' à scacchi*, fol. 51ʳ.

Chapter XIV. Magic squares

A *magic square* is a square divided into a square number of cells in which natural numbers, all different, are arranged in such a way that the same sum is found in each horizontal row, each vertical row, and each of the two main diagonals. The constant sum to be found in each row is called the *magic sum* of this square.

What is usually considered are squares filled with the first natural numbers. If the square is of order n, that is, has n cells on each side, the sum of all these numbers will equal

$$\frac{n^2 (n^2 + 1)}{2};$$

therefore the sum in each of the rows (lines, columns, main diagonals), thus the magic sum for such a square, will be

$$M_n = \frac{n(n^2 + 1)}{2}.$$

A magic square of order 2 is not possible. There is a single arrangement for the square of order 3 (seen above, p. 2), while there are 880 for that of order 4, after which their number increases rapidly. There are general methods for constructing magic squares of whatever size, but their application is restricted to specific forms of orders. Thus we have such methods for constructing squares of orders 3, 5, 7, ..., generally $n = 2k + 1$ ('odd' orders or squares), others for orders 4, 8, 12, ..., generally $n = 4k$ ('evenly-even' orders or squares), others for the remaining orders, thus 6, 10, 14, ..., generally $n = 4k + 2$ ('evenly-odd' orders or squares).

§ 1. Early history

It is commonly said that magic squares appeared in China at the beginning of our era. Now, first, this means the square of order 3, and, second, higher-order squares involving general construction methods do not occur in China before the 12th century, and are clearly of Arabic or Persian origin.

As a matter of fact, the earliest studies on magic squares go back to ancient Greece. True, a single anonymous text has survived, but it is quite elaborate and written for advanced readers, without the basic theory. There must have been more elementary treatises, but none has come to light nor, even, any allusion to magic squares in ancient texts. As to this single testimony, we do not have the original text, for it is preserved in an Arabic translation of the early 10th century. Even the survival of this ancient text at the time was some kind of miracle, as we gather from the introductory words of the translator into Arabic, Mufaḍḍal ibn Thābit ibn Qurra: *I found in the Library, among the books of the caliphs' collection, two books, for the greater part damaged by termites, so that one could understand just little of them* (...). *Then I examined them, found elucidating them very arduous, (but) it occurred to me that it might be possible to make sense of those parts which had been damaged in one by what had been preserved in the other and to restore the proper meaning by replacing a word by another until the account was correct.*[409]

This translation gave an impulse to Arabic studies on magic squares. The first was Abū'l-Wafā' Būzjānī's (940-997/8; above, p. 63). Actually, it was an attempt —only partly successful despite the mathematical competence of its author— to find the background of the basic, general constructions: indeed, the placing of each number is merely recorded in the ancient text, presumably because it could be learned in detail elsewhere.[410] We then see continuous development in the 11th and early 12th centuries, with the discovery of various general construction methods. Here the contribution of Ibn al-Haytham (*c.* 965-1041) proved to be essential; whereas Būzjānī constructed the squares individually, order by order, Ibn al-Haytham approached the problem generally. Considering the properties of the *natural squares*, that is, of squares filled with the natural numbers taken in succession, he noted that, first, all their diagonals, main and broken, always contain the magic sum while, second, pairs of symmetrically placed lines and columns differ from the magic sum by the same amount but with a different sign. The first property gave him the clue to one general construction for odd orders and the second was applied to even orders. But Ibn al-Haytham was only partly successful for the less simple case of evenly-odd orders, which was finally solved towards the end of the 11th century.[411]

---

[409] Mufaḍḍal ibn Thābit ibn Qurra was possibly the son of one of the most known translators of Greek works, Thābit ibn Qurra (836-901). Both were Sabeans, that is, belonged to a sect worshipping Greek deities, thus polytheists, but tolerated in early Islamic times because of their exceptional abilities as scholars and translators. The 'Library' in question may be the House of Wisdom (*bayt al-ḥikma*), known to have preserved books collected in early Islamic times.

[410] Recording these placings was a preliminary to the two main purposes of the treatise, namely the construction of squares displaying other configurations, expounded in its second and fourth parts. For further details, see our recent edition of Mufaḍḍal's Arabic translation.

[411] For more about how general methods were discovered, and Ibn al-Haytham's and his successors' contributions, see our *Magic squares, their history and construction* (hereafter '*Magic squares*'), pp. 25–29, 51–56, 88–93; or earlier editions: *Les carrés magiques*, pp. 25–28, 49–51, 85–89; *Маг. квадраты*, pp. 33–37, 58–61, 96–100.

Squares filled with non-consecutive numbers also began to appear at that time. Whether original or inspired by earlier texts, quite an elaborate study on the construction of such squares with a set of given numbers not in arithmetical progression appeared already in the early 11th century.[412] The origin of that has to do with the association of Arabic letters with numerical values —an adaptation of the Greek numerical system (Fig. 190); this adaptation appeared in early Islamic times, before the adoption of Indian numerals, but remained in use later. Thus to the letters of a word or of a phrase can be associated a set of numbers. Then, with an n-letter word, or a phrase comprising n words, written in one row of a square (thus determining its order), the task was to complete the square numerically so that it would display in each row the sum in question — a mathematically interesting problem since this is not always possible. Although the author of this treatise makes no mention of such squares being used as amulets, the possibility would certainly have occurred to the contemporary reader.

ᾱ	β̄	γ̄	δ̄	ε̄	ϝ̄	ζ̄	η̄	θ̄
ا	ب	ج	د	ه	و	ز	ح	ط
1	2	3	4	5	6	7	8	9
ῑ	κ̄	λ̄	μ̄	ν̄	ξ̄	ο̄	π̄	ϙ̄
ى	ك	ل	م	ن	س	ع	ف	ص
10	20	30	40	50	60	70	80	90
ρ̄	σ̄	τ̄	ῡ	φ̄	χ̄	ψ̄	ω̄	ϡ̄
ق	ر	ش	ت	ث	خ	ذ	ض	ظ
100	200	300	400	500	600	700	800	900
͵ᾱ								
غ								
1000								

Fig. 190

Indeed, the construction of such squares paved the way to the *practical* use of magic squares, from the 11th century on. Since this use aroused much interest, for wearing such a square was supposed to bring good luck, many 'popular' works merely taught readers how to complete the rest of a square after filling one row displaying the letters of some holy name, or the words of some Koranic sentence, and left out any mathematical justification. One example of such a square is given below (Fig. 191-192; the numbers corresponding to the given words are underlined).

Readers without any knowledge of reckoning were not forgotten: some treatises just gave a set of seven magic squares of orders 3 to 9, each of

---

[412] See our *Un traité médiéval*, pp. 84–132 & (Arabic text) 164–135.

which was associated with one of the seven then known planets (including Moon and Sun) and was supposed to embody its qualities, good or evil. These qualities were described and commented in detail in the text, and the reader was even told on what material and when to draw each of such squares; for both the nature of the material and the astrologically predetermined time of drawing were supposed to increase the square's efficacy (see below). The square was then ready to be placed in the vicinity of the beneficiary or victim. This must have been of great help in solving personal or business problems.

هو   اَلله   الرحمن   الرحيم   الملك القدوس

201	121	289	329	66	11
60	170	256	137	191	203
196	254	146	139	189	93
32	276	144	141	187	237
172	176	42	143	185	299
356	20	140	128	199	174

٢٠١	١٢١	٢٨٩	٣٢٩	٦٦	١١
٦٠	١٧٠	٢٥٦	١٣٧	١٩١	٢٠٣
١٩٦	٢٥٤	١٤٦	١٣٩	١٨٩	٩٣
٣٢	٢٧٦	١٤٤	١٤١	١٨٧	٢٣٧
١٧٢	١٧٦	٤٢	١٤٣	١٨٥	٢٩٩
٣٥٦	٢٠	١٤٠	١٢٨	١٩٩	١٧٤

Fig. 191   Fig. 192

§2. Arrival of magic squares in late mediaeval Europe

Of such kind were the Arabic texts on magic squares translated into Latin in 14th-century Spain. Characteristic examples of theory and application are the following 4 × 4 and 5 × 5 squares, attributed to Jupiter and Mars, with their respective properties, mainly favourable for the first and unfavourable for the second.[413]

[174] *The figure of Jupiter is square, four by four, with 34 on each side* (Fig. 193). *If you wish to operate with it: make a silver plate in the day and the hour of Jupiter, provided Jupiter is propitious,*[414] *and engrave upon it the figure; you will fumigate it with aloes wood and amber. When you carry it with you, people who see you will love you and you will obtain from them whatever you request. If you place it in the store-house of a merchant, his trade will increase. If you place it in a dovecote or in a hive, a flock of birds or a swarm of bees will gather there. If someone unlucky carries it, he will prosper and be more and more successful. If you place it in the seat of a prelate, he will enjoy a long prelature and will not fear his enemies, but be successful among them.*

---

[413] This and other examples, all taken from the text *Incipiunt figure 7 planetarum*, in our *Magic squares for daily life*.

[414] Thus on Thursday, 1st and 8th hours of the day and 3rd and 10th of the night, and with Jupiter in direct motion and increasing in brightness.

14	10	1	22	18
20	11	7	3	24
21	17	13	9	5
2	23	19	15	6
8	4	25	16	12

16	3	2	13
5	10	11	8
9	6	7	12
4	15	14	1

11	24	7	20	3
4	12	25	8	16
17	5	13	21	9
10	18	1	14	22
23	6	19	2	15

Fig. 193   Fig. 194   Fig. 195

The figure of Mars when unfavourable means war and exactions. It is a square figure, five by five, with 65 on each side (Fig. 194 —note the separation of odd and even numbers, with the latter in the corners; another square transmitted in similar texts is that of Fig. 195, different but having the same effect). *If you wish to operate with it, take a copper plate in the day and hour of Mars,*[415] *when Mars is decreasing in size and brightness, or malefic and retrograding, or in any way unfavourable, and engrave the plate with this figure; and you will fumigate it with the excrement of mice or cats. If you place it in an unfinished building, it will never be completed. If you place it in the seat of a prelate, he will suffer daily harm and misfortune. If you place it in the shop of a merchant, it will be wholly destroyed. If you make this plate with the names of two merchants and bury it in the house of one of them, hatred and hostility will come between them. If you happen to fear the king or some powerful person, or enemies, or have to appear before a judge or a court of justice, engrave this figure as said above, when Mars is favourable, in direct motion, increasing in size and brightness; fumigate it with one drachma* ($= \frac{1}{8}$ ounce) *of carnelian stone. If you put this plate in a piece of red silk and carry it with you, you will win in court and against your enemies in war, for they will flee at the sight of you, fear you and treat you with deference. If you place it upon the leg of a woman, she will suffer from a continuous blood flow. If you write it on parchment on the day and the hour of Mars and fumigate it with birthwort and place it in a hive, the bees will all fly away.*

It was thus the arrival of such texts in late mediaeval Europe which first aroused interest in, and later led to the study of, such squares there. This incidentally explains the use of the term 'magic' to qualify them —formerly also 'planetary', which we find still employed by Fermat.[416] One 4 × 4 square thus transmitted was used by Dürer in his *Melencolia* in order to date it (Fig. 193: '1514'). As to the mediaeval Arabic (perhaps originally Greek) denomination, 'Harmonious arrangement of numbers'

---

[415] On Tuesday, same times as before.

[416] *Varia opera mathematica*, p. 176; or *Œuvres complètes*, II, p. 194.

(*wafq al-a'dād*), which had a more mathematical connotation, it remained unknown, just as did the various general constructions described in Arabic and Persian manuscripts.

The earlier transmission towards the East was more fruitful. India and China received many more examples of squares, sometimes also construction methods. In Byzantium, a treatise was written at the very beginning of the 14th century by Manuel Moschopoulos who, taking a set of examples surely received from some Persian or Arabic source, reconstructed a few methods for odd and evenly-even squares.[417] Obviously, Moschopoulos did not know anything about the existence of magic squares in ancient Greece.

§3. First attempts at reconstruction

European scholars of the fifteenth and sixteenth centuries attempted, with the squares received, to do just the same as Moschopoulos, but these initial studies were not altogether successful.

— Odd-order squares

In addition to that of order 3 (particular case) and that of order 5 in Fig. 195, those of orders 7 and 9 were transmitted (Fig. 196 & 197).[418]

37	78	29	70	21	62	13	54	5
6	38	79	30	71	22	63	14	46
47	7	39	80	31	72	23	55	15
16	48	8	40	81	32	64	24	56
57	17	49	9	41	73	33	65	25
26	58	18	50	1	42	74	34	66
67	27	59	10	51	2	43	75	35
36	68	19	60	11	52	3	44	76
77	28	69	20	61	12	53	4	45

22	47	16	41	10	35	4
5	23	48	17	42	11	29
30	6	24	49	18	36	12
13	31	7	25	43	19	37
38	14	32	1	26	44	20
21	39	8	33	2	27	45
46	15	40	9	34	3	28

Fig. 196   Fig. 197

All these squares display a regularity in their arrangement which led to the following construction rule explained by Bachet.[419] We start by adding on each side of the empty square of the desired order a stair-like figure (Fig. 198, order 5); we then write in the alternate (oblique) rows

---

[417] Edited and translated by Tannery; on its origin, see our study of it.

[418] Some of these squares were transmitted in the right-to-left form.

[419] Ed. 1624 (not in the 1612 ed.), pp. 161–166.

the natural sequence; after that, each of the stair-like parts is transposed, with its numbers, to where it fits inside, on the opposite side (Fig. 199). That is how all these squares may be obtained. Curiously, here Bachet, who says that he has seen these squares in many authors, attributes that rule to himself, although it appears (but very concisely expressed) in Cardan's *Practica arithmetice*.[420]

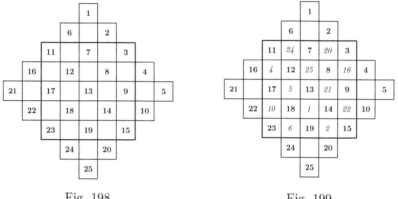

Fig. 198   Fig. 199

Now from a practical point of view this auxiliary construction is superfluous, for the sequence of natural numbers may be placed directly, as follows. We first write 1 in one of the four cells adjacent to the central cell, say that below. We move diagonally away from it, cell by cell, writing the sequence of numbers; when we reach a side, we move to the opposite one as if to continue the diagonal move. (In order to determine the cell and the subsequent diagonal moves, we may draw on each side auxiliary rows, as in Fig. 200, with $n = 5$.) When a sequence of n numbers has been thus placed, the next cell is occupied; we then move, whatever the (odd) order, two cells down in the same column, and resume the diagonal placing.

The square of Fig. 194, which separates the numbers by parity, may also (like any such square of odd order) be obtained using an auxiliary figure (Fig. 201, order 7). Writing first the natural sequence of odd numbers in the largest oblique central square within that which is to be constructed, then the even ones in the auxiliary figure, we shall just move its three outer parts, with their numbers, to where they fit in the empty

[420] Ed. 1539, Ch. XLII, No. 39 (squares) & Ch. LXVI, No. 72 (rule). By the way, in a letter to Mersenne, Fermat notes, about the method expounded by Bachet: *pour la regle des quarrez impairs, je dis premierement qu'elle n'est pas de son invention, car elle est dans l'Arithmetique de Cardan* (see his *Varia opera mathematica*, p. 174).

corners. This is a well-attested construction in Arabic mediaeval trea-tises.[421]

11	24	7	20	3	
4	12	25	8	16	4
17	5	13	21	9	17
10	18	1	14	22	10
23	6	19	2	15	23
11	24	7	20	3	
				16	

Fig. 200

		1				
	15	9	3			
29	23	17	11	5		
43	37	31	25	19	13	7
8	2	45	39	33	27	21
22	16	10	4	47	41	35
36	30	24	18	12	6	49
44	38	32	26	20	14	
46	40	34	28			
48	42					

Fig. 201

— Evenly-even orders

Besides the 4×4 ('Dürer') square already seen, Europe received an-other form of this square, in which the two median vertical rows are inverted (Fig. 202),[422] and two 8×8 squares (Fig. 203 & 204).

1	15	14	4
12	6	7	9
8	10	11	5
13	3	2	16

Fig. 202

1	63	62	4	5	59	58	8
56	10	11	53	52	14	15	49
48	18	19	45	44	22	23	41
25	39	38	28	29	35	34	32
33	31	30	36	37	27	26	40
24	42	43	21	20	46	47	17
16	50	51	13	12	54	55	9
57	7	6	60	61	3	2	64

Fig. 203

1	2	62	61	60	59	7	8
16	10	51	53	12	54	15	49
48	47	19	20	21	22	42	41
25	39	38	28	29	35	34	32
33	31	30	36	37	27	26	40
24	23	43	44	45	46	18	17
56	50	11	13	52	14	55	9
57	58	6	5	4	3	63	64

Fig. 204

There were attempts in the 16th century to construct the squares of order 4, 6 and 8 by drawing auxiliary figures in the manner seen for odd-order squares. At best, such a construction produced only the square considered.[423] These attempts could only fail for two reasons: first,

---

[421] See our *Magic squares*, pp. 34–36 (or earlier edition: *Les carrés magiques*, pp. 33–35; *Маг. квадраты*, pp. 42–44).

[422] In this particular case such a transformation again gives a magic square.

[423] Attempts of Spinula (16th century), see *Récréations*, pp. 215–216.

each order was considered independently, and second, since no distinction was made between evenly-even and evenly-odd orders, no relatedness was sought between the squares of order 4 and 8 —in any event, it is unlikely that a general theory could have been established from single examples. There had to be a fresh start, disregarding the existing squares, as mathematicians were to make from Fermat's time on, and as had been done before in Arabic times (above, p. 190).

As a matter of fact a general method for evenly-even squares could have been found by considering the two squares of Fig. 202 and 203, namely by observing how each one's numbers were placed relative to those of the corresponding natural square. Indeed, let us mark, (say) by writing dots, the diagonals of an empty 4×4 square (Fig. 205), and then reproduce this structure in an empty 8×8 square (Fig. 206). We shall next enumerate the cells from two opposite corners. By filling first the dotted cells and then the blank ones (thereby equalizing opposite rows of the natural square, see p. 190), we shall obtain the magic squares of Fig. 202 & 203.[424]

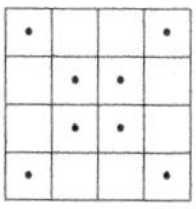

Fig. 205 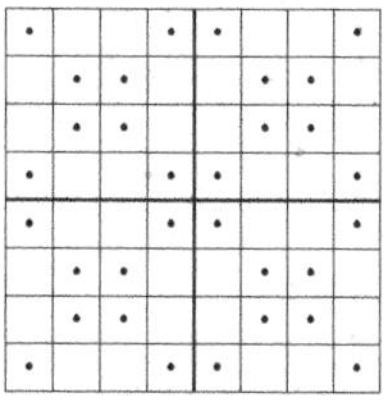   Fig. 206

— Evenly-odd orders

The two following 6×6 squares were transmitted:

1	35	34	3	32	6
30	8	28	27	11	7
24	23	15	16	14	19
13	17	21	22	20	18
12	26	9	10	29	25
31	2	4	33	5	36

Fig. 207

1	32	34	3	35	6
30	8	27	28	11	7
20	24	15	16	13	23
19	17	21	22	18	14
10	26	12	9	29	25
31	4	2	33	5	36

Fig. 208

---

[424] This method is applicable to the construction of all evenly-even squares. See *Magic squares*, pp. 45–47 (*Les carrés magiques*, pp. 43–45; *Маг. квадраты*, pp. 53–54). But no true method will produce the second square of order 8 (Fig. 204), which was found by trial and error; see *Les carrés magiques*, p. 261, or *Маг. квадраты*, p. 268.

Whereas one (Fig. 207) displays a reasoned construction, the other (Fig. 208) seems to have been obtained, like the second square of order 8, empirically.[425]

One exception to this failure to develop general methods is found in a work by Michael Stifel (1487-1567), already encountered (pp. 94, 106): it concerns *bordered* squares, that is, ones where the removal of successive borders leaves each time a magic square. Indeed, Stifel has general methods for all three types of orders (Fig. 209, odd square, & 210, even square with, alternately, evenly-even and evenly-odd borders).[426] He does not mention any source, nor does he give himself any credit, and his methods are not exactly like the ones we know from Arabic sources. We may only observe that the construction of successive bordered squares is somewhat easier; for here it makes sense to start from one magic square and add a border —which seems indeed to be how the general constructions of such magic squares were found.[427]

16	81	79	77	75	11	13	15	2
78	28	65	63	61	25	27	18	4
76	62	36	53	51	35	30	20	6
74	60	50	40	45	38	32	22	8
9	23	33	39	41	43	49	59	73
10	24	34	44	37	42	48	58	72
12	26	52	29	31	47	46	56	70
14	64	17	19	21	57	55	54	68
80	1	3	5	7	71	69	67	66

Fig. 209

256	9	247	246	12	13	243	242	16	17	239	238	20	21	235	2
3	226	213	45	46	210	209	49	50	206	205	53	54	201	32	254
4	33	200	63	193	192	66	67	198	188	70	71	185	58	224	253
252	34	59	178	169	89	90	166	165	93	94	161	80	198	223	5
251	222	60	81	160	101	155	154	104	105	151	98	176	197	35	6
7	221	196	82	99	146	141	117	118	137	112	158	175	61	36	250
8	37	62	174	100	113	136	123	122	133	144	157	83	195	220	249
23	38	73	173	107	114	129	126	127	132	143	150	84	184	219	234
24	218	183	85	108	115	125	130	131	128	142	149	172	74	39	233
232	217	75	86	148	138	124	135	134	121	119	109	171	182	40	25
231	41	76	87	147	145	116	140	139	120	111	110	170	181	216	26
27	42	180	162	159	156	102	103	153	152	106	97	95	77	215	230
28	43	179	177	88	168	167	91	92	164	163	96	79	78	214	229
228	202	199	194	64	55	191	190	68	69	187	186	72	57	55	29
227	225	44	212	211	47	48	208	207	51	52	204	203	56	31	30
255	148	10	11	245	244	14	15	241	240	18	19	237	236	22	1

Fig. 210

[425] See *Les carrés magiques*, pp. 257–258; *Маг. квадраты*, pp. 265–267.

[426] See his *Arithmetica integra*, fol. $24^{v}-30^{r}$; or *Magic squares*, pp. 149–150 & 154, 161–163 & 166–167, 170–171 & 173 (*Les carrés magiques*, pp. 125–126 & 130, 137–139 & 141–142, 146 & 148–149; *Маг. квадраты*, pp. 136 & 141, 148–150 & 153–154, 157–158 & 160–161).

[427] See *Magic squares*, p. 141 *seqq.* (*Les carrés magiques*, p. 117 *seqq.*; *Маг. квадраты*, p. 127 *seqq.*).

Chapter XV. Infinite sets

The consideration of infinite sets does not usually appear in connection with mathematical recreations. Two reasons have led us to include the subject here. First, as with many of the recreational problems seen above, there was then no general solution to the questions thus raised. Second, their treatment by mediaeval scholars brings clearly to light the apparent contradictions resulting from the comparison of infinite sets.

§1. Natural numbers and rational fractions

We have already seen a Provençal text which admitted for the first time a negative number as solution (p. 61). This same text also shows that two consecutive natural numbers enclose an infinite quantity of fractional numbers. Since, from Greek times, the set of natural numbers was recognized as infinite, its author was abruptly confronted with the question of the relative size of two infinite sets in one of which there are 'infinitely more' elements than in the other.

He was led to these considerations whilst constructing rational approximations of square and cube roots of natural numbers not themselves being squares or cubes. So let us consider such an integer N; furthermore, since the author's reasoning is general, we may describe it in general terms. To find then a sequence of increasingly close rational approximations of $\sqrt[m]{N}$ when N is not a mth power.

Since the integer N is not itself a mth power, it must be comprised between the mth powers of two consecutive natural numbers, and so we shall have

$$a^m < N < (a+1)^m,$$

with a an integer.

The author sets out two assertions as preliminaries.[428]

(1) An exact value (that is, according to ancient and mediaeval concepts, a rational value) can never be found.

Indeed, the author notes, if that were the case, then $\sqrt[m]{N}$, being between the two consecutive integers a and $a+1$, would have to be the sum of a and a fraction. But it is not possible for the power of a fraction (which we may suppose to be reduced) to give an integer.

(2) An approximate value may always be ameliorated.

Indeed, the author observes, the sequence of aliquot fractions (thus with numerator 1) is endless, just as the sequence of natural numbers is;

---

[428] *L'Arithmétique de Pamiers*, pp. 113–115 & (text) pp. 277–278.

consequently, it will always be possible to increase or reduce, if need be, an approximate value using small fractions.

It is at this point that the author shows how an infinite sequence of rational approximations may be set out.[429] For that purpose, he considers two sequences of fractions, one ascending and the other descending —related to one another since corresponding terms add up to unit. Let the first, the so-called 'natural progression in augmentation' (*natural progressio en augmentatio*) be

$$\frac{1}{2}, \frac{2}{3}, \frac{3}{4}, \dots, \frac{k}{k+1}, \dots$$

and the second, the 'natural progression in diminution' (*natural progressio en diminutio*) be

$$\frac{1}{2}, \frac{1}{3}, \frac{1}{4}, \dots, \frac{1}{k+1}, \dots ;$$

thus (for us) the first approaches 1 and the second, 0.

Since $a^m < N < (a+1)^m$, let us start our calculation by computing $(a+\frac{1}{2})^m$. If it turns out that

$$\left(a+\frac{1}{2}\right)^m < N,$$

we shall replace $\frac{1}{2}$ by the terms, taken successively, of the ascending sequence; computing *each time* the mth powers, we shall finally attain the two quantities enclosing N; their roots are such that

$$a + \frac{k-1}{k} < \sqrt[m]{N} < a + \frac{k}{k+1},$$

which can also be written as

$$\frac{ak(k+1) + k^2 - 1}{k(k+1)} < \sqrt[m]{N} < \frac{ak(k+1) + k^2}{k(k+1)}. \quad (*)$$

Likewise, if

$$\left(a+\frac{1}{2}\right)^m > N,$$

we are to replace $\frac{1}{2}$ by the terms, taken in succession, of the descending sequence until we arrive at the two approximations enclosing N, thus

$$a + \frac{1}{k+1} < \sqrt[m]{N} < a + \frac{1}{k}$$

and therefore

---

[429] *L'Arithmétique de Pamiers*, pp. 118–120 & (text) pp. 280–282.

$$\frac{ak(k+1)+k}{k(k+1)} < \sqrt[m]{N} < \frac{ak(k+1)+k+1}{k(k+1)} . \quad (**)$$

We see that in either case the limits obtained are of the form

$$\frac{s}{t} < \sqrt[m]{N} < \frac{s+1}{t} .$$

Now between two such fractions can be inserted an arbitrary quantity of rational fractions: choosing any natural number h and writing the previous expression as

$$\frac{sh}{th} < \sqrt[m]{N} < \frac{(s+1)h}{th} ,$$

it appears that there will be between the two given fractions $h-1$ intermediate ones, for

$$\frac{sh}{th} < \frac{sh+1}{th} < \frac{sh+2}{th} < \ldots < \frac{sh+h-2}{th} < \frac{sh+h-1}{th} < \frac{sh+h}{th} ,$$

and we shall find out, by computation, which pair, raised to the mth power, will enclose N. Since the numerators of this pair of fractions differ once again by 1, we may choose a new h (the larger the better, the text says) and thus find $h-1$ closer values of which two will once again contain the required value. Continuing in this way, we can obtain any number of pairs of fractions enclosing $\sqrt[m]{N}$, each pair a better approximation than the one before.

This method certainly appeals. It is, however, inapplicable because of the disproportionate number of computations involved; furthermore, there existed from antiquity various convenient ways of attaining values close to quadratic and cubic roots (see below, pp. 211–214). Our author's has, though, the advantage of showing clearly that between two consecutive squares or cubes, thus a^2 and $(a+1)^2$ and a^3 and $(a+1)^3$, there is an infinite quantity of square or cubic fractions; for, as the author says, *all square or cubic fractional numbers are smaller than 1 or comprised between two consecutive integral square or cubic numbers.*[430] Therefore, *although the ascending (sequence of) squares and cubes of integral numbers does not come to an end, there are more fractional square and cubic fractional numbers.*[431] Consequently, considering now their roots, there will be an infinity of fractions between two consecutive (positive) integers. That is, as said above, although the sequence of natural numbers is infinite, there is an infinity of fractions between any two consecutive ones. The fractions have therefore, as the

---

[430] *L'Arithmétique de Pamiers*, pp. 116 & 279. The author distinguishes here between proper and improper fractions.

[431] *Op. cit., ibid.*

author puts it, *many more reasons* to be in infinite quantity than the natural integers; but, he adds, this does not mean that, considering the infinity of natural numbers, *there would be more fractional numbers, for this is against reason.*[432]

It fell to Georg Cantor (1845-1918) to show that the set of all positive rational fractions and that of all natural numbers are 'equivalent', which is to say that the first cannot be considered to contain more elements. A particular arrangement enabled him to make that easier to grasp. Suppose first the set of all such fractions less than 1 to be lined up; this is obtained by considering the quantities $\frac{k}{l+1}$, with $k < l+1$, and letting l run through the set of all natural numbers while k, for each l, runs through the finite set $1, \ldots, l$; we are further to suppose the fractions reduced in order to avoid repetition. We shall then have, in the first line, *all* fractions between 0 and 1, and each appearing only once, as desired. All other (improper) fractions will be obtained by adding the successive natural numbers to the first line (Fig. 211).

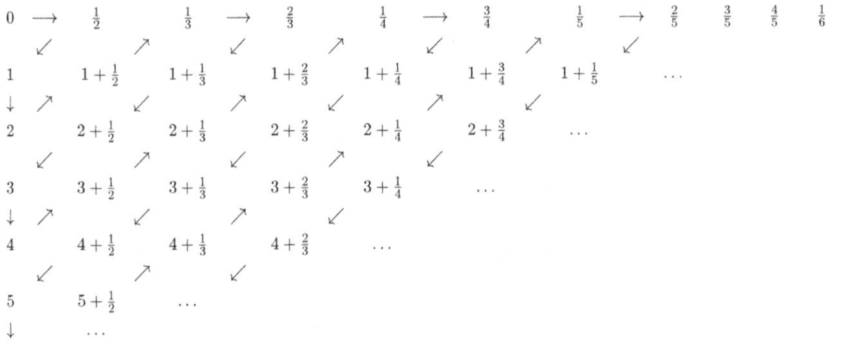

0 →	$\frac{1}{2}$		$\frac{1}{3}$ →	$\frac{2}{3}$	$\frac{1}{4}$ →	$\frac{3}{4}$	$\frac{1}{5}$ →	$\frac{2}{5}$	$\frac{3}{5}$	$\frac{4}{5}$	$\frac{1}{6}$	...
1	$1+\frac{1}{2}$	$1+\frac{1}{3}$	$1+\frac{2}{3}$	$1+\frac{1}{4}$	$1+\frac{3}{4}$	$1+\frac{1}{5}$	...
2	$2+\frac{1}{2}$	$2+\frac{1}{3}$	$2+\frac{2}{3}$	$2+\frac{1}{4}$	$2+\frac{3}{4}$	...
3	$3+\frac{1}{2}$	$3+\frac{1}{3}$	$3+\frac{2}{3}$	$3+\frac{1}{4}$	...
4	$4+\frac{1}{2}$	$4+\frac{1}{3}$	$4+\frac{2}{3}$	...
5	$5+\frac{1}{2}$	...

Fig. 211

The resulting table now displays all positive rational fractions, and each one only once. We can pass through all of them, and each one only once, by following the diagonal path indicated by the arrows; that is, we can enumerate these fractions by means of the set of natural numbers. The one-to-one correspondence between the two sets then becomes clear. The set of positive fractions is thus 'equivalent' to that of the natural numbers, and it is therefore, according to the usual terminology, 'denumerable' (*gleichmächtig* and *abzählbar* were Cantor's terms). This thus answers the question put, some five centuries earlier, by the Provençal author.

---

[432] *L'Arithmétique de Pamiers*, pp. 117 & 279–280.

Remark. This diagonal way of enumerating is essential: were we to go along the first row, we would never reach the second. (We would only have shown that the set of fractions between 0 and 1 is denumerable.)

The above reasoning may be generalized to a sequence of infinite sets provided the elements of each may likewise be set out in a row. In other words, an (in)finite denumerable sequence of infinite denumerable sets $\{M_i\}$ is itself denumerable (Fig. 212): as for the above fractions, we shall enumerate their elements diagonally, starting from the corner in the finite and taking in succession $m_{11}, m_{12}, m_{21}, m_{31}, m_{22}, m_{13}, \ldots$.

m_{11}	m_{12}	m_{13}	m_{14}	m_{15}	m_{16}	m_{17}	$\ldots$		$= \{M_1\}$
m_{21}	m_{22}	m_{23}	m_{24}	m_{25}	m_{26}	m_{27}	$\ldots$		$= \{M_2\}$
m_{31}	m_{32}	m_{33}	m_{34}	m_{35}	m_{36}	m_{37}	$\ldots$		$= \{M_3\}$
m_{41}	m_{42}	m_{43}	m_{44}	m_{45}	m_{46}	m_{47}	$\ldots$		$= \{M_4\}$
m_{51}	m_{52}	m_{53}	m_{54}	m_{55}	m_{56}	m_{57}	$\ldots$		$= \{M_5\}$
m_{61}	m_{62}	m_{63}	m_{64}	m_{65}	m_{66}	m_{67}	$\ldots$		$= \{M_6\}$
m_{71}	m_{72}	m_{73}	m_{74}	m_{75}	m_{76}	m_{77}	$\ldots$		$= \{M_7\}$
m_{81}	m_{82}	m_{83}	m_{84}	m_{85}	m_{86}	m_{87}	$\ldots$		$= \{M_8\}$
$\ldots$									$\ldots$

Fig. 212

Since in Fig. 211 the first column contains the set of natural numbers, we see how for infinite sets the whole may become equivalent to one of its parts: for even though the quantities of their respective elements would seem *a priori* to be different, each element of the first can be seen to be in a one-to-one correspondence with each element of the second. Whereas pairing the elements of two *finite* sets will immediately determine if their quantities are different or equal. It is precisely by applying this pairing to infinite sets that 14th-century scholars will render the paradox evident. We shall see their examples after examining the basis for their comparisons.

§2. Finite and infinite in Greece and the Middle Ages

1. Greece

Greek scholars did not venture, *as far as we know*, into such considerations. This is not to say that they refused to consider the existence of infinites; for they recognized three:

— the sequence of natural numbers: whatever the number N, there is a number $N + 1$;

— the two continuous magnitudes time and space are seen to be infinite;

— a continuous finite quantity may be divided indefinitely (as had been mostly admitted from classical times).

The characteristic of these infinite sets is that they are 'potentially' infinite: they consist of an endless succession, and their last constituent part can never be reached. Thus, in order to be able to study infinite sets and compare them, mediaeval scholars had to create in a somewhat artificial manner 'actual' infinites, that is, infinites which could be considered as complete. They reached this virtual creation and artificial comparison by adapting or extending Greek concepts, namely series and Euclid's axioms.

SUMMING SERIES. Certain formulae of summing series were known in Greece: summing consecutive natural numbers, their squares, their cubes; summing an arithmetical progression; summing a geometrical progression. For this last case, the Greeks knew that

$$ a + a \cdot r + a \cdot r^2 + a \cdot r^3 + \ldots + a \cdot r^n = a \frac{r^{n+1} - 1}{r - 1} \quad \left(= a \frac{1 - r^{n+1}}{1 - r} \right), $$

as demonstrated in Euclid's *Elements of geometry* (proposition IX, 35).

But this was restricted to a finite number of terms, whether they were increasing or $(r < 1)$ decreasing, with each term added contributing to increase the sum. But certain ancient results may have been obtained in the case $r < 1$ by having $r \to 0$ in an infinite series and then demonstrating these results in a classical way; such a procedure is almost explicit in Archimedes' *Quadratura parabolae*.

SETTING AXIOMS. Besides fundamental definitions, Euclid put at the beginning of his *Elements* postulates and axioms. There are five postulates, all of geometrical nature, which had to be introduced as a basis for demonstrating the theorems to come; for, when writing a textbook, it is necessary to start with a limited number of undemonstrated propositions. There are also five axioms or, to remain faithful to the Greek, 'common notions' (κοιναὶ ἔννοιαι), which are not specifically geometrical but, rather, represent something one would intuitively accept as being true. They are the following:

I. Those equal to a same are equal to one another.

II. If equals be added to equals, the wholes will be equal.

III. If equals be subtracted from equals, the remainders will be equal.

III′. If equals be added to unequals, the wholes will be unequal.[433]

---

[433] The number of axioms was increased in late antiquity. This is just one example, mentioned here because it will play a rôle in what follows.

IV. Those which coincide with one another are equal to one another.

V. The whole is greater than the part.

Today's mathematical reader will see at once the difficulties involved in any comparison of sets differing by (relatively) negligible quantities, such as finite sets differing by infinitely small quantities, but also infinite sets differing by finite quantities. Thus it was necessary to complete these axioms with certain restrictions or specifications, and that was what mediaeval scholars were to do.

2. The Middle Ages

For the reasons set forth above, the idea of actual infinites could no longer be *a priori* rejected in late mediaeval times.

— First, theologians observed that if actual infinites do not exist, this could not be for *material* reasons, such as availability of space and duration of creation, for God's omnipotence could overcome these limitations; thus, if actual infinites do not exist, it must be for *logical* reasons. This led to the search for inconsistencies inherent in the very notion of actual infinite.

— But in the search for such inconsistencies and to avoid any *a priori* rejection of their existence by opponents, such infinites must have been supposed to exist, and thus some theoretical means of creating them was to be found. Here mathematics, in particular the convergence of some infinite series, was to provide the necessary tool.

SUMMING SERIES. Scholars at the time of Nicolas Oresme ($c.$ 1320–1382), bishop of Lisieux, had noticed that series with an infinite number of terms could have not only infinite sums but also finite ones.[434]

(i) As an example of an infinite sum, Oresme proved the divergence of the harmonic series, that is, of the sum of the natural numbers' inverses. For, considering

$$\frac{1}{2} + \frac{1}{3} + \frac{1}{4} + \frac{1}{5} + \dots,$$

and grouping together the quantities of successive terms determined by the sequence of powers of 2,

$$\frac{1}{2} + \left(\frac{1}{3} + \frac{1}{4}\right) + \left(\frac{1}{5} + \frac{1}{6} + \frac{1}{7} + \frac{1}{8}\right) + \dots,$$

he observed that this is larger than

$$\frac{1}{2} + 2 \cdot \frac{1}{4} + 4 \cdot \frac{1}{8} + \dots = \frac{1}{2} + \frac{1}{2} + \frac{1}{2} + \dots,$$

---

[434] See Wieleitner's study, or the second of Oresme's *Quaestiones super geometriam Euclidis*.

whereby the harmonic series must *a fortiori* have an infinite sum and thus be divergent.

(*ii*) To find an example of a finite sum, late mediaeval scholars started with the Greek formula seen above, namely

$$a + a \cdot r + a \cdot r^2 + a \cdot r^3 + \ldots + a \cdot r^n = a\,\frac{1 - r^{n+1}}{1 - r},$$

and applied it to the infinite sum of the inverses of the powers of 2, thus with $a = \frac{1}{2} = r$, and thus $r < 1$. Less rigorous than the Greeks, they envisaged considering that

$$\frac{1}{2} + \frac{1}{4} + \frac{1}{8} + \ldots = \sum_{k=1}^{\infty} 2^{-k} = \frac{1}{2}\,\frac{1}{1 - \frac{1}{2}} = 1$$

since the term subtracted in the numerator was becoming negligible.

Thus emerged a possible means of constructing actual infinites and thus challenging any *a priori* rejection. For one could then envisage the step-by-step construction of such an infinite with an infinite number of finite increments over infinite parts of an arbitrary but finite length of time, say one hour. Indeed, using the above sequence, we shall take a finite magnitude $\mathfrak{a}$ and start by doubling it in half an hour; then we double the result in half this time, thus a quarter of an hour, then double the new result in an eighth of an hour, and continue doubling the successive results in the successive fractions 2^{-k} of an hour. This will produce the quantities $2\mathfrak{a}$, $4\mathfrak{a}$, $8\mathfrak{a}$, ..., and finally, at the end of the hour, the infinite quantity $\mathfrak{A}$. The convergence of the series 2^{-k} has thus enabled us to complete the construction of an actual infinite in a finite time.

Such was the procedure used in particular by Oresme and his student Albert of Saxony in the middle of the 14th century in order to study the properties of actual infinites (as a matter of fact they aimed at asserting their non-existence by exposing the inconsistencies their existence would imply). Oresme's arguments are found in some places of his commentary on Aristotle's *Physics*, and the same arguments are repeated, but more clearly expressed, in Albert of Saxony's commentary on Aristotle's treatise *On the heavens*.[435]

SETTING AXIOMS. The basis of mathematical knowledge in the Middle Ages, in both the Christian and Moslem worlds, was Euclid's *Elements*.

---

[435] Arguments and quotations in our *Vergleiche zwischen unendlichen Mengen* or pp. 40–48 of our *On an algorithm* (both quote the original texts). On the mediaeval considerations of infinites in mediaeval times, see the works of Duhem, Anneliese Maier and Murdoch.

Since his axioms were proper to finite magnitudes, they had, if one were to consider infinite magnitudes, to be adapted. This is what those two scholars did. For the first axiom, which asserts the transitivity of equality, they considered as equivalent two infinites of which one arises from the second by merely rearranging its parts. Disregarding the second axiom, here irrelevant, they modified the last three by taking into account the existence of imperceptible differences.

I. *Nothing can become larger than it is by merely changing the place of its parts, (thus) without external addition or rarefaction, nor can it become less without (external) withdrawal or condensation. Likewise, no set may become larger without addition of elements or less without subtraction of elements.*

Note the two facets of this assertion: first are considered continuous sets (the substance of which remains unalterable in quantity and volume if not subject to physical or chemical alteration), then the discrete sets, constituted by separate elements. The same will be done in the fourth axiom.

III. *If equals be subtracted from equals, the remainders will be equal, whether the parts removed have to the remainders a (determined) ratio or not.*

Indeed, a ratio can be expressed only between magnitudes of the same kind, not between a finite magnitude and one infinitely large or infinitely small; for, if that were the case, taking a finite a finite number of times would produce an infinite.

III'. *If equals be added to unequals, the wholes will be unequal, whether there is between them, as before, a (determined) ratio or not.*

IV. *If, for those which are superposed or, in thought, applied one to another, one does not exceed the other nor is exceeded by it, then none is larger or smaller than the other. The same holds for sets such that to each element of one corresponds one (and only one) element of the other.*

Note that the term 'equality' of the fourth Euclidean axiom has disappeared: since the magnitudes are infinite, equality cannot be ascertained. There is also, for discrete sets, the principle of one-to-one correspondence between their elements, which is exactly what Cantor was to use some four centuries later for ascertaining equivalence of sets.

V. *The part of a whole exceeding it and nowhere exceeded by it is not equal to the whole from which it is a part.*

This new form of Euclid's fifth axiom is thus adapted to sets which cannot be estimated in their entirety.

§3. Examples of comparing infinites in the Middle Ages

Here are given three 'assertions' by Oresme; they will not be demonstrated but intuitively verified by means of examples.[436]

First assertion: *No infinite is larger or smaller than another infinite.*

This is the modification of the concept of equality, already seen in Axiom IV, now to be applied to examples of infinites, first continuous then discrete.

Example 1: Let $\mathfrak{A}$ be a body infinite from all sides (*sit a unum corpus infinitum omniquaque*). Let us disregard it for a while. Let then (Fig. 213) $\mathfrak{B}$ be a bar with a section of one square foot and infinitely long in one direction, thus with one of its ends in the finite; we imagine it divided into cubes with side one foot. Let us form from $\mathfrak{B}$ a new body in the following way. We take the first cube and give it, in half an hour, a spherical shape. In the following quarter of an hour, we make another sphere from the first one and the second cube. We do the same with the third cube in the next fraction of time, namely half a quarter of an hour. We continue in the same way along the bar, each time taking half the previous time, thus for the successive portions 2^{-k} of an hour. At the end of the hour, we shall have obtained the body $\mathfrak{C}$. Now this $\mathfrak{C}$ is an infinite from all sides, for it has resulted from infinitely many additions of (equal and finite) quantities.

By the fourth axiom, $\mathfrak{C}$ cannot be said to be larger or smaller than $\mathfrak{A}$. But $\mathfrak{C}$ resulted from $\mathfrak{B}$ and, by the first axiom, it cannot have increased or decreased during the transformation. Consequently, as asserted, $\mathfrak{B}$ cannot be larger or smaller than $\mathfrak{A}$.

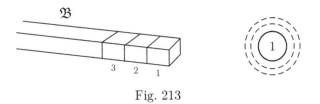

Fig. 213

It is even remarked that $\mathfrak{B}$ can produce an arbitrary number of infinites. Indeed (Fig. 214), we may first take every other element and thus have, at the end of the first hour, the infinite I_1, consisting of all the elements of even rank ($\{2k\} = 2, 4, 6, \ldots$). After joining together

---

[436] Details: see above references (p. 206n).

the remaining cubes, we shall once again take every other one and create, in another hour, the infinite I_2 (thus formed from the initial elements $3, 7, 11, \ldots$, that is, from the initial set of elements $\{2^2 \cdot k - 1\}$). We likewise construct I_3 from the elements initially numbered 5, 13, 21, $\ldots (\{2^3 \cdot k - 3\})$. And so on: the infinite I_n will consist of the elements initially bearing the numbers $\{2^n \cdot k - (2^{n-1} - 1)\}$. This procedure can be continued as far as desired since the bar still has the same capacity of creating infinites as it did initially.

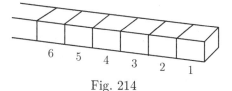

Fig. 214

This first mediaeval example illustrates for us in the best possible way how the infinite sequence of natural numbers remains an infinite set when from it is removed an infinite sequence, or infinite sequences, of elements in arithmetical progression with a constant difference other than unit, and how the remainder will have the same potential as the initial set.

Example 2. Oresme considers a bar, of finite length, divided into the parts 2^{-k} (Fig. 215). Let us put on these parts alternately 'something white' and 'something black'. Let us remove the first black thing and put in its place the second white one, with this taking half an hour to be done. Next, in the parts 2^{-k} of an hour, we remove each black and replace it with the next white (which we shall have to take each time further along, but will always find). At the end of the hour, there will remain only whites. Now, initially, the whole set of blacks and whites was neither larger nor smaller than the set of parts, according to the fourth axiom; by the same, and also Axiom I, the whole set of whites at the end is neither larger nor smaller than the set of parts. Consequently, the whole (initial) set of whites and blacks is neither larger nor smaller than the whole (final) set of whites.

The same reasoning can be applied, it is added, if the first white were initially put on the first part and the others on each thousandth counted from the first one on.

Fig. 215

This again illustrates the equivalence of the set of natural numbers and that of even (or odd) numbers, or that of integers in arithmetical progression.

Second assertion: *There exist pairs of infinites which are not equal.*

Oresme's illustrations will appear to the reader to be less impressive than the previous ones; this is because they rely on the more familiar notions of whole and part, clearly unequal for finite quantities, and of finite differences between two infinite quantities, today immediately felt by us to be negligible.

Example 3. Let there be a body infinite from all sides. We take from it a bar infinitely long in one direction. According to the fifth axiom, these two bodies will not be equal.

Example 4. Let (Fig. 216) $\mathfrak{A}$ and $\mathfrak{B}$ be two infinite bars as above, with their extremities, in the finite, aligned, and let $\mathfrak{C}$ and $\mathfrak{D}$ be two finite bars with the same cross-section and lengths (say) four and two feet. We shall then have two unequal infinite bodies. Indeed, either $\mathfrak{A}$ and $\mathfrak{B}$ are unequal to begin with (we cannot judge of the situation at the other end), or they are equal but then $\mathfrak{A} + \mathfrak{C}$ and $\mathfrak{B} + \mathfrak{D}$ will be unequal according to Axiom III$'$.

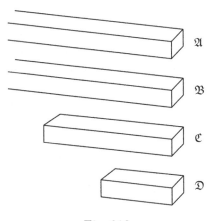

Fig. 216

Third assertion: *No infinites are comparable one to another.*

This third assertion results from the other two. The first led us to conclude that no infinite can be said to be larger or smaller than another; consequently, they must all be equal or incomparable. But the second assertion has shown that there exist unequal infinites. They must there-

fore all be incomparable. It is on this basis that Oresme and Albert of
Saxony will later assert the non-existence of actual infinites.

As seen, Oresme and Albert of Saxony had shown the possibility of
establishing a one-to-one correspondence between the elements of an infinite set and those of one of its subsets (whence a somewhat unclear notion
of equality between two infinite sets). It was then easy to show that this
would contradict the intuitive notion that the whole is larger than the
part. Two and a half centuries later, Galileo will adopt a more cautious
approach: considering in his *Discorsi e dimostrationi matematiche* (1638)
the problem of comparing the infinites of dots forming two segments of
straight line unequal in length, both of which are continuous infinite sets
but admit of a bijection between their elements (Fig. 217), he resorts to
a comparable situation for discrete infinite sets: pairing the squares of
whole numbers with their roots (thus $\{n^2\} \longleftrightarrow \{n\}$) suggests that the
quantities of their elements are the same, although the first set is included
in the second. We may at least conclude, he then writes, that the concepts of 'larger than', 'equal to' and 'smaller than' should be applied to
finite sets only.[437]

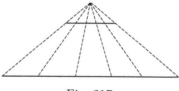

Fig. 217

§4. Ancient and mediaeval root approximations

As observed above, the root approximation in the Pamiers arithmetic,
although very attractive in theory, is of no use in practice. On the other
hand, with the exact procedure for extracting square and cube roots
being too long, most treatises, from ancient times on, restrict themselves
to teaching only approximation formulae, which reach a satisfying value

---

[437] Salviati (Galileo), asked by Sagredo about the conclusions to be drawn, answers:
*Io non veggo che ad altra decisione si possa venire, che à dire infiniti essere tutti i
numeri, infiniti i quadrati, infinite le loro radici; nè la moltitudine dei quadrati esser
minore di quella di tutti i numeri, nè questa maggior di quella; & in ultima conclusione
gli attributi di eguale, maggiore, e minore non haver luogo ne gl'infiniti, mà solo nelle
quantità terminate. E però quando il Sig. Simplicio mi propone più linee diseguali, e mi
domanda come possa essere, che nelle maggiori non siano più punti, che nelle minori,
io gli rispondo, che non ve ne sono nè più, nè manco, nè altrettanti; mà in ciascheduna
infiniti* (ed. 1655, p. 25).

using a small number of computations. All these methods basically rely on the simple binomial formulae for $(u + v)^2$ or $(u + v)^3$.[438]

1. Square roots

Consider the integer N, with $a^2 < N < (a+1)^2$. Since a is the largest integer contained in $\sqrt{N}$, we shall have $\sqrt{N} = a + x$, with $0 < x < 1$, and therefore $N = a^2 + r$, with $r = 2ax + x^2$. Now since $2ax < r < 2ax + x$, we have

$$\frac{r}{2a + 1} < x < \frac{r}{2a},$$

whence

$$a + \frac{r}{2a + 1} < a + x = \sqrt{N} < a + \frac{r}{2a}. \qquad (*)$$

a. Use of the two limits

By adding together the numerators and denominators of the fractions in $(*)$ we shall obtain the better value

$$\sqrt{N} \approx a + \frac{2r}{4a + 1}.$$

Indeed, if $\frac{p}{q} < \frac{r}{s}$, the fraction $\frac{p+r}{q+s}$ will fall between them (as, by the way, does $\frac{mp+nr}{mq+ns}$ with m, n natural numbers). This property was known in antiquity (and proved by Pappos of Alexandria, c. 350, in his *Collectio*, VII, theorem 8), then used by Uqlīdisī, c. 950, for the above formula.[439]
As a particular case $(\frac{r}{2a+1} < x < 1)$, we have

$$\sqrt{N} \approx a + \frac{r + 1}{2a + 2},$$

which is also encountered in Arabic texts.[440]

b. Use of the upper limit

We have, from $(*)$,

$$\sqrt{N} < a + \frac{r}{2a} = a + \frac{N - a^2}{2a} = \frac{a^2 + N}{2a} = \frac{1}{2}\left(a + \frac{N}{a}\right).$$

This procedure can be repeated. For from the above value $a_1 = a + \frac{r}{2a}$ we can obtain a second, better one, namely $a_2 = a_1 + \frac{r_1}{2a_1}$ ($r_1 = N - a_1^2 < 0$), and so on, whence, generally,

$$a_i = a_{i-1} + \frac{r_{i-1}}{2a_{i-1}} = a_{i-1} - \frac{|r_{i-1}|}{2a_{i-1}}.$$

---

[438] What follows is taken from our *On an algorithm*, pp. 48–52.

[439] See Saidan, *The Arithmetic of al-Uqlīdisī*, pp. 164, 441. This property is also mentioned by Chuquet (*Triparty*, p. 653).

[440] See Günther, p. 45; Hunrath, pp. 21, 27; Saidan, pp. 445, 448.

Since

$$r_{i-1} \equiv N - a_{i-1}^2 = -\left(\frac{r_{i-2}}{2a_{i-2}}\right)^2,$$

and $r_1 < r_2 < \ldots$, these values approach indefinitely close to $\sqrt{N}$ from the right side, that is, they are all by excess.

The basic approximation, first found in Mesopotamia, is mentioned in Greece by Heron of Alexandria (*Metrica* I.8) —who points out the possibility of its being iterated (τάξομεν τὰ νῦν εὑρεθέντα (...) καὶ ταῦτα ποιήσαντες εὑρήσομεν πολλῷ ἐλάττονα (...) τὴν διαφορὰν γιγνομένην).[441] It occurs throughout the Middle Ages, beginning with Johannes Hispalensis and Fibonacci. Cataldi (1552-1626) found the relation between the sequence of approximations and the partial quotients of the development of $\sqrt{a^2 + r}$ in a continued fraction (the terms of the approximations are the terms with indices 2^n in the development).[442] We may also observe that this approximation method corresponds to Newton's procedure for $f(x) = x^2 - N$.

c. Use of the lower limit

Consider this time $a_1' = a + \frac{r}{2a+1} < \sqrt{a^2 + r}$, and, generally,

$$a_i' = a_{i-1}' + \frac{r_{i-1}'}{2a_{i-1}' + 1}.$$

Since

$$r_{i-1}' \equiv N - a_{i-1}'^2 = \frac{r_{i-2}'}{2a_{i-2}' + 1}\left[1 - \frac{r_{i-2}'}{2a_{i-2}' + 1}\right] > 0,$$

and $a_1' < a_2' < \ldots$, the a_i' form a sequence of increasingly better approximations for $\sqrt{N}$, this time from the left-hand side on.

The basic approximation, namely a_1', was probably known in Greece. In any event, we encounter it in the tenth century in Islamic countries. In the West, it is attested *c.* 1300;[443] it was, however, not widely used

---

[441] Heron uses the equivalent form $\sqrt{N} \approx a_i \equiv \frac{1}{2}\left(a_{i-1} + \frac{N}{a_{i-1}}\right)$, in which $\sqrt{N}$, the geometric mean of a_{i-1} and $\frac{N}{a_{i-1}}$, is seen to be approximated by their arithmetic mean.

[442] See Treutlein, pp. 68–69; Günther, pp. 29–30, 45–46, 57–58; Hunrath, pp. 27–30, 37–43; Tropfke, pp. 264–265, 267, 272, 277, 289; Saidan, p. 441; Maracchia, pp. 67 *seqq.*, 105.

[443] MS. Colmar 414 (365, cat.), fol. 9^r, 12–25 (among various additions on mean proportionals, roots and recreational problems): *Posito quod unum castrum distet ab alio per 4 leucas et aliud ab alio per 4 et distent ad angulum rectum, si vis scire quantum distent a se remotiora, facias de numeris propositis quadratos, et illos addas sibi adinvicem; deinde, si possis, invenias unum numerum qui ductus in se quadrate equivaleat illis quadratis, et ille numerus ostendet distantiam iam dictam. Si vero nullus numerus*

until the 16th century. The iterative process was studied at the end of that century by Cataldi.[444]

2. Cube roots

Consider the integer N, with $a^3 < N < (a+1)^3$, thus a the largest integer contained in $\sqrt[3]{N}$. Then we shall have $\sqrt[3]{N} = a + x$, with $0 < x < 1$, and therefore $N = a^3 + r$, with $r = 3a^2x + 3ax^2 + x^3$. Now since $3a^2x < r < (3a^2 + 3a + 1)x$, we have

$$\frac{r}{3a^2 + 3a + 1} < x < \frac{r}{3a^2},$$

and consequently

$$a + \frac{r}{3a^2 + 3a + 1} < a + \frac{r}{3a^2 + 3a} < a + \frac{r}{3a^2 + 1} < a + \frac{r}{3a^2},$$

where the first approximation on the left is always by defect and that on the far right by excess (the two medians may be on either side). All these approximations are attested in mediaeval and Renaissance writings:

— $a + \dfrac{r}{3a^2 + 3a + 1}$ occurs in the works of Arabic mathematicians, in

     Fibonacci's and later;[445]

— $a + \dfrac{r}{3a^2 + 3a}$ is used by Tartaglia and Buteo;[446]

— $a + \dfrac{r}{3a^2 + 1}$ is found in Arabic texts;[447]

— $a + \dfrac{r}{3a^2}$ is used in Arabic texts;[448] it is employed iteratively by Car-

     dan. It is Newton's algorithm for $f(x) = x^3 - N$.

---

valeat inveniri qui precise reddat, in se ductus, illos duos quadratos, sume numerum qui reddat quadratum minorem proximum tantum, et vide quot unitates desint in illo quadrato, quod non comprehendit alios duos quadratos; et tot partes appones, de illis, dico, que provenirent per appositionem gnomonis. Verbi gratia. Bis quater 4 sunt triginta duo. Nullus quadratus constituit ipsum. Accipe ergo minorem, scilicet 25, et constat quod 7 deficiunt. Si ipsi apponeremus gnomonem $(2 \cdot 5 + 1)$, resultarent per appositionem gnomonis undecim. Ergo 7 undecime defuerunt.

[444] See Treutlein, pp. 67–69; Perott, p. 168; Günther, pp. 44–45; Hunrath, pp. 21, 26, 27, 35, 37; Carruccio, p. 127, or Maracchia, p. 75 *seqq.*; Tropfke, p. 277.

[445] Treutlein, p. 76; Hunrath, pp. 35, 38, 47; Tropfke, p. 278.

[446] Hunrath, p. 38; Buteo, *Logistica*, Ch. II, p. 83, writes about this approximation: *Ad hoc autem falsas quidem regulas commenti sunt. Vera autem sic habet.*

[447] Tropfke, p. 278.

[448] Treutlein, p. 76; Saidan, p. 463.

Chapter XVI. Geometrical recreations

The present work would be incomplete without some of the geometrical problems which recurrently find a place among other mathematical recreations. Now these problems are all based on elementary geometrical formulae, and thus do not require elaborate reasonings as do some of the previous cases. That is why, for each topic, we shall just start with the formula and then give some examples. The mathematical notions are indeed simple: Pythagorean theorem, length of the circumference of a circle, volumes of cubes or parallellipipeds. That may also explain why such recreational problems are seen to occur among the oldest. A characteristic example is that of a ladder leaning against a wall, the foot of which is pulled away and the top thus moves down. It is already found in early Mesopotamian texts (*c.* 1800 BC), and various forms of it then occur in ancient and mediaeval times. It must also be the longest-living of such problems, for it is still encountered by modern students, who might either be asked about the arc described by a given point on the ladder when the foot is moved away (elliptical —or circular for the mid-point) or about the relative speeds displayed by top and foot (one constant, the other increasing or decreasing).

§ 1. The tower and the river

[175] *There is a tower 40 cubits high, and at its foot passes a river 30 cubits wide. I want to know how long will be a rope joining the bank of the river to the top of the tower.* [449]

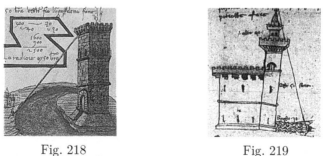

Fig. 218   Fig. 219

Given the height of the tower l and the distance on the ground d, the length of the rope is immediately determined since we have a simple application of the Pythagorean theorem. The data chosen here are

---

[449] F. Calandri, *Opusculum*; same problem in his manuscript, Riccardiana 2669, fol. 85^v, with illustration (left; text ed. p. 171) and in MS. Rome Accad. naz. dei Lincei Cors. 1875, fol. 52^r, with illustration (right).

multiples of the sides of the simplest right-angled triangle, namely 3, 4, 5. Otherwise, we shall have to use an approximation, as in the second triangle of Fig. 220.[450]

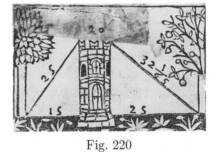

Fig. 220

A variant of this problem is that of the tree which, after breaking at a certain height with its upper part remaining attached to the trunk, has its top touching the ground; here the initial height of the tree is the sum of two sides of the triangle. Indeed, if (Fig. 221) l is the original height of the tree, d the horizontal distance between top and root, and g the vertical distance between root and break, then $l - g$ is the length of the fallen part. Thus three quantities are involved, and we may be given l and g, or l and d, or g and d, or, finally, $g : l = k$ and d. The relation

$$d^2 + g^2 = (l - g)^2, \quad \text{or} \quad d^2 = l^2 - 2\,l \cdot g,$$

underlies the four solving formulae:

$$d = \sqrt{(l - g)^2 - g^2},$$

$$g = \frac{1}{l}\left(\frac{l^2 - d^2}{2}\right),$$

$$l = \sqrt{d^2 + g^2} + g,$$

$$l = \frac{d}{\sqrt{1 - 2k}}.$$

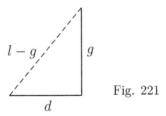

Fig. 221

---

[450] From F. Pellos' *Compendion de lo abaco*, fol. 88^{r}. Approximation formula $a + \frac{r}{2a+1}$, see p. 213.

Examples of these four cases appear together in the *Liber mahameleth*.[451] Particularly common is the second problem, an example of which just follows the previous one in Calandri's *Opusculum*, with a 50-cubit tree, broken by the wind, and a 30-cubit river. The lengths of the two parts of the tree are found to be $g = 16$, $l - g = 34$. Earlier, the 11th-century Persian scholar Bīrūnī mentions the occurrence of problems of this type in algebra treatises;[452] such a problem is also found in Indian and Chinese sources.[453]

Fig. 222   Fig. 223

§ 2. Measuring an inaccessible height

This is not recreational, but since it also involves right-angled triangles it often finds a place among these recreational problems in mediaeval treatises (Fig. 224). Let h be the unknown height of the tree or tower, and d the horizontal length of its shadow (measured); let then a stick be planted upright, in the same vertical plane, and let h' be its height over the ground and d' the measured length of its shadow. Then, since $h : h' = d : d'$, we can calculate h.[454]

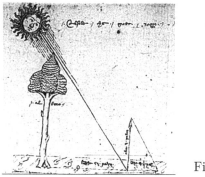

Fig. 224

[451] Problems B.350–B.353.

[452] Edition of the *Liber mahameleth*, p. 1670.

[453] Tropfke, *Geschichte*, p. 620. Illustrations from F. Calandri's *Opusculum* and from MS. Florence BNC Magl. XI 86 (Paolo dell'Abbaco, ed. p. 139).

[454] Illustration from MS. Rome Accad. naz. dei Lincei Cors. 1875, fol. 53^r ($d = 53$, $h' = 3$, $d' = 4 + \frac{1}{3}$, whence $h = 36 + \frac{9}{13}$; top: *è 'l sole che gietta raggi*).

The *Liber mahameleth* considers a similar case, but with two sticks, for then we do not need the sun.[455] Let (Fig. 225) h be the unknown height, l_1 and l_2 the known (visible) heights of two sticks planted upright on the ground, in the same vertical plane as the object and in such a way that all three upper ends lie in a straight line; let finally d be the distance between the base of the shorter stick and that of the object, and d_1, d_2 be the distances between the lower ends of the sticks and the observer. Since, by similar triangles, $(h - l_2) : d = (l_1 - l_2) : (d_1 - d_2)$, we have

$$h = l_2 + \frac{d\,(l_1 - l_2)}{d_1 - d_2}.$$

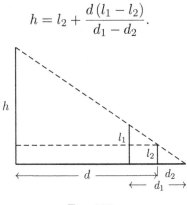

Fig. 225

This can also be done with a single stick and no sun, as explained by Chuquet —though the measuring may be somewhat awkward (Fig. 226).[456] The unknown height ed is to the length of the stick bc as the horizontal distance ae is to ab. Now the place of the observer a, which is the lower end of the 'visual ray' ad (*radius visualis* in the picture, *ray visual* in the text), is obtained by having the eye 'as close as possible to the ground' (*metz ton oeil au plus pres de terre*) and moving back and forth until the end of the stick and the top of the tower are aligned.

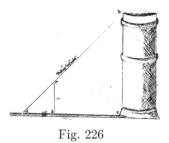

Fig. 226

---

[455] Problem B.362, thus found among the previous geometrical recreations —obviously by analogy of the method.

[456] Illustration from Chuquet's manuscript, MS. Paris BNF fr. 1346, fol. 225^r.

§3. The ladder leaning against a wall

As said above (p. 215), this has been known since Mesopotamian times.[457] Its twofold aspect, recreational and computational, became evident in Arabic times, for then it was found in both collections of mathematical recreations and algebra handbooks, as mentioned by Bīrūnī (above, p. 217) in his treatise *On Shadows*.[458] Here again, there are examples in Indian and Chinese texts.

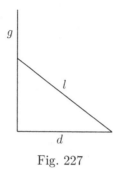

Fig. 227

A ladder is standing against an equally high wall. The foot being moved away from the wall by a certain length, the top of the ladder will descend by some other length. Let (Fig. 227) l be the length of the ladder and the height of the wall, g the distance the upper end of the ladder drops from the top of the wall after the lower end is drawn away by d. There are then three quantities, which are linked by the relation $(l - g)^2 + d^2 = l^2$, with $l > g$. Accordingly, each can be determined from the other two; but other relations between the three may also be given. All the cases below are treated, and their formulae used and carefully demonstrated geometrically, in the *Liber mahameleth*.[459]

1. Given l, d, required g.

$$g = l - \sqrt{l^2 - d^2}.$$

2. Given l, g, required d.

$$d = \sqrt{l^2 - (l - g)^2}.$$

---

[457] Two examples reported in Vogel's *Vorgriechische Mathematik*, II, p. 67. Eight such problems also found in Demotic mathematical papyri from the third century BC, see Parker, pp. 35–40. See also below, *Remark*.

[458] *Mas'ala al-sullam al-mudawwana fī ḥisābāt al-muṭāraḥa wa-furū' al-jabr wa'l-muqābala*; see edition of the *Liber mahameleth*, p. 1670.

[459] Problems B.342 to B.349 (each involving the same values: $l = 10$, $d = 6$, $g = 2$).

3. Given g, d, required l.[460]

$$l = \frac{d^2 + g^2}{2g}.$$

Fig. 228

Remark. There is one early Mesopotamian example of case 2, and, among the eight Demotic problems, there are three examples of each of cases 1 and 2 and two of case 3; a further example of case 3 occurs in a contemporary Mesopotamian text.

4. Given l, $d + g$, required d and g.

$$d = \sqrt{\left[\frac{l - (d + g)}{2}\right]^2 + \frac{l^2 - [l - (d + g)]^2}{2}} - \frac{l - (d + g)}{2},$$

whereby we infer $g = (d + g) - d$.

5. Given l, $d - g$, required d and g.

$$d = l - \left(\sqrt{\left[\frac{l - (d - g)}{2}\right]^2 - \frac{(d - g)^2}{2}} + \frac{l - (d - g)}{2}\right),$$

whereby we infer $g = d - (d - g)$.

6. Given l, $\dfrac{d}{g} = k$, required d and g.

$$g = \frac{2l}{k^2 + 1}.$$

---

[460] The illustration is from Pacioli's *Summa* (geometrical part, fol. 54ᵛ, No. 25 —but solved by algebra). Pacioli's problems 25–30 are in fact the same as in the *Liber mahameleth*, with the same data. Since in his No. 31 (B.351 of the *Liber mahameleth*) Pacioli refers to *maestro Gratia*, who is Grazia de' Castellani, a 14th-century commentator of the *Liber mahameleth*, this must be his source —for Pacioli himself does not seem to have had any direct knowledge of that work. On all that, see our edition of the *Liber mahameleth*, p. lxi.

7. Given $d + g$, $d \cdot g$, required l, d and g.

$$d, \; g = \frac{d + g}{2} \pm \sqrt{\left(\frac{d + g}{2}\right)^2 - d \cdot g},$$

whence l.

8. Given $d - g$, $d \cdot g$, required l, d and g.

$$d, \; g = \sqrt{\left(\frac{d - g}{2}\right)^2 + d \cdot g} \pm \frac{d - g}{2},$$

whence l.

§4. The two towers

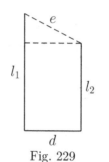

Fig. 229

Let (Fig. 229) l_1, l_2 be the respective heights of two towers ($l_1 \neq l_2$), d the distance between their bases and e the distance between their tops. These four quantities thus obey the relation

$$\left(l_1 - l_2\right)^2 + d^2 = e^2.$$

Three of them being known, we are to calculate the fourth. This gives rise to three cases:[461] given first l_1, l_2, d, then l_1, l_2, e, and finally l_1 or l_2, e, d, which are solved with the respective formulae

$$e = \sqrt{\left(l_1 - l_2\right)^2 + d^2}, \quad d = \sqrt{e^2 - \left(l_1 - l_2\right)^2}, \quad l_1 - l_2 = \sqrt{e^2 - d^2}.$$

§5. Two towers and a fountain

[176] *There are two towers on a flat ground, one 80 cubits high and the other, 90 cubits, and distant by 100 cubits. Between these two towers there is a fountain in*

---

[461] *Liber mahameleth*, B.359–B.361.

such a place that two birds flying from each of them in different ways arrive at the fountain directly. I want to know how far the fountain is from each tower.[462]

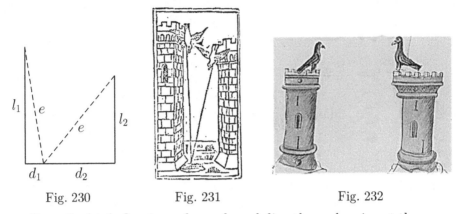

Fig. 230   Fig. 231   Fig. 232

Since the birds fly at equal speeds and directly, and arrive at the same time, they cover an equal distance e (Fig. 230); the fountain is thus closer to the higher tower. The bird flying from this higher tower arrives at a horizontal distance from the tower equal to $d_1^2 = e^2 - l_1^2$, the other at the horizontal distance $d_2^2 = e^2 - l_2^2$, so that $d_2^2 - d_1^2 = l_1^2 - l_2^2 = 90^2 - 80^2 = 1700$. But this is $(d_1 + d_2)(d_1 - d_2)$, and $d_1 + d_2 = 100$, so $d_1 - d_2 = 17$, and therefore $d_1 = 58 + \frac{1}{2}$, $l_2 = 41 + \frac{1}{2}$.

Fig. 233

§6. Further problem of two towers

Pacioli proposes the problem of two towers distant by $d = 30$, with the height of the taller one $l_1 = 50$, and a rope joining the top of each to

---

[462] F. Calandri, *Opusculum* (from which the illustration of Fig. 231); Fig. 232 (*quisti palumbi se parteno per andare a bere*, but no fountain drawn) is from Muscarello's *Algorismus*, fol. 99[v] (ed., II, pp. 227–228). Fig. 233: Measuring the difference in height between two towers (from: Oronce Finé, *Bref et singulier traicté*, MS. Paris BNF fr. AF 7481, dated 1538).

the foot of the other; then, there is a plumb-line, $h = 15$ in length, falling from their intersection to the ground, at the respective distance d_1, d_2 from l_1, l_2. Required the height l_2 of the lower tower (Fig. 234).[463]

Since, by similar triangles, $l_1 : d = h : d_2$ ($bc : ab = on : ao$), we find first $d_2 = 9$ for the shorter horizontal distance, thus $d_1 = 21$ for the other; since likewise $l_2 : d = h : d_1$, we find $l_2 = 21 + \frac{3}{7}$.

There is a simpler form of such a problem, for it involves only the plumb-line and the two adjacent oblique sides. We are told that two trees of given heights l_1, l_2 and horizontal distance d have fallen, with the result that their tops meet (Fig. 235). Required at what distance from their roots a plumb-line hung from that point will reach the ground.[464]

With $l_1 = 40$, $l_2 = 30$, $d = 50$, we have as before $40^2 - d_1^2 = 30^2 - d_2^2$, thus $d_1^2 - d_2^2 = (d_1 + d_2)(d_1 - d_2) = 1600 - 900 = 700$, so $d_1 - d_2 = 14$, whence $d_1 = 32$, $d_2 = 18$.

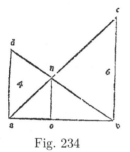

Fig. 234

Fig. 235

§ 7. The falling tree

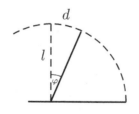

Fig. 236

An initially vertical tree falls towards the ground at a constant daily angular velocity. If (Fig. 236) l is the height of the tree, its top will describe in the course of time an arc of length $d = l \cdot \varphi$, which in our case must be the arc of a quadrant. Given thus d_0, the arc-length covered

---

[463] In his arithmetical manuscript, MS. Vat. lat. (in Italian) 3129, fol. 298^r. Similar problem (with l_1, l_2, d given and h required) in Pacioli's *Summa* (geometrical part, fol. 56^r, No. 48, above illustration, $l_1 = 6$, $l_2 = 4$, $d = 8$, $d_2 = 3 + \frac{1}{5}$, $h = 2 + \frac{2}{5}$).

[464] Widman's *Behende und hubsche Rechenung*, with illustration (ed. 1508, fol. 83^r).

daily, and dividing $l \cdot \frac{\pi}{2}$ by d_0, we shall know the number of days t taken by the tree to be on the ground, or its length if the time of fall is given.[465] Indeed, with the usual mediaeval approximation for π,

$$t = \frac{l \cdot \frac{\pi}{2}}{d_0} \cong \frac{\frac{1}{2} l \cdot (3 + \frac{1}{7})}{d_0}, \qquad l \cong 2 \frac{d_0 \cdot t}{3 + \frac{1}{7}}.$$

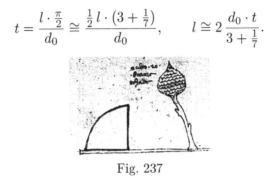

Fig. 237

[177] *A tree the trunk of which is rooted in the ground is 20 cubits high, and (its top) bends each day one cubit, no more and no less. Tell me in how many days the top will be on the ground and (the trunk) lying down.*[466]

Since $l = 20$, $d_0 = 1$,

$$t \cong 10 \cdot \left(3 + \frac{1}{7}\right) = 31 + \frac{3}{7}.$$

In the *Liber mahameleth*, with $l = 30$ and $d_0 = 1$, the result is $47 + \frac{1}{7}$.

[178] *A peasant wants to cut down a tree 60 cubits high. He strikes it hard and with each blow the top of the tree bends towards the ground by one cubit. I ask after how many blows said tree will be on the ground.*[467]

Since $l = 60$, $d_0 = 1$, the number of blows is twice the last result above, thus $94 + \frac{2}{7}$.

Fig. 238

In the *Liber mahameleth* the downward movement is then partly com-

---

[465] *Liber mahameleth*, B.354 & B.356.

[466] MS. Rome Accad. naz. dei Lincei Cors. 1875, fol. 58$^\mathrm{v}$, with illustration (and comments: *è alto 20 braccia, e chade*).

[467] F. Calandri, *Opusculum*, with 'illustration'.

pensated by an upward one.[468] Let d be the distance which must be covered by the top, d_0 the arc-length covered daily, δ_0 the daily retrogression partly compensating the downward movement ($\delta_0 < d_0$); assuming that once on the ground the tree will not rise again and that both movements take place at the same uniform angular velocity, the number of days to reach the ground is calculated as

$$t = \left[\frac{d - \delta_0}{d_0 - \delta_0}\right] + \frac{\delta_0 + r}{d_0 + \delta_0},$$

with our square brackets indicating that the integral part of the quotient must be taken and r being the remainder. In the *Liber mahameleth*'s example, $d = 70$, $d_0 = 3$, $\delta_0 = 1$, so $d = \frac{1}{2} l \cdot (3 + \frac{1}{7}) = 110$, and therefore

$$t = \left[\frac{110 - 1}{3 - 1}\right] + \frac{1 + 1}{3 + 1} = 54 + \frac{1}{2}.$$

§ 8. The stone thrown into a cistern

Into a cistern of given volume filled with a given quantity of water a stone is thrown; knowing the volume of the stone we can determine the quantity of water overflowing; or, conversely, we may determine the volume (or one unknown dimension) of the stone when the quantity of water overflowing is given; furthermore, knowing the depth of the cistern, its content and the quantity overflowing enables us also to calculate the drop in water level when the stone is removed.[469]

Indeed, if v_1 is the volume of the cistern, v_2 the volume of the stone, with l_i, r_i, h_i the respective dimensions, q the quantity of water contained in the full cistern, q' the quantity overflowing and h'_1 the drop in water level, then

$$\frac{q'}{q} = \frac{v_2}{v_1} = \frac{l_2 \cdot r_2 \cdot h_2}{l_1 \cdot r_1 \cdot h_1} = \frac{h'_1}{h_1}.$$

Here is the *Liber mahameleth*'s first example.

[179] *Into a cistern ten cubits long, eight wide, six deep, containing one thousand measures of water, a stone four cubits long, three wide, five thick is thrown; how much water will overflow?*

Given thus $l_1 = 10$, $r_1 = 8$, $h_1 = 6$, $q = 1000$, $l_2 = 4$, $r_2 = 3$, $h_2 = 5$, required q'. (Each dimension of the stone being half one of the dimensions of the cistern, the two bodies are similar in shape.)

---

[468] Problem B.355. We have already encountered a similar case and solving formula in the problems of movement (pp. 100–101).

[469] The *Liber mahameleth* has examples of all these cases, see its B.334 & B.337, B.335–336 & B.338, B.339–340.

$$q' = \frac{v_2 \cdot q}{v_1} = \frac{4 \cdot 3 \cdot 5 \cdot 1000}{10 \cdot 8 \cdot 6} = \frac{60\,000}{480} = 125.$$

In his *Opusculum*, F. Calandri gives the following example:

[**180**] *There is a fish-tank with length 12 cubits, width 10 cubits and height 8 cubits, into which falls a spherical stone with diameter 3 cubits. I want to know by how much the water of said fish-tank will rise* (quanto alzerà l'acqua nel decto vivaio) *in consequence of the fall of this stone.*

$$h_1' = h_1 \cdot \frac{v_2}{v_1} = 8 \cdot \frac{\frac{4}{3}\frac{22}{7}\left(\frac{3}{2}\right)^3}{12 \cdot 10 \cdot 8} = \frac{33}{280}.$$

h_1' is thus the (theoretical) increase in height —in his illustration, Fig. 239, the water overflows.

Fig. 239

Chapter XVII. Hidden numbers

In the simplest form of this game, there are two players: the asker, who wishes to find out the number kept secret, and the one who knows it. The latter is told to mentally perform a series of computations and then give the final result, after which the asker will tell him the hidden number.

In his revised edition of Ozanam's *Récréations*, Montucla observed about its chapter X, on finding such numbers, that he will make a selection of the problems presented in the original edition. For, he says, most are banal and the underlying trick obvious. He is certainly right in that respect, for the solving formula is indeed often banal. That, however, is clear only to us: the person thinking of a number will be distracted answering the questions, or making the computations he is told to perform, to be aware of what is going on.

§ 1. The walnuts

The following game is played with walnuts.[470] The asker tells his companion to take as many walnuts in one hand as in the other, then to transfer a number of them, chosen by the asker, to the first hand ($\mathfrak{a}$), then to transfer from this ($\mathfrak{a}$) to the other hand as many as are left (in $\mathfrak{b}$). The number of walnuts thus remaining (in $\mathfrak{a}$) will be twice the number chosen by the asker.

With x the initial number of walnuts in each hand, the successive steps are

$\mathfrak{a}$	$\mathfrak{b}$
x	x
$x + m$	$x - m$
$2m$	$2x - 2m$

Fig. 240

with m known by the asker, thus also $2m$; x itself remains undetermined.

At the end of his *Triparty* Chuquet has a whole section about such games. He has in particular the same problem, which he calls 'the game of equal things'.[471]

[181] *If a person has as many coins, or other things, in one hand as in the other and you want to know it whilst giving the impression that you guess it, (proceed in*

---

[470] MS. Tours 399, fol. 138$^\mathrm{r}$ – 138$^\mathrm{v}$. Original text in *Récréations*, p. 239.

[471] *Appendice*, p. 456; MS., fol. 206$^\mathrm{r}$ – 206$^\mathrm{v}$.

this way): Tell him to transfer from one hand to the other some number, whatever you wish to tell him, provided he can do it —for if he has not enough, you must reduce this number. This being done, tell him to put back from the hand where he has put said number to the other, as much as remains. You are to know that in the hand into which you first had the number put there is (now) the double of this (number).

Example. Let us put that in each hand he had 12 denarii without your know-ing it. You may tell him to put from the right hand into the left 7 —if this is possible.[472] This being done, you will tell him to put back from the left to the right, as much as there is presently. You are to know that in the left there are 14, which is twice the number 7 you told him.

If you want to know how much there is in the right one, you may know it by the novenary (novenaire). After that, you will be able to tell him that he has in both hands 24, namely 14 in the left and 10 in the right.

With this game, we can indeed determine, as seen, only the quantity in the hand first receiving, for the content of the other serves merely as an auxiliary set. In order to determine the content of the other hand, and thus the whole, Chuquet suggests using the 'novenary', which we shall treat in the next section. Before closing this one, we may mention a case of cheating taught unscrupulously by Trenchant:[473]

[182] *If you want to subtlely know the number of coins your companion holds in his hand or in his purse: take an arbitrary number of small stones, or other things, of which you know the number although you pretend not to. Suppose he has 14 coins and you, 10 stones or counters. Putting these stones in front of him, you will say a number which you think exceeds the number of coins he has, for instance you will tell him that he has 18 coins. Then, if he holds less, he must take as many of these stones as needed. Then, you close your eyes and, apparently unseen by you, he will take 4 of these stones. As a matter of fact, you open your eyes and take a look at these stones; or when you take them back, you must secretly see how many he took. Knowing that he has taken 4, subtract these 4 from 18, there remain 14. Consequently, you will declare that he holds 14 stones.*

§ 2. The novenary

This is no doubt one of the most known games, in view of both its simplicity and the easiness of the computations to be performed. It is found both East and West, and must be of Greek origin, for we find it in two Arabic authors known to have been strongly influenced by Greek

---

[472] If he had enough to begin with.

[473] *Arithmetique*, III, Ch. XI, No. 4 (ed. 1561, p. 265; ed. 1602, p. 336).

science: Abū Yūsuf al-Kindī (d. around 870) and ʿAlī ibn Aḥmad al-Anṭākī (d. 987). The latter, who in his Commentary on the Arithmetic of Nicomachos (*c.* 100) reproduces mainly translations of Greek texts, explains it as follows.[474]

[**183**] *You tell the person to multiply what he keeps hidden by 3, then to take half the result, to multiply what remains by 3, and to take half the result. Then he eliminates the result with 9s. For your part you take, for each of their subtractions, 4, and, for what remains according to this computation which does not make 9, you will take one for each two and a fourth. The result will be what was hidden.*

The hidden number x has become, after the transformations, $\frac{9x}{4}$. In order to know it, the asker will take 4 for each possible subtraction of 9. If a quantity remains, he will add 1 for each possible subtraction of $2 + \frac{1}{4}$ —since each unit of x turns out to have been multiplied by $2 + \frac{1}{4}$.[475]

Fig. 241

A variant is to multiply x twice by 3 but to take half the result only once. This will ease the task of the asker: each subtraction of 9 counts as 2 and the remainder, if any, as 1 (there will be a remainder of $\frac{9}{2}$ if x is odd). Here is how Chuquet expresses it in the first problem of his section on hidden numbers:[476]

[**184**] *If you want to know the number someone has thought of and keeps in mind, tell him to triple that and to take half of the result, and have him triple this again. From this triple, have him remove one or more nines, like 18, 27, 36, etc., and this be*

---

[474] See our *Magic squares in the tenth century*, pp. 186 & 320 (*A*.III.15); for al-Kindī, MS. Istanbul Aya Sofya 4830, fol. 81ᵛ.

[475] Fig. 241: finding out a number, from F. Calandri's manuscript, Riccardiana 2669, fol. 109ᵛ —text, ed. p. 219: *Diciamo che l'amico si ponessi 10 (in chuore).*

[476] *Appendice*, p. 456; MS., fol. 206ʳ. Same in MS. Paris BNF fr. 1339, fol. 72ʳ (without example) and in the *Subtilitates*, Nos. 3 (with mnemonic verses) and 22.

continued and repeated several times until he can no longer remove integral nines. Then tell him to remove from what remains to him 1, or 2, or 3, or some other number which is less than 9; then if he can remove only 1, 2, or some number, count it as 1, and count secretly 2 for each nine you have made him remove from his number. In this way you will know what he had thought of.

 Example. Let him have thought of 5. Taking the triple makes 15. Of which he retains half, which is $7\frac{1}{2}$. Which being tripled again makes $22\frac{1}{2}$. Then, according to the size of the number thought of, he is told to remove a greater or smaller quantity of nines.[477] *Now tell him to subtract 18 from this and to keep in mind the remainder; then, from this remainder, 9, if possible. He answers that it is not. So tell him to subtract 1 or 2, whichever he likes, provided it can be done; for be it 1, or 5, or 8,*[478] *what he subtracts (which is) less than 9 will mean for you 1. And the 2 nines mean 4. This with 1 makes 5, which is the number he had thought of.*

Chuquet also mentions another application of this game: *by means of the novenary one may also say which day of the week or of the month the lover has kissed his love.* This determination of days leads us directly to the next section.

§3. The unknown date

[185] *If you want to know on which day someone has kissed his love:*[479] *Tell him to double the day, to add 1, to multiply the whole by 5, then the result by 10, and to subtract 50 from the whole sum. Ask then how many times 100 can be subtracted from the whole sum. If once, it will be Sunday, if twice, Monday, if thrice, Tuesday, and so on.*

With x the day, beginning with Sunday, the calculation is

$$10\,[\,5(2x+1)\,] - 50 = 100x + 50 - 50.$$

More frequently, the underlying formula is

$$10\,[\,5(2x+5)\,] - 250 = 100x + 250 - 250,$$

which comes to the same: the addend has been changed and the subtracted term accordingly. This must be of ancient origin: it occurs among problems attributed to the Venerable Bede (7th c.). There is a similar late mediaeval example:[480]

---

[477] In order to shorten the operations, he may subtract directly a multiple of 9.

[478] In fact, these last two values are inappropriate.

[479] *Subtilitates*, No. 10. Some texts are more explicit, as is the MS. Montpellier 323, fol. 237vb: *Si vis scire quota feria iacuit cum amica sua socius tuus.*

[480] MS. Tours 399, fol. 143^{v} – 144^{r}; or MS. Paris BNF fr. 1339, fol. 69^{v}.

[186] *If you want to guess on which day of the week someone has done something, whatever it is, which he claimed several times without specifying the day, tell him to think about the day's number —that is, 1 for Sunday, 2 for Monday, 3 for Tuesday, 4 for Wednesday, 5 for Thursday, 6 for Friday, 7 for Saturday. Then tell him to mentally double the number of the day he has done this thing, then add to this doubled number 5, all that mentally, without telling you. Tell him to multiply the sum obtained by 5, then again the whole by 10, and next to subtract from the whole 250. The remainder (after dividing by 100) you will take as the day he has done said thing.*

Example. Suppose he has done the thing on Sunday; this is 1. It must be doubled, this gives 2. He must add 5 to this 2, which makes 7. These 7 he must multiply by 5, which gives 35, then by 10, which is 350. There remains 100, which you are to count as 1. He did the thing on Sunday.

This way you must always keep (in mind): subtract from the whole sum 250, and take 1 for each hundred. If there remains 1, this is for Sunday; if there remain two hundred, this is for Monday; and so on for the others.

After giving a table of the solutions for each day of the week, the text (and that related, see footnote) extends this method to determining the year, the month, the week, the day, the hour on which something has been done or a person born. The example given is that of a person having in mind ten years; this will give 3652 days (for in ten years there are two intercalary years), and so the computation will be $10 \, [\, 5(2 \cdot 3652 + 5)] - 250 = 365\,450$, which, reduced by 250, gives 365 200, the division of which by 100 indeed gives ten years. The same can be applied, the text continues, to the month (12 in a year), the week (4 in a month), the hour (24 in a day), with operations and rule of determination remaining exactly the same; indeed, the result is $100 \cdot t$.

§ 4. The unknown sum

The question is again about determining a sum —but within an interval of a hundred consecutive integers— according to its remainders relative to 3, 5, 7.[481]

[187] *If a man puts on a table some number of denarii he chooses, provided it is over 7 and not more than 106, and you want to know this number: Tell him to remove 3 repeatedly, and, if nothing remains, to tell you; then again 5, then 7. For each coin remaining from the 3s, count 70, for each from the 5s, 21, from each of*

---

[481] MS. Paris BNF fr. 1339, fol. 72^r–72^v. Same problem found earlier in Fibonacci's *Liber abaci*, p. 304 (with $n = 53$).

the 7s, 15; subtract, from the sum thus made, 105 (repeatedly[482]). *The remaining number of denarii will be that put on the table.*

Let indeed the number of denarii be n, with $7 \leq n \leq 104$ (and not, as in the text, $8 \leq n \leq 106$); then, if $n = 3q_1 + r_1 = 5q_2 + r_2 = 7q_3 + r_3$, we shall have

$$70r_1 + 21r_2 + 15r_3 = (70n - 210q_1) + (21n - 105q_2) + (15n - 105q_3)$$
$$= 105(n - 2q_1 - q_2 - q_3) + n.$$

After 105, which is the lowest common multiple of the divisors considered, the same sequence of remainders will recur, in the same order, and the number hidden will no longer be univocally determined.

Fig. 242

Of course, there is the possibility of determining a sum in an easier way, of the kind seen before for determining dates:

[188] *If you want to know how much someone has in his purse, say so: Double, and add 5; multiply the whole by 5; add 10; take ten times the whole you have, and have him tell you the sum; but you are to subtract from it 350. For each remaining hundred he had one denarius.*[483]

We are thus to calculate

$$10\,[\,5(2x + 5) + 10\,] - 350 = 100x.$$

We also find in this text the same sequence of computations, but this time in order to determine a fraction which the person has in mind. As we are told, according to whether the result found is 50, 25, 20, $12 + \frac{1}{2}$, 10, the fraction thought of will be $\frac{1}{2}$, $\frac{1}{4}$, $\frac{1}{5}$, $\frac{1}{8}$, $\frac{1}{10}$. Normally, the numbers chosen are integers; indeed, with fractions, the person asked is unlikely to want to continue playing.

---

[482] This specification, omitted in this MS., is added in another (MS. Rouen I 58, fol. 87$^\mathrm{v}$: *gettes* (...) *par tant de foys comme vous pourres C et V*).

[483] *Rascioni d'algorismo*, No. 119. Illustration from F. Calandri's manuscript, Riccardiana 2669, fol. 107$^\mathrm{r}$ (text, ed. p. 214: *Quando uno havessi posto in sur una tavola 20 grossi et 2 compagni fra loro se gli dividono*).

§5. The hidden ring

In the next problems we are to find out who, among a group of people, is holding a ring. In the first, determination is limited to the person; then it specifies the finger ([190]–[191]), next ([193]) the part of the body —which for that purpose is divided into a hundred numbered parts. The purpose now being to determine a person, the asker addresses some participant who knows who has the ring.

[189] *You tell someone: 'Take what there is between you and the ring'. Then tell him: 'Double what you have'. Then tell him: 'Add 5 to it'. Then say: 'Multiply it by 5'. Then tell him: 'Add 10 to it'. Then say: 'Multiply what you have by 10'. Then say: 'Subtract 400 from what you have'. After he subtracts it, take 1 for the 400, and keep it in mind. (...) Then tell him to subtract from what he has as many times 100 as he can, and take yourself 1 for each subtracted hundred. Then, when he is left with less than 100, consider what you have. Indeed, the number reached will indicate who has taken the ring.*[484]

Put x the number of people counted from the calculator to the holder of the ring (both included). The sequence of computations,

$$10[5(2x + 5) + 10] - 400,$$

reduces to

$$100x + 350 - 400 = 100(x - 1) + 50,$$

therefore the subtraction of the hundreds will reveal the value of $x - 1$, and adding the unit kept in mind after the subtraction of 400 gives x. Note that, as seen several times above, the multiplications performed are limited to those by 2, 5 or 10, and thus always to products that are mentally feasible.

Remark. It would have been simpler to calculate

$$10[5(2x + 5) + 10] - 350,$$

which would leave $100x$, as in the previous problem; but adding useless complications is part of the game, in order to distract from what is actually being computed.

So far the addends involved have been given quantities; in the following problems they must be determined as well.

[190] *A certain number of persons are gathered and one of them hides a ring on some phalanx of the finger of one hand, and you want to determine who and where (...).*[485]

---

[484] *Liber augmenti et diminutionis*, p. 369; or p. 83 of our *Recueil*.

[485] *Liber abaci*, p. 305. The illustration is taken from F. Calandri's manuscript, fol. 99ᵛ, text ed. p. 199 —other problem: *maestro* with his *scholari*.

Fig. 243

Have them, to begin with, sit in an orderly manner. Tell one of them, the most able for reckoning, to count starting from himself to the one holding the ring. Have him double this number, add 5, multiply by 5, then add the number of the finger; that is, he is to add 1 if the holder has it on the little finger of its left hand, 2 if on the annular, 3 if on the middle finger, 4 if on the forefinger, 5 if on the thumb of this same hand; if he has it on the little finger of its right hand, 6, if on the annular, 7, if on the middle finger, 8, if on the forefinger, 9, if on the thumb, 10. Let him multiply the whole sum by 10. He will then add to the product the number of the phalanx: 1 if the holder has it between the first and the second joint, 2 if between the second and the third, 3 if between the third and the end of the nail. Have him tell you the sum. After removing from it 250, the number of remaining hundreds will give you the number of persons, counted from the calculator to the holder of the ring, including them both; the remaining tens will give you the number of the finger, counted from the little finger of the left hand, as said above; the units will give you the number of the phalanx where the ring is found.

Let x be the number of people, f that of the finger, p that of the phalanx. We shall thus calculate

$$10 \left[5 \left(2x + 5 \right) + f \right] + p - 250.$$

Since this reduces to

$$100x + 250 + 10f + p - 250 = 100x + 10f + p,$$

the determination is unambiguous, except in the case $f = 10$.

Let us illustrate that with two examples (Fibonacci has none, nor for the subsequent problem [193]). Take first $x = 7$, $f = 3$, $p = 2$. We obtain $982 - 250 = 732$. Therefore it is indeed the seventh from the calculator (included) who has the ring, namely on the second phalanx of the middle finger of his left hand. Take next $x = 7$, $f = 10$, $p = 2$. We obtain $1052 - 250 = 802$. Since there is no indication of finger, it will be that corresponding to 10, and this will remove a hundred.

Remark. A 14th-century Italian text calculates

$$10 \left[10 \left[5 \left(2x + 5 \right) + 10 + m \right] + f \right] + p - 3500,$$

where $m = 1$ or $m = 2$ according to whether the ring is in the left or right hand.[486] Since the above expression reduces to $1000x + 100m + 10f + p$, the number obtained will indicate the place directly. In the example given, the calculation gives 9751, and 6251 after subtraction. The ring is then on the first phalanx of the sixth person's right-hand thumb. Since now f runs from 1 to 5, there is no ambiguity this time.[487]

Fig. 244

The same game as [190] above occurs in Chuquet's *Triparty*, which explicitly mentions the ambiguous case.[488]

[191] *In a gathering of many persons, someone has a ring or a piece of jewelry of gold or silver. If you want to know who has it, and in which hand, on which finger, and on which phalanx, have the persons sitting together so as to be numbered, with one of them being the first, the next the second, and so on.*[489] *Likewise, let the fingers be numbered to 10. This being done, you are to be just outside the group. Tell one of them to double the number of the person having the ring; to this double, have 5 added and the sum multiplied by 5. Have him adding to that product the number of the finger with the ring. In front of this number, have the number of the phalanx of the finger put in such a way that it occupies the first order of the number — as if one were to put 2 in front of 7, he would have 72.*[490] *Ask him then what number he holds. Subtract from it 250, for the remainder will indicate what*

---

[486] *Rascioni d'algorismo*, No. 115.

[487] Illustration again from F. Calandri's manuscript, fol. 101^r (text, ed. p. 202: *6 giuchatori a una taverna*).

[488] *Appendice*, p. 458; MS., fol. 208^r–208^v. Same game again in the MS. Paris BNF fr. 1339, fol. 71^r, with a man throwing a ring among a group of ladies, then going away, and later returning to find out which woman has it, and on which phalanx and finger of which hand.

[489] Here the numbering starts with any one of the players, known to the asker.

[490] Either a trick to complicate the reasoning or an operation for a player unable to multiply by 10.

you want to know; you must know that the hundreds represent the person with the ring, the tens the number of the finger, and the first figure[491] *indicates on which phalanx.*

Example. Let us put that the number indicated by the person who performed the prescribed multiplications and additions be 932. You secretly subtract 250 from 932, it leaves 682. Then you can say that the sixth person has the ring, on the second phalanx of the 8th finger. Should it happen that after your subtraction of 250 from the number given to you there is 0 in the place of the tens, you should remove a hundred and count it as ten tens, saying that the ring is on the tenth finger. As if the multiplications and additions had given 951: remove 250, there remains 701, which means that the sixth person of the group has the ring on the first phalanx of the tenth finger.

Fig. 245   Fig. 246

The expression

$$10 \left[5 \left(2x + 5 \right) + f \right] + p - 250 = 100x + 10f + p$$

is also used by Chuquet in order to determine the respective results from three dice being thrown. This time there is no ambiguity since all three numbers are less than 10:[492]

[192] *A man has thrown three dice, and you want to know the marks of each and all together. Tell him to double the mark of one of them, whichever he wishes. Have him add 5 to the result and multiply the sum by 5. Have the mark of one of the other two dice added, and those of the third put 'before' this sum.*[493] *Then ask him the resulting number and remove from it 250. The three remaining figures will be those of their marks.*

---

[491] That of the units.

[492] *Appendice*, pp. 458–459; MS., fol. 208^v. The illustrations are from F. Calandri's manuscript, fol. 105^v (text, ed. pp. 210–211: *Uno che trae 3 dadi in su una tavola*), and from Muscarello's, fol. 78^v (ed., II, pp. 193–194, *4 giucatori e dadi*). Tiny player? No, tiny notion of the rules of perspective; same in Fig. 67.

[493] This corresponds to a multiplication by 10 of the previous sum.

Let us now consider Fibonacci's determination of the part of the body where the ring's holder has hidden it.[494]

[193] *If he wants to hide this ring on some part of the body and you wish to know who has it and where:*

We are first to distinguish in the body 100 parts. Of these, ten are the fingers, with the same numbering as previously (see [190]). The ten toes are, likewise, ten others. There are thus 20. The twenty-first is the instep of the left foot, near the toes, its sole is the twenty-second part; the instep of the right foot, near the toes, is the twenty-third, its sole the twenty-fourth. The instep of the left foot, the ankle, the twenty-fifth, and underneath the heel of the same foot, the twenty-sixth; above the right foot, where the foot meets the leg, the twenty-seventh, and underneath the right heel, the twenty-eighth. The outer joint of the left foot, the twenty-ninth, the inner one, the thirtieth; the outer joint of the right foot, the thirty-first, the inner, the thirty-second. The outer part of the left leg, the thirty-third, the inner one, the thirty-fourth; the outer part of the right leg, 35, the inner one, 36. The outer part of the left knee, 37, the inner one, 38; the outer of the right, 39, the inner one, 40. The outer part of the left thigh, 41, the inner one, 42; the outer part of the right, 43, the inner one, 44. The left hip above or below the posterior, 45, above the pubis, 46. On the right hip towards the posterior, 47. On the waist near the posterior, 48. Near the anus, 49. Near the manly parts, 50.[495] On the left hip, underneath the loins or near the loins, 51. The navel, 52. The right hip, underneath or near the loins, 53. On the waist near the loins, 54. Underneath the left armpit, 55, the breast, 56, the right armpit, 57. Between the shoulder blades, 58. Between the neck and the left shoulder, or near the neck on the left side, 59. The clavicle underneath the throat, 60. On the right side near the neck or the shoulder, 61. Behind, on the joint of the neck, 62. Between the elbow and the shoulder on the outer part of the left arm, 63, on the inner, 64; the outer of the right, 65, the inner, 66. The outer part of the left elbow, 67, inner, 68; outer of the right, 69, inner, 70. Between the elbow and the left hand, on the outer part, 71, on the inner, 72, outer of the right, 73, inner, 74. The outer left hand knuckle, 75, inner, 76; outer of the right, 77, inner, 78. On the back of the left hand, 79, in the hand itself, 80; on the back of the right hand, 81, in this same hand, 82. In the mouth, 83. The left nostril, 84, the right, 85. The left ear, 86, the right, 87. Behind the left ear, 88, behind the right ear, 89. On the brow near the hair, 90. The curvature of the neck, 91. The top of the head, 92. Behind the left ankle, 93, behind the right, 94. On the left knee, 95, underneath, 96; on the

---

[494] *Liber abaci*, pp. 305–306.

[495] This median location seems to have been chosen intentionally; otherwise, it is not clear why, when numbering the upper parts of the body, we return to lower ones (93–98).

right, 97, underneath, 98. Underneath the left nostril, 99, underneath the right, 100.

These parts being thus determined, have these (participants) all sitting in or-dered manner. Tell one of them, the one who best knows how to compute (qui magis de abbaco sciverit) and who knows the aforesaid attributions (by heart!), to count from himself to the holder of the ring, including himself in this number. This counting being done, tell him to double this number, add to the resulting quan-tity 10, multiply the whole by 10 and add 5. Have him multiply the whole by 5. Let him add then the number of the place where the ring is, according to what has been indicated above. Let him multiply the sum by 10. When he gives you the result, subtract from it 5250. The number of remaining thousands will be the number of the man. Divide the remainder by 10; the quotient will be the number of the place where the ring is hidden.

By means of the same rule we could find out who has picked up one thing among a hundred things, and which one, if they have been given numbers in succession from one to a hundred.

Let once again x be the number of people counted from the calculator to the holder of the ring, both included, and l the part of the body. Calculating then

$$10\left[5\left\{10\left(2x+10\right)+5\right\}+l\right]-5250,$$

which is

$$1000x+5250+10l-5250=1000x+10l,$$

the factor of 1000 will indicate who has the ring, and the remainder divided by 10 the part of the body where he hides it. As in the previous case, if the number found is divisible exactly by 1000, we shall remove once 1000, for there must be a remainder.

Example (not in Fibonacci). Take $x=11$, $l=100$; the formula gives 12 000, which is $x=11$ and $l=100$.

§6. Two people and two things

The asker wishes to know which of two different things the player is hiding in each hand. To do so, he attributes two numbers of different parity to the things, and he does the same for each hand. Then he asks the player to compute the product of the number of each thing by the number of the hand holding it, and to tell him the sum of these two products. The asker will then be able to say which thing is in which hand.

Indeed, there are four possible situations, according to the parity (e, o) of the numbers attributed to the two things, one thing being thus t^e and the other t^o, and to the parity attributed to the left and right hands l, r: see Fig. 247 (in I & II the right hand has an even number and in III & IV, an odd one).

Consider now the parity of the product of the numbers of each thing and its hand. Only one such product can be odd —namely if it results from odd by odd, which occurs in II and III— and since then the other product must be even, their sum will be odd; in that case thing and hand must be of the same parity and the asker will know which thing is in which hand. But in I and IV their respective products, being odd by even, are both even, making therefore an even sum, which will mean that thing and hand are of a different parity.

	I		II		III		IV	
thing	t^e	t^o	t^o	t^e	t^e	t^o	t^o	t^e
hand	l^o	r^e	l^o	r^e	l^e	r^o	l^e	r^o
product	e	e	o	e	e	o	e	e
sum		e		o		o		e

Fig. 247

The four cases of the above figure are considered successively in the next two mediaeval examples; in the first ([194], [194′]), to the right hand is attributed an even number (= I, II), in the second ([195]) an odd one (= III, IV).

[194] *If you want to guess if someone has gold in one hand and silver in the other, or a gold or silver ring, or some other thing: Say that you want the gold to be worth 5 denarii (= t^o) and the silver 4 (= t^e), or any number you wish, provided one is even and the other odd and the gold is worth more than the silver; and you do not know in which hand he holds the gold or the silver. Tell him that you want the value of the piece of jewelry which is in the right hand to be worth twice (= r^e) its value, and the piece of jewelry which is in the left hand to be thrice (= l^o). Then ask if the number (sum of) the triple and the double is even or odd; if it is even (I), the gold is in the right hand, if odd (II) the silver is in the right hand.*[496]

To the gold is thus attributed an odd number (t^o, which is expected to be larger than the even since the money unit is for both the denarius; but their relative values are irrelevant) and to the right hand an even number (r^e) since one is to double what is in the right hand. As said,

---

[496] MS. Tours 399, fol. 140^r – 140^v; also in the *Subtilitates*, No. 23.

this is the situation of I and II, for each of which there is an example:

[194'] *Example. The gold ring is worth 5 denarii ($= t^o$), and the silver ring, 4. Now it is so that he will put the gold ring in his right hand or in the left.*

Suppose he puts it in the right. It is worth 5 denarii, and you have said that what he has in his right hand is to be doubled, which is 10 ($= t^o \cdot r^e$), and what is in his left hand, tripled, which is the silver ring worth only 4 denarii. Triple 4, it gives 12 denarii ($= t^e \cdot l^o$). Add the whole. The (resulting) number is even, for 10 and 12 make 22.

If it happens that he has put the gold in the left and the silver in the right: he is to double that of the right, and the silver is worth 4 denarii, it gives 8 denarii ($= t^e \cdot r^e$), and to triple that of the left, and the gold is worth 5 denarii, it gives 15 ($= t^o \cdot l^o$). (Summing,) it is an odd number, therefore the gold is in the left and the silver, in the right.

A similar problem is proposed by Chuquet.[497]

[195] *If a man has two different things, say gold and silver, and he holds in one hand the gold and in the other the silver, by cunning you will know (the answer) while giving the impression that you guess: Attribute to the gold a certain price, and also to the silver, another price, in such a way that one be even and the other odd; for instance, tell him that the gold is worth 4 ($= t^e$) and the silver, 3 ($= t^o$). The same with any other two provided one is even and the other odd.[498] Then tell him to multiply by the odd number what he holds in the right hand, and what he holds in the left hand by the even number. Ask him whether the sum of these two products is even or odd; if odd (case III, $t^e \cdot l^e + t^o \cdot r^o$), it indicates that the silver is in the right hand and the gold in the left, if even (IV, $t^o \cdot l^e + t^e \cdot r^o$) it indicates that the gold is in the right and the silver in the left.*

Chuquet adds that this could also be determined by a single multiplication, though that would make the trick too obvious. Indeed, one might just multiply the odd hand (l^o or r^o) by its object and ask for the parity of the result: it will be odd only if this object is also odd (t^o). The same remark about simplifying the computation is made by J. Trenchant.[499] His question here is about two friends, 'Jaques' and 'Pierre', one of whom would like to be a king with an income of 4 millions and the other an emperor with an income of 5 millions, and we are to determine who wishes what.

---

[497] *Appendice*, p. 457; MS., fol. 207^r – 207^v (text: *Récréations*, p. 253).

[498] This should refer to the numbers now attributed to the hands.

[499] *Aritmetique*, III, Ch. XI, No. 5 (ed. 1561, pp. 266–267; ed. 1602, p. 337).

§ 7. Three people and three things

[196] *If three men hold three things of different kinds and you want to know
which holds each one, you must first know each thing and say to yourself that one
is 'the first', the other, 'the second' and the other, 'the third'; call, once again to
yourself, the first 'two', the second '9' and the third '10'. Then, call one of the men
'2', another '3', the third '5'. Instruct one of them, who must know what each
one holds, to multiply the name of the holder of the first thing by 2, the name of
the holder of the second by 9, the name of the holder of the third by 10, to add the
results of the multiplications and to give you the result. Subtract it from 100 and
divide the remainder by 8. The quotient of the division will be the name of that who
holds the first thing, the remainder will be the name of that who holds the second,
while the other holds the third.*[500]

Fig. 248   Fig. 249

In other words, the asker attributes to each of the three men a number
(here: 2, 3, 5), then to each of the three things a number (here: 2, 9, 10).
He communicates these numbers to the calculator, who then declares the
sum of the products of each man by his object. The asker will then be
able to determinate what each possesses.

Let us, for instance, take the things in direct order and the men in
inverse order. We shall then have

$$\frac{100 - (5 \cdot 2 + 3 \cdot 9 + 2 \cdot 10)}{8} = \frac{43}{8} = 5 + \frac{3}{8},$$

whereby we may say that the third man has the first thing, the second
man, the second, and (therefore) the first man, the third.

Fig. 250   Fig. 251

---

[500] *Liber augmenti et diminutionis*, p. 371 or (better text) our *Recueil*, pp. 84–85.
Illustrations from Tagliente's *Componimento*, ed. 1525, 1547 and (below) 1554, and F.
Calandri's manuscript, fol. 106[r] (ed. p. 212: *anello, fiorino, groso in su una tavola*).

Consider generally that n_1, n_2, n_3, with $n_1 + n_2 + n_3 = t$, are the numbers attributed to the men and m_1, m_2, m_3 those of the things; but we are to take $m_1 = m_3 - s$ (with $s > n_i > 1$) and $m_2 = m_3 - 1$; then we shall have $m_3 > m_2 > m_1$. The sum given to the asker is then

$$n_1 m_1 + n_2 m_2 + n_3 m_3 = n_1 \left(m_3 - s\right) + n_2 \left(m_3 - 1\right) + n_3 m_3.$$

The asker then subtracts this sum from $m_3 t$, which he knows, then divides the remainder by s, which he also knows. He then obtains

$$\frac{m_3 t - \left[n_1 \left(m_3 - s\right) + n_2 \left(m_3 - 1\right) + n_3 m_3\right]}{s} = \frac{n_1 s + n_2}{s} = n_1 + \frac{n_2}{s}.$$

The integral part thus gives n_1, the possessor of the first thing, while the remainder gives n_2. From this determination we see why we had to take $s > n_i$.

In our text, $m_3 = 10$, $s = 8$, then 2, 9, 10 are the numbers attributed to the objects. To the n_i are attributed the three values 2, 3, 5, so that $t = 10$; but we might just as well have taken 3, 6, 1 or 4, 1, 5. As regards the lowest possible choice for t, it would be $6 = 1 + 2 + 3$. Fibonacci uses it in one of his examples, with the same m_3 and s (and therefore subtracting from 60).[501]

[197] Another way to determine the object held by each of three people, but without any computation or external help, is the following, where the asker gives them tokens.[502]

Fig. 252

Let there be three people, A, B, C, three objects, say a, e, i, and 24 tokens. Before leaving the room, the asker gives A one token, B two,

[501] *Liber abaci*, pp. 307–308; Fibonacci mentions (p. 308) the condition for s: *Et nota quod unumquemque numerorum datorum* (the n_i's) *oportet esse minus quam 8, cum in 8 dividere oporteat*. In the *Subtilitates*, No. 19, these three numbers are employed to determine, in three couples, who is the wife of whom.

[502] Added by a 14th-century reader to the previous problem, see *Recueil*, p. 85 (MS. Paris BNF lat. 15120, fol. 57ᵛ). The illustration is from F. Calandri's manuscript, fol. 108ʳ (ed. p. 216; *Essendo tre a dividere 30 grossi o altra moneta*).

C three, thus leaving 18, and he requires each participant to take, after choosing one object, a given number of the remaining tokens, namely
— the holder of a as many tokens as already received;
— the holder of e twice the number of tokens received;
— the holder of i four times the number of tokens received.

Upon returning, the observer will be able, knowing the number of tokens remaining, to determine who has what.

Indeed, the six possible combinations lead each to a different remainder, as indicated by Fig. 253, where we find how many tokens each participant had after the initial distribution (second row), then what he receives according to the object he has (combinations I to VI). The last two columns indicate the number of tokens distributed and the number of remaining ones, and the latter determines who has what.

	A	B	C	total	remainder
	1	2	3	6	18
I	(a) 1	(e) 4	(i) 12	17	1
II	(a) 1	(i) 8	(e) 6	15	3
III	(e) 2	(a) 2	(i) 12	16	2
IV	(e) 2	(i) 8	(a) 3	13	5
V	(i) 4	(a) 2	(e) 6	12	6
VI	(i) 4	(e) 4	(a) 3	11	7

Fig. 253

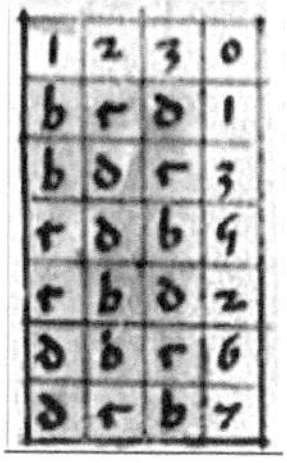

Fig. 254

At the end, a sequence of trisyllabic words serves as a mnemonic for the solution:

aperi, prelati, flamine, riome, preias, inager, bifena.

The place of the word corresponds to the remainder, the three vowels indicate the holders, and their order who has what. Thus to the remainder 7 corresponds *bifena*, which indicates that the third participant has the first thing, the second participant, the second thing, while the third holds the first. As to *riome*, which contains the missing vowel o, it corresponds to the missing remainder 4.[503]

Chuquet has the same problem, distributes the tokens in the same way, and gives (instead of a mnemonic) a table indicating the respective holder according to the remainder, with the three things designated by b, c, d (Fig. 254).[504]

---

[503] Other occurrences and other mnemonics: *Récréations*, p. 256.
[504] *Appendice*, pp. 457–458; MS., fol. 207^{v} – 208^{r} (with above table).

§8. The game with coins

This game may not be unknown to today's reader, for it is still in use, but with playing cards. It is again found as a marginal addition by a (French) reader in the same manuscrit as above (p. 242n); it also occurs among Pacioli's recreations.[505]

[198] We suppose that there are sixteen coins m_i and that a participant has chosen one, say m_4; the asker is to determine which one.

(1) The asker has the coins placed in two rows:

$$m_1 \quad m_2 \quad m_3 \quad \boldsymbol{m_4} \quad m_5 \quad m_6 \quad m_7 \quad m_8$$
$$m_9 \quad m_{10} \quad m_{11} \quad m_{12} \quad m_{13} \quad m_{14} \quad m_{15} \quad m_{16}$$

He asks in which row the required coin is.

(2) Then he makes two other rows by taking from the two previous ones, alternately and in order, each time one coin, beginning with the row not containing the required coin:

$$m_9 \quad m_1 \quad m_{10} \quad m_2 \quad m_{11} \quad m_3 \quad m_{12} \quad \boldsymbol{m_4}$$
$$m_{13} \quad m_5 \quad m_{14} \quad m_6 \quad m_{15} \quad m_7 \quad m_{16} \quad m_8$$

(3) After asking where the coin is, he makes two rows as before, but beginning with the row containing the coin:

$$m_9 \quad m_{13} \quad m_1 \quad m_5 \quad m_{10} \quad m_{14} \quad m_2 \quad m_6$$
$$m_{11} \quad m_{15} \quad m_3 \quad m_7 \quad m_{12} \quad m_{16} \quad \boldsymbol{m_4} \quad m_8$$

(4) He does the same again, but beginning this time with the row not containing the coin:

$$m_9 \quad m_{11} \quad m_{13} \quad m_{15} \quad m_1 \quad m_3 \quad m_5 \quad m_7$$
$$m_{10} \quad m_{12} \quad m_{14} \quad m_{16} \quad m_2 \quad \boldsymbol{m_4} \quad m_6 \quad m_8$$

After being told which row contains the required coin, he will know that it is in the antepenultimate place.

---

[505] MS. Paris BNF lat. 15120, fol. 57^{v}–58^{v} (*Recueil*, p. 74); *De viribus quantitatis*, MS. Bologna BU it. 250, fol. 114^{r}–115^{r}; see Agostini's summary, p. 190.

Appendices

A. Brief outline of the history of mathematics

1. Arithmetic and its applications

(a) Denominations

In Greek times, 'arithmetic' was used to designate the properties of natural numbers (p. 131n), while reckoning itself was called 'logistic' (λο-γιστική; see p. 8n —whence the title of Buteo's work, see below B.2s). In the Middle Ages the term *arithmetica* had the same (Greek) sense, before gradually acquiring, from the 15th century on, its modern one (see below). As to the decimal base, it was always in current use (the ten fingers); but other bases were also occasionally used, in particular the duodecimal (pp. 113–114) and the vigesimal; the Mesopotamian sexagesimal base has since antiquity remained in use for astronomy.

A major change in the expression of numbers came from the use by the Arabs of the Indian system representing the numbers with nine symbols, the value of each of which depended on its place within the number, while a further symbol, the zero, marked an empty place. Such a practice of linking value to position had, necessarily, already been in existence, as well as that of leaving a place empty, or even using a sign for it, but the systematic use and application of these ten symbols to all arithmetical operations was of lasting importance. It thus came into current use in Islamic countries in the 9th century, and spread in the Christian world from the 13th century on.

Indeed, around 820, the Persian residing in Baghdad al-Khwārizmī wrote an arithmetical treatise which taught, first, the nature of the decimal positional system, then how to transcribe in figures numbers expressed verbally, next how to perform with figures the four arithmetical operations, and finally the (exact) extraction of square roots. This arithmetical work played an essential part in propagating the use of the new numerals in the East. It was translated into Latin in Spain in the 12th century and played once again an essential part, this time in the West, mainly on account of the works it inspired in the 13th century. A testimony to its influence is seen in the various terms used at that time to mean 'arithmetic', all erroneous Latin transcriptions of al-Khwārizmī's name, such as *algorismus, algorizmus, algoritmus, argori(s)me*, which thus came to designate the subject and no longer the author.[506] Furthermore, this designation soon came to apply not only to operations but

---

[506] The Arabic letters *kh* and *g* (or *j*) may easily be confused.

also to their applications to trade and daily life (see below, B, Nos 2*i*, 2
l, 2*o*, 2*p*, 3*c*, 4*b*, 4*g*, 4*n*, 7). The word *algoritmus* survives nowadays in
the word 'algorithm', the *h* having been introduced in the belief that the
term was of Greek origin; only with the study of mediaeval arithmetical
manuscripts in modern times did its true origin come to light.

Algorismus and the like were not the only particular names employed
in mediaeval times in connection with arithmetic and its practical applications. Fibonacci had called his main work '*Liber abaci*', a book on *abacus*,
the meaning being not the reckoning table (ἄβαξ) but arithmetic (see p.
238) and algebra and their applications to trade and daily life. Due to
Fibonacci's predominant influence in late mediaeval times in Italy, mathematical treatises in Italian were from then on mainly called books on
abaco or some orthographical variant of it (see below B, 4*a*, 4*c*, 4*d*, 4*e*, 4
i, 4*j*, 4*o*, 6*b*); likewise, schools teaching this science were *botteghe d'abaco*
and their teachers *maestri d'abaco*. Before that, in 12th-century Spain,
the word *mahameleth* was used in exactly the same sense, as appears
from the title of the main work on the subject, the *Liber mahameleth*
by Johannes Hispalensis (John of Seville); the word itself was a rough
transcription of the Arabic معاملات (*mu'āmalāt*) meaning 'commercial
transactions' but was actually used to designate the science of mathematics with its applications to trade and daily life, just as *abaco* was in
Italy. By the end of the 15th century, the earlier denominations came to
be replaced by 'arithmetic', but then still including the applications of
the arithmetical operations to trade and daily life (see below B, 2*u*, 2*w*,
3*h*, 3*j*, 4*a*, 4*h*, 4*k*, 4*l*, 4*o*, 4*p*, 4*q*, 4*r*, 4*s*).

The Latin *cautelae* (written *cautele* in mediaeval Latin), or more precisely *cautele algorismi* ('cunning arithmetical questions'), came to designate recreational problems; whence its presence in the titles of some of
our sources, either Latin or French (see below B, 2*i*, 2*l*, 3*c*).

(*b*) Writing integers

In Greek, the numbers from 1 to 9, 10 to 90, 100 to 900 were expressed
in letters (often overlined, so as not to be mistaken for words), while the
nine thousands were the unit signs, but with a further distinctive sign.
See pp. 46*n*, 191. The period is 10 000; the same signs are used, preceded by the word 'myriad', or an abbreviation of it, which is repeated
for each further period (*k* times for a multiple of 10^{4k}). The simplifi-
cation brought about by the new system of Indian numerals is evident:
however large the number, these ten symbols sufficed (mediaeval arithmetical treatises emphasize this point). In Latin mediaeval mathematical

texts Roman numerals, unsuitable for reckoning, were only exceptionally used for numbers (pp. 97n, 140n, 232n —same MS., Fig. 233), which were either expressed in words (systematically in the *Liber mahameleth*, as was also usual in Arabic texts of the 9th to 12th centuries) or with the new numerals (which later Arabic texts used too).

(*c*) Writing fractions

In Greece, $\frac{1}{2}$ has a symbol of its own. For aliquot fractions, an apostrophe is added to the integer of the denominator; for non-aliquot fractions, one way is to write the denominator as an exponent. In mediaeval times, both Arabic and Latin texts adopt the fraction bar along with the writing of numbers in figures (before that, the numerator, in words, was said to be 'parts of' the denominator, in words). To facilitate oral expression, a fraction is often expressed as the sum of simple unit fractions, in antiquity as well as in mediaeval Arabic and Latin texts; this sum is then written by juxtaposing them (pp. 91, 97, 173). Likewise, an integer followed by a fraction means, implicitly, that they are added (pp. 39, 66, 76, 86, 87, 98, 99, 104–106, 110, 166, 174, 216, reproducing the original writing —Fig. 56 is an exception, for typographical reasons).

The decimal notation of fractions will become common towards the end of the 16th century, after the publication by Stevin of his 'decimal arithmetic' (*De Thiende*, Leiden 1585). There are, though, isolated occurrences before, the earliest found is in the work of al-Uqlīdisī, about 950 (above, p. 118) —it must be even earlier, for this author neither claims it as his invention nor explains it. The fraction bar is then used systematically: see the figures on pp. 45, 70, 72, 81, 169 (but see below p. 252, bottom), 251. This was particular true for approximations of π (or, as it was then called, 'the ratio of the circumference to the diameter'): in order to express it, one took as the diameter a power of 10 having as many 0 as the required number of decimals, and this became the denominator (see p. 168n, analogous situation). As to the symbol π, the initial letter of περίμετρον, 'circumference', it came into current use in the 18th century.

(*d*) Logarithms

While in Prague, the Swiss Jost Bürgi invented logarithms to simplify Kepler's astronomical calculations; his *Progress-Tabulen*, however, published later (1620), left the priority to John Napier's *Mirifici logarithmorum canonis descriptio* (1614). Logarithms came into common use in the 18th century, and were praised for their convenience (p. 167n).

2. Progressions and series

(a) Arithmetical progressions

If a_1 is the first term and r the common difference, then $a_2 = a_1 + r$, $a_3 = a_1 + 2r$, and the nth term will be $a_n = a_1 + (n-1)r$; the sum of the progression will be

$$a_1 + \cdots + a_n = \frac{n}{2}[a_1 + a_n] = na_1 + r\frac{(n-1)n}{2}.$$

See pp. 66, 111, 115n, and (problems) Nos. [60], [61], [63], [68]–[72], [80], [81], [116]–[119].

(b) Geometrical progressions

Towards 300 BC, Euclid demonstrated (*Elements* IX.35) that, for such a progression to $n+1$ terms, we have, in our writing, the proportion

$$\frac{a_{n+1} - a_1}{a_1 + a_2 + \cdots + a_n} = \frac{a_2 - a_1}{a_1},$$

whence, with $a_1 = a$ the first term and r the common ratio,

$$a_1 + a_2 + \cdots + a_n \left(= a + ra + \cdots + r^{n-1}a \right) = a\frac{ar^n - a}{ar - a} = a\frac{r^n - 1}{r - 1}.$$

Several of the problems seen above involved geometrical progressions ([15], [16], [64], [73]–[75], [120], [121]). The extension to an infinity of terms (with $|r| < 1$) appears in the 14th century (pp. 204, 206).

(c) Series

Summing the cubes of consecutive natural numbers —for which the formula had been known since antiquity— occurs in [81] & [82]. Infinite series with a convergent sum (other than the above geometrical progression) are considered at the end of the Middle Ages (pp. 205–206).

3. Algebra

(a) Denomination

In the introduction to his *Arithmetica*, Diophant explains how to reduce an equation to its final form, containing only one term of each of the unknown's powers (see also his introduction to Book IV, p. 88 of the edition of the Arabic extant Books; also *ibid.*, p. 77 *seqq.*). The first operation consists in adding to both sides the terms, taken positively, which are subtracted; the equation will then contain only additive terms. The second operation consists in removing from both sides the common quantities; the equation will then be of the desired form. For instance, if the equation considered is $5x^2 + 18 - 11x = 16$, it will be transformed first into $5x^2 + 18 = 11x + 16$, then into $5x^2 + 2 = 11x$, which is the final form. These two operations have no known specific denominations in

Greek (Diophant uses, if anything, κοινὴ προσκείσθω ἡ λεῖψις, thus 'add in common what is subtracted', and ἀπὸ ὁμοίων ὁμοια, '(subtract) like from like') but they do in Arabic: the first is al-jabr (الجبر, pronounced in certain regions al-gabr), 'restoration', the second al-muqābala (المقابلة), 'opposition' or 'comparison'. Since they were considered to be character-istic of algebraic reckoning, they came to designate it; thus the algebraic work by Abū Kāmil bears the title 'Book on al-jabr and al-muqābala' (p. 269); later, the name of this science was shortened to 'al-jabr'. In the Christian world, in the 12th century, at the time of translation from the Arabic, these names were either translated (into *restauratio* and *oppositio*, respectively) or just transcribed. Here too, the first of these two transcriptions came to designate the science; thus the *Liber mahameleth* already uses *algebra* (indeclinable word) in this sense.

(b) Indeterminate equation of the first degree

The indeterminate equation of the first degree $ax - by = c$ (with a, b, c natural numbers) is always solvable in natural numbers if a and b are relatively prime, and with an infinite number of solutions; furthermore, from one pair of solutions any other can be obtained since their general form is $x = x_0 + b \cdot t$, $y = y_0 + a \cdot t$ with t natural and x_0, y_0 the smallest positive solution. See pp. 42–43, 140.

(c) Systems of linear equations

We have seen such systems applied to exchanges of money between partners (Ch. V). They are a prominent feature of mediaeval mathemati-cal treatises, mainly as mathematical recreations on account of implausi-ble contexts. Their importance in mediaeval mathematics merely reflects how much space is devoted to them in Fibonacci's works. Even though their forms are often complicated, the solving principle is fairly simple and goes back to Greek algebra (p. 52).

(d) Quadratic equations

The solution of trinomial quadratic equations is attested from early Mesopotamian times, about 2000 BC. Until the end of the 16th century, they were divided into three types, namely $ax^2 + bx = c$, $ax^2 = bx + c$, $ax^2 + c = bx$, or, in reduced form, $x^2 + px = q$, $x^2 = px + q$, $x^2 + q = px$, with the coefficients a, b, c, p, q always positive; their solving formulae are then, respectively,

$$x = \frac{-\frac{b}{2} + \sqrt{\left(\frac{b}{2}\right)^2 + ac}}{a}, \qquad x = -\frac{p}{2} + \sqrt{\left(\frac{p}{2}\right)^2 + q},$$

$$x = \frac{\frac{b}{2} + \sqrt{\left(\frac{b}{2}\right)^2 + ac}}{a}, \qquad x = \frac{p}{2} + \sqrt{\left(\frac{p}{2}\right)^2 + q},$$

$$x = \frac{\frac{b}{2} \pm \sqrt{\left(\frac{b}{2}\right)^2 - ac}}{a}, \qquad x = \frac{p}{2} \pm \sqrt{\left(\frac{p}{2}\right)^2 - q},$$

for that is how they were expressed and calculated, but still always verbally. Note that the first two types have only one positive solution whereas the third has two provided the discriminant is positive.

The reason why these three types are considered, and not the remaining one $ax^2 + bx + c = 0$ with positive coefficients, is that only positive solutions were admitted. For this same reason, the first-degree equation $ax + b = 0$, with positive a & b, does not occur. By the way, it is extremely rare for algebraic expressions to be set equal to zero (see p. 42n). As we have seen, negative solutions are first considered at the end of the Middle Ages (pp. 60–62).

Examples of quadratic equations, sometimes with only the answer given, are found in problems [32], [60], [62], [63], [71], [80]–[82]. It is hardly surprising to find relatively few quadratic equations here: mathematical recreations, being mainly intended for a larger audience, rarely display them.

Cubic and quartic equations will be solved algebraically in the 16th century. Thus they do not occur here, unless reducible to a lower degree (p. 88); Diophant also uses higher powers of the unknown, but in a fictitious way since dividing by the unknown with the lowest exponent reduces the problem.[507]

(e) *Exponential equations*

A few cases of exponential equations occur, but are then reduced to solving equations of the first two degrees (see pp. 43–44, 83–84).

4. Symbolism

(a) *Use of abbreviations*

Mediaeval algebra, Arabic as well as Latin, is initially verbal, and even the numbers are written in words (above, p. 247); when number symbols do occur, it is mainly in tables or, if in the text, on a copyist's initiative. (The use of numerals is more common in arithmetical treatises, the operations being summarized in the margins.) This verbal form of

---

[507] That is in fact his third law of equation reduction (above, a). This occurs in the intermediate chapters of his *Arithmetica*, preserved in Arabic, where higher powers of the unknown occur; see their edition, p. 179 ('Def. XIII').

numbers, which may seem rather laborious, has at least the advantage of limiting the impact of any copyists' mistakes or omissions; furthermore, with inflected languages like Latin, the grammatical ending may confirm the rôle of a number in a sentence. From the 13th century on, however, the use of numerals in mathematical texts becomes the rule (above, p. 245).

Algebraic symbolism appears already in antiquity: there are in Diophant's *Arithmetica* signs for equality and subtraction (though none for addition since terms added were just juxtaposed); the subtractive terms followed the positive ones, separated from them by the minus sign.[508] Any kind of symbolism then disappears, and reappears only towards the end of the Middle Ages. As in Diophant's case, the purpose is clearly to avoid having to rewrite in full recurring mathematical terms by abbreviating the corresponding words. Therefore the Latin *plus* and the Italian *più* become p, $\bar{p}$ or $\tilde{p}$ (pp. 44, 81) and *minus* and *meno* accordingly m, $\bar{m}$ or $\tilde{m}$ (pp. 44, 72, 75–76, 89),[509] while *radix* and *radice* become $\mathcal{R}$ (pp. 44, 89). Chuquet has instead $\mathcal{R}^2$ (pp. 72, 75–76, 81) since he also uses higher roots. If the root comprises several terms, Chuquet underlines them, thus making the extent of the root clear (Fig. 255 below, $\sqrt[2]{5\frac{2}{5} - \sqrt[2]{22\frac{2}{5}}}$);[510] in this respect, he was a forerunner, preceding by a century Bombelli, who encloses the expression in a kind of square brackets, namely L and ⌐ (Fig. 256, cube root of $49 - \sqrt{2400}$).[511] Before, people tried with more or less success to express this inclusion verbally (p. 89n).[512] There is no doubt that the solution of higher-degree equations, involving many computations with roots, largely contributed to the development of algebraic symbolism.

Fig. 255   Fig. 256

In 12th-century texts the Latin *et*, 'and', links addends, and remained in use in later texts (with *plus* also being used to eliminate any possible ambiguity). Our sign + in fact arose from one of the (numerous) scribal abbreviations of *et* in 15th-century manuscripts (mathematical or not).

---

[508] See our *Introduction to the history of algebra*, pp. 32–33 (French edition, p. 33).

[509] Abbreviations were in fact in general use and not specific to mathematics: in [69], $\bar{m}$ stands for *meyl*, 'mile'.

[510] MS. Paris BNF 1346, fol. 127$^\mathrm{r}$. Here there is a single underlining, but there may be another one if the second root covers a longer expression.

[511] *L'algebra* ed. 1579, p. 155.

[512] A characteristic example in the *Liber mahameleth*, pp. 1373–1374; see also our *Introduction to the history of algebra*, pp. 71–72, 123 (French edition, pp. 75, 127–128).

It clearly has the algebraic meaning in Germany towards the end of the 15th century (Widman, below p. 266) and comes into common use in the 16th (pp. 77, 93).

(b) *Designation of unknowns*

Diophant has specific 'signs' to designate the unknown and its powers to the ninth; here again, they are in fact scribal abbreviations. In mediaeval (verbal) algebra, the unknown of an algebraic problem, our x, is only given a specific name. In Latin, it is the word *res*, 'thing', translating the equivalent Arabic word شيء, perhaps from the Greek τί. In 14th-century Italy, when mathematics began to be written in the vernacular, *res* was translated as *cosa* (pp. 44, 74, and below p. 262, 4a), then that was transcribed as *Coß* in Germany, which became the usual designation there for 'algebra' in the 16th century (whence the title of Rudolff's book, see p. 275), with the users of algebra becoming *Cossisten*.

Because of the need for abbreviations already discussed, *res* and *cosa* were reduced, from the end of the 15th century on, to *r.* and *co* (see pp. 77, 93; 44, 74). Chuquet has a designation of his own: the unknown is expressed by its degree put as an exponent to the coefficient; thus, 2^1 is $2x$, 35^5 is $35x^5$ (pp. 81, 174); as a rule, it is appended to the integer, which may cause some ambiguity when fractions occur (p. 81). All this is applicable as long as there is a single unknown. There is in our selection of problems one case of verbal denomination of a second unknown (namely 'sum', see p. 42). This is not exceptional.[513]

(c) *Further remarks*

At the very beginning of the 17th century capital letters began to be used to designate quantities, vowels for the unknowns and consonants for the knowns. Descartes started by using capitals for the unknowns and small letters for the knowns; in his *Geometrie* of 1637, however, he adopted the usage of minuscules throughout: for the knowns, from the beginning of the alphabet, for the unknowns, from the end. Accordingly, his main unknown was z and not x.

In our problem [152], p. 169, a dot is written between numbers to be multiplied; but it is merely a separation and by no means a multiplication sign. Using a dot for multiplying two quantities does not occur before the end of the 17th century. It was proposed by Leibniz, in his letter of July 19, 1698 to Johann Bernoulli; the purpose was to avoid confusion between ×, first used in 1631 by Oughtred, and the unknown x.

---

[513] See our *Introduction to the history of algebra*, pp. 76, 105 (French ed., pp. 80, 109).

B. Sources used

1. Sources in Greek

(**1a**) Eratosthenes (3rd c. BC), one of the greatest Alexandrian scholars, is mainly known for his 'sieve', a simple way to separate prime and composite numbers. We have seen his measurement of the earth (pp. 122–124, 126).

(**1b**) Poseidonios (*c.* 100 BC), known cosmographer and geographer, teacher in particular of Cicero. His measurement of the earth, less precise than Eratosthenes', was largely adopted later (pp. 122, 124–125, 126).

(**1c**) Nicomachos, *c.* 100, is a (second-rate) Greek mathematician, whose *Introduction to arithmetic* played an essential rôle in the transmission of Greek elementary number theory, in both the Latin and Arabic Middle Ages (adaptation by Boëtius, in the early 6th century, and translation by Thābit ibn Qurra, in the 9th century). See pp. 53, 131, 132, 135–138, 229. Iamblichos (*c.* 320) wrote a commentary on it, to which we have alluded (p. 53).

(**1d**) The greatest astronomer of antiquity, Ptolemy, living in Alexandria *c.* 150, is mentioned on pp. 121, 125, 126n, 130. His main work, the *Almagest*, or Μαθηματικὴ σύνταξις ('mathematical compilation'), remains the classical astronomical treatise and basis of later studies in both Latin and Arabic times.[514]

(**1e**) The mathematician Diophant, *c.* 250 AD, is our only source for higher Greek algebra; as with Euclid for geometry and Apollonios for conic sections, his work has supplanted (if not condemned to oblivion) earlier ones. His *Arithmetica* originally comprised thirteen 'Books' (= chapters), of which six are extant in Greek (namely I-III and VIII-X, ed. Tannery) and four in an Arabic translation (namely IV-VII), this latter part from a late Greek commentary. The last three Books are lost (on their presumable content, see the edition of the extant Arabic part, pp. 76–84). From Diophant is taken problem [41]; see also pp. 53, 54.

(**1f**) Book XIV of the Greek Anthology (*Anthologia Graeca*) contains various epigrams compiled in late antiquity, and represents the earliest surviving collection of puzzle problems, including recreational problems of the type examined in this study. See problem [44] and pp. 33n, 68n, 111, 173n.

(**1g**) Moschos is a Byzantine mathematician of *c.* 1200, known to us only from a reference by Fibonacci. He was one of the sources of Fibonacci's

---

[514] 'Almagest' is an Arabic adaptation, with the article *al*, of the more popular Greek designation μεγίστη σύνταξις.

interest in linear systems of equations —one of the latter's domain of predilection. See pp. 51–53.

(1*h*) The Byzantine monk Maximos Planudes, *c.* 1290, is mainly known in mathematics for collecting manuscripts of Diophant's *Arithmetica* and writing a treatise on arithmetic with the nine signs 'according to the Indians' (κατ' Ἰνδούς). A minor mathematician, he is nevertheless a prominent figure of the mathematical revival there. See pp. 45–46 (problem [34]).

(1*i*) His friend Manuel Moschopoulos is the author of a treatise on magic squares, where he attempted, starting from examples of squares probably taken from Persian manuscripts, to find construction methods. See p. 194.

(1*j*) The MS. Paris BNF Suppl. Gr. 387 contains a collection of 119 problems written in the early 14th century, which Vogel edited. See our [12], [35], [79], [85], [119].

(1*k*) Among the Greek manuscripts bought in Constantinople in the 16th century by the ambassador of Ferdinand I, Archduke of Austria, and now in the Vienna National Library is the MS. phil. gr. 65, containing various mathematical treatises. The first, written in the mid-15th century, is in two parts, the first (arithmetical) one of which has recently been edited by Deschauer. It includes the problem mentioned on p. 102*n*.

(1*l*) Further on in the same manuscript, by another copyist (and in non-classical Greek), we find a collection of problems, edited by Hunger and Vogel, from which are taken our [47] and [54].

2. Sources in Latin

(2*a*) Bede (the Venerable Bede, *Beda Venerabilis*), an English Benedictine monk living around 700, author of historical and theological works. With him originated the expression 'the year of our Lord' (*anno Domini*), calculated according to the number of years since Christ's birth. Scientific writings include calculation of Easter and a small collection of mathematical problems. See p. 230.

(2*b*) Alcuin of York is an English ecclesiastic of the 8th century (d. 804), educated at the cathedral school of York, which preserved Bede's legacy. In 781 Charlemagne met him in Parma, and asked him to reorganize education in his empire. Among Alcuin's writings of an educational nature there is the collection of 'Propositions to sharpen the minds of young people' (*Propositiones ad acuendos iuvenes*), which contains 53 problems, most of them doubtless of ancient origin (some even involving camels, which clearly indicates an Alexandrian Greek origin). Certain recreational ones have been reported here, as well their successors. See [11], [59], [83], [120], [140], [141], [162], [163], [171], and pp. 115n, 170, 177.

(2c) The *Liber augmenti et diminutionis* (Book on increasing and decreasing) by Ibrahim or Abraham is a collection of some 30 problems translated from Arabic into Latin in 12th-century Spain. For the problems presented here ([27], [189], [196]), we used the 19th-century edition of the Latin text by G. Libri (which has minor, easily detected faults).

(2d) Johannes Hispalensis (John of Seville) wrote towards the middle of the 12th century in Spain a vast treatise called 'Book on commercial transactions' (*Liber mahameleth*, see above, p. 246), explaining in the first part arithmetical operations using the Indian number system and in the second part their application to daily and trade problems. It obviously follows Arabic-Hispanic treatises intended for merchants, but suffers from a certain inconsistency: sometimes elementary computations or reasonings are expounded at length, thus of little interest to a mathematician, whereas elsewhere highly specialized mathematical topics are treated, such as extracting the square root of the sum of rational and irrational terms, which is certainly out of reach of the average merchant. Like many Arabic mathematical treatises of the time, it concludes with a set of recreational problems, some of which are included here [45], [58], [65], [96], [156], [157], [179]; see also pp. 67, 71n, 101–102, 161, 213, 217–221, 224–225.

(2e) The most important mathematician in mediaeval Europe is doubtless the Pisan Leonardo Fibonacci (d. after 1240). Of his five works, the main one is the *Liber abaci*, written around 1202 and revised later, which teaches the elements of mathematics with applications to commerce and daily life, thus covering the same topics as the *Liber mahameleth*. But being a merchant Fibonacci travelled in both eastern Moslem countries and the Byzantine Empire where, as a mathematician, he sought contact with others and gathered all the information he could. This gave him a much wider mathematical education than was available at the time, with works such as the *Liber mahameleth* relying solely on texts available in Spain, either Arabic or translated from the Greek. The 12th and 13th chapters of the *Liber abaci* offer a wide variety of recreational problems, including many linear systems of equations, in which Fibonacci changes the data so as to obtain, at times, a zero or even negative solution among a set of positive ones. This latter point is illustrated in [39]–[40]. But there are numerous other problems of his: [17], [26], [30], [31], [48], [52],

[84], [91], [99], [103], [107], [112], [124], [125], [136], [145], [146], [150], [158], [190], [193]; see too pp. 29–30, 34, 42–45, $50n$, 51-54, 66, 79, 119, 120, 135, 136, 141, 161–164, 213, 214, $231n$, 236, $242n$.

(**2f**) The *Annales Stadenses* is a wide-ranging chronicle written in the middle of the 13th century by the monk Albert; the work is particularly informative about more recent events. It is a valuable early source for some mathematical problems. See [1], [25], [164], [167], [173] and p. 154.

(**2g**) The Dominican friar Jacobus de Cessolis was the author, in the second half of the 13th century, of the *Liber de moribus hominum et officiis nobilium*, one of the many mediaeval works drawing a parallel between the chess game and social conditions of the time. This one was translated into nine languages, sometimes several times (from the French into English by W. Caxton in 1474, and printed a few years later). The manuscripts of the *Liber de moribus* are often beautifully illustrated with scenes of daily life. This is the case of the MS. Paris Bibliothèque de l'Arsenal fr. 5107 rés., from the end of the 15th century, which includes the French translation of de Cessolis' book by the friar Jehan de Vignay. See pp. $70n$, $120n$, $183n$.

(**2h**) MS. Paris Bibliothèque Nationale de France lat. 15120 contains mathematical treatises copied by various hands, apparently all from the 13th century. One set of problems is taken from the *Liber augmenti et diminutionis*, another from the *Liber mahameleth* (2c, 2d). It includes (added by two readers of the 14th and early 15th century) our problems [197] and [198].

(**2i**) The *Subtilitates enigmatum*, also called *Cautele algorismorum*, is an anonymous collection of 34 problems (some repeated in a different form). Most of the common recreational types appear in it, and the numerous extant manuscript copies from the 14th century on attest to its wide use. Our intention being to draw from various sources there is only one of its problems here ([185]); but references are numerous: pp. $3n$, $5n$, $41n$, $80n$, $113n$, $115n$, $154n$, $229n$, $239n$, $242n$.

(**2j**) Nicolas Oresme first studied in Paris together with Albert of Saxony, last became Bishop of Lisieux (1377-1382). He translated from Latin into French and commented various Aristotelian texts. In mathematics, he was the first to introduce diagrams representing, at each point of the base line and perpendicular to it, the variation of intensity, thus becoming a precursor of coordinate geometry. In his *Questiones super geometriam Euclidis* he noted the existence of convergent and divergent series. For some of his arguments on infinites, see pp. 205–211.

(**2k**) Albert of Saxony, German theologian, mathematician and physicist, educated at the University of Paris, became its rector (1353), then first rector of the University of Vienna (1365), before being appointed bishop of Halberstadt (1366-1390). While in Paris he worked with Oresme. Author of various 'questions' on treatises by Aristotle. Less original than Oresme's, his treatises are nevertheless of remarkable clarity. See pp. 206, 207, 211.

(**2l**) MS. Montpellier Ecole de Médecine 323, copied in the 13th and 14th centuries, comprises various texts, mainly astronomical and on the calendar, but also on arithmetic; three leaves, headed *Cautele algorismi*, contain recreational problems (fol. $235^v - 237^v$). See p. 230n.

(**2m**) MS. Reims G.559 (cat. 696) of the 14th century, contains theological writings; a 14th-century marginal gloss of it is our [131].

(**2n**) The 14th-century anonymous translation from the Arabic headed *Incipiunt figure 7 planetarum* represents seven magic squares, from order 3 to 9, attributed to the seven planets. It describes the attributes, favourable or unfavourable, of each square according to its planet and the efficacy of their action depending upon the planet's position. Our [174] is a good instance of that interaction.

(**2o**) MS. Lyons 59 (cat. 127) includes, along with lexicons of names and verbs, a compilation dated 1400 (fol. $53^r - 68^v$) which explains how to write numbers using the numerical symbols and perform the arithmetical operations (*species algorismi*); finally there are thirteen 'rules' teaching how to solve problems of specific kinds. Instances illustrating the application of these rules are our [110], [111], [116], [117]; see also p. 91n.

(**2p**) The *Algorismus Ratisbonensis* is a collection of problems, in Latin or German, copied towards the middle of the 15th century by the monk Frederic in the Benedictine cloister St. Emmeram in Regensburg. Italian influence is clear. These problems are obviously intended for merchants, but many are of recreational nature. See [92] and pp. 3n, 47n, 147n.

(**2q**) The MS. Vienna Palat. 5203, of the 15th century, written by the astronomer Johannes Müller (Regiomontanus, 'from Königsberg'), contains astronomical and mathematical treatises. On fol. 167^r a table of perfect numbers, our [130].

(**2r**) MS. Dijon 268 (cat. 447), of the 15th century, contains texts on astronomy (mainly in Latin) and, towards the end (fol. $112^r - 136^v$), an arithmetical treatise explaining the rule of three and then various solving rules for different problems; this is followed by instructions for measuring pieces of land and receptacles. Said rules (*particulares regule*) concern problems of a recreational nature. See [22], [23].

(**2s**) Johannes Buteo (Jehan Borrel), b. 1492, entered the abbey of St. Anthony in 1508, and on account of his outstanding abilities was sent by his superiors to study in Paris. There he pursued his mathematical education. In spite of Cardan's somewhat subjective judgment (see below, 2*w*), he has generally been considered quite an able scholar. See, from his *Logistica*, our problems [7], [18], [56], [72], [132], [148], [154], and p. 214.

(**2t**) Jean Fernel (1497-1558) was a physician, author of *Universa Medicina* printed 1554 and reprinted some thirty times. Earlier, he was interested in mathematics and, in particular, in geodesy, which led him to make a new measurement of the meridian degree, reported on p. 126.

(**2u**) Gemma Reiner/Regnier (1508-1555), better known on account of his origin as Gemma Frisius, was an eminent Dutch mathematician of the first half of the 16th century. His *Arithmeticae practicae methodus facilis*, combining theory and applications, was reprinted some sixty times, first in 1540. From it is our problem [57].

(**2v**) Michael Stifel (d. 1567) wrote the *Arithmetica integra*, a very original scholarly work printed 1544. He also had reprinted (1553/4) an algebraical work by Christoff Rudolff, the *Coß* (see pp. 252, 266), to the original edition of which (1525) he added various comments. See, from that second edition, pp. 94, 106, and, from the *Arithmetica integra*, p. 198.

(**2w**) The name of Girolamo Cardano (Cardan) is associated with the algebraical solution of the cubic and quartic equations. Before dealing with that, he published in 1539 a well-written treatise, *Practica arithmetice*. Following general theory are application problems, from which our [5], [38], [70], [75], [81], [89], [127], [147]; see also pp. 31, 73, 120, 126*n*, 129, 195, 214. The strangeness of Cardan's character is well documented, in particular his attempts to learn from Tartaglia (our 4*t*) how to solve algebraically the third-degree equation.[515] Of notable interest is his autobiography, at the end of which he mentions all those who cited his mathematical works. Those who criticized him receive rough treatment: Buteo (our 2*s*) is a 'grindstone impervious to any knowledge and teaching', and he wonders how Buteo, Tartaglia and a few others dare

---

[515] See our *Introduction to the history of algebra*, pp. 129–134 (French ed., pp. 134–138).

call themselves scholars since their knowledge does not go beyond the elementary.[516]

(**2x**) Pietro Bongo was an ecclesiastic of the 16th century (d. 1601), the author of a book reprinted several times on the mystery, magic and metaphysical signification of numbers. See pp. 133, 134, 136.

(**2y**) Marin Mersenne (1588-1648) was a theologian, and also a scientist, to whom we owe editions of earlier mathematical works. He was a correspondent of Fermat, with whom he shared an interest in number theory (seen on p. 133).

(**2z**) Pierre (de) Fermat (1608-1665) was a counsellor of the parliament of Toulouse who became interested in mathematics. Although particularly known for his assertions in number theory (still the object of studies), he played a great part in the birth of analytical geometry and infinitesimal calculus. The beginning of the study of magic squares in Europe is also associated with him; see pp. 193, 195n, 197.

3. Sources in French

(**3a**) MS. Paris Bibliothèque Nationale de France AF fr. 2050, from the 15th century (as are the following, also anonymous MSS.), is an arithmetical manual for traders. The arithmetical part proper (fol. $1^r - 120^r$) is followed by a description of the various coins minted in France and other countries, with their values. See problems [68], [97], [152] and pp. 33n, 64n, 67n, 70n, 96n, 170n.

(**3b**) MS. Paris Bibliothèque Nationale de France AF fr. 1339 contains arithmetic and geometry, followed by a treatise on the use of the astrolabe. We have mentioned it several times, but generally as a secondary source, for it treats problems examined in other manuscripts. See [23], [187] and pp. 3n, 33n, 38n, 41n, 47n, 140n, 151n, 154n, 229n, 230n, 235n.

(**3c**) MS. Bibliothèque de Tours 399 is a set of three quite different manuscripts, the second of which (fol. $125^r - 147^r$) contains, as its title (*Ce sont lez cautelez d'argorime*) indicates, a collection of recreational problems, some of which go back to Alcuin's collection. It was one of our main sources. See [21], [113], [114], [115], [118], [121], [122], [186], [194], [194′] and pp. 3n, 17n, 33n, 41n, 64n, 79n, 91n, 151n, 154n, 227n.

(**3d**) MS. Nantes 456 (*olim* fr. 290) teaches arithmetic and its commercial applications, then (fol. 86^r) how to perform arithmetical operations using

---

[516] *Buteo lapis molaris qui nec scit nec doceri potest (...). Ex his qui male dixerunt neminem agnovi qui Grammaticam excesserit, nescio qua audacia eruditorum ordini se intitulerint* (*De vita propria liber*, XLVIII).

counters (*getz*), next geometry (that is, as in other such treatises: calculation of surfaces and volumes of common geometrical figures). From the first part are our [73], [98]; see also pp. 3*n*, 170*n*.

(**3e**) MS. Rouen Bibliothèque municipale I 58 (cat. 1006), of the 15th century, is clearly intended for merchants, since it gives a detailed account of coins and change. At the end a collection of recreational problems (with use of Roman numerals). See pp. 97*n*, 140*n*, 146, 232*n*.

(**3f**) The main mediaeval mathematical work in French is Nicolas Chuquet's *Triparty en la science des nombres*, thus, as indicated by the title, a treatise in three parts on the science of numbers (calculating with rational numbers, with roots, solving equations). To it is appended, first, a collection of (mainly) recreational problems, now commonly called *Appendice au Triparty* (MS., fol. 148^r), then a part on geometry and measurements (211^r), and a further one on commercial applications (264^r). About its author we know what he himself tells us: a bachelor in medicine, he was from Paris but lived in Lyons where he wrote the above work in 1484.[517] The *Triparty* was published by A. Marre in its entirety, but not the *Appendice*, Marre giving mainly just enunciation and answer. Since it is the *Appendice* which contains recreational problems, we have drawn from it extensively, completing at times Marre's extracts using the MS. Paris BNF AF fr. 1346. See problems [2], [15], [29], [37], [43], [49], [60], [64], [71], [77], [80], [93], [128], [138], [151], [153], [160], [165], [181], [184], [191], [192], [195], and pp. 23*n*, 28–29, 30, 47, 48, 141, 146, 154, 177, 218, 243.

(**3g**) The anonymous *Livre de chiffres et de getz* (Book on figures and counters) is one of the earliest mathematical works printed in French, namely in Lyons in the very beginning of the 16th century. From the first, 1501 edition, a single exemplar has survived (Staats- und Stadtbibliothek Augsburg); from the second, printed a few years later, again a single exemplar is preserved (Bibliothèque Méjanes, Aix-en-Provence). The difference between the two editions is seen only in details, with minor changes in type setting and spelling. As we are told in the introduction, it was intended for merchants not knowing Latin; accordingly, the level is not very high and the solutions sometimes erroneous. It first teaches how to perform arithmetical operations with counters (*getz*), next with figures, then gives commercial problems followed by thirty-two 'rules and problems' (fol. 48^v – 64^r). Some of the latter problems have been reproduced here: [24], [36], [67], [76], [90], [133]; see also pp. 33*n*, 146, 170*n*.

---

[517] MS. Paris BNF fr. 1346, fol. 147^r (ed., p. 814): *il a este fait par Nicolas Chuquet parisien, bachelier en medecine; je le nomme 'le triparty de Nicolas en la science des nombres'. Lequel fut commancé, medié et finy a Lyon sus le Rosne l'an de salut 1484.*

(**3h**) In 1520 there was published in Lyons a lengthy arithmetic by Estienne de la Roche, *L'arismethique novellement composee*, teaching arithmetic and its applications, with numerous indications on the coins in use, then a few elements of geometry (reprinted 1538). In the introduction, de la Roche mentions among his sources Chuquet and Pacioli, and speaks of his own 'minor additions' (*quelque petite addicion de ce que j'ay peu* (= pu) *inventé et experimenté en mon temps en la pratique*). As a matter of fact, much is drawn from Chuquet's handwritten work, which thus became indirectly known before its 19th-century edition. See pp. 23n, 73n, 87n.

(**3i**) *S'ensuit jeux partis*, also called 'The game of princes and young ladies' is an anonymous text of 12 leaves printed around 1530. Its purpose is to teach chess to 'all noble hearts' wishing to know it for the purpose of entertainment and to avoid idleness. See p. 187.

(**3j**) Pierre Forcadel was born in Béziers in southern France and, in 1560, appointed professor of mathematics at the Collège Royal (now Collège de France), where he remained until his death (1576). Before that, he had stayed in various Italian towns. At the beginning of his *Arithmetique* (1557), he writes that his purpose was to record his 'infinite discoveries in arithmetic'. One problem of his is our [8].

(**3k**) Jean Trenchant, living in Lyons, is the author of an Arithmetic published in 1558, later revised and reprinted several times. Indeed, in the later editions the author tells that it was well received, and that he who understands and puts into practice its teaching will have no need of other works. We have reported a few of his recreations, most taken from earlier authors: [9], [20], [182], and p. 240.

(**3l**) The Frisian Edouard Leon Elcius Mellema, professor of Latin, French and mathematics, wrote an arithmetical treatise in two volumes, published respectively in 1582 and 1586, partly presented as a dialogue between master and pupil, and embellished here and there with verses in Greek, Latin, French, German and Dutch. Although not specifically written for merchants, there are numerous applications to trade. Problem [155] is taken from that work.

(**3m**) Bachet de Méziriac, who first edited the Greek text of Diophant, is commonly known for his *Problemes plaisans et delectables qui se font par les nombres* ('Amusing and delightful problems solved by means of numbers'), which is often considered as the ancestor of modern works on recreational mathematics. The second edition (1624) was longer than the first (1612). To the title, Bachet added that these problems were taken from 'various authors' and commented by him. It appears that the work was still in keeping with the mediaeval tradition of recreational problems, which is why it is cited here. See, for its problems, [10], [137], [142], and various remarks pp. 19–20, 23, 31, 151, 194, 195, 195n. Bachet's second edition was reprinted in 1874 by Labosne, whose instructive comments supplement Bachet's (see pp. 10, 21–22, 154n, 157n, 158n). Further observations were made by Schubert (see pp. 10n, 13) and by Lucas (p. 154n).

(**3n**) A later collection of recreational problems, this time from the end of the 17th century, is by Jacques Ozanam. During the 18th century, it was successively revised and augmented, first by Martin Grandin and then by Jean Etienne Montucla, the author of two renowned works on the history of mathematics (1754, 1758). From it are taken our [139], [149]; see also p. 227.

4. Sources in Italian[518]

(**4a**) The purpose of an anonymous *Libro d'abaco* written in the 14th century was to teach *molte ragioni d'abaco, cioè d'arismetricha e di giometria*; it also gives indications on the coins and units of measurement of various Mediterranean countries. It is thus a typical commercial arithmetic, while also containing some algebra (problems solved *per la chosa*). See p. 171n.

(**4b**) *Rascioni d'algorismo* is a 14th-century collection of some 140 problems (arithmetical operations with integers and fractions, commercial applications, recreational problems, measuring plane geometrical figures) followed by indications on coins; here again, clearly a treatise for merchants. See our [28], [188], and pp. 5n, 30n, 97n, 235n. From its manuscript (MS. New York Columbia University Library X511.AL3) and edition are reproduced two illustrations, pp. 151, 152.

(**4c**) To Paolo dell'Abbaco (d. 1374), astronomer and mathematician, is attributed a *Trattato di tutta l'arte dell'abacho* (below, 4d, 4i) and various other works, of which he is perhaps not always the author. MS. Florence Biblioteca Nazionale Centrale Magliabechiano XI 86 contains an *Istratto di ragioni*, written, we are told, *per lo venerabile strolagho Maestro Pagholo*. Our [19], [94] are taken from it; see also pp. 5n, 56n, 85n, 93n, 162n.

(**4d**) MS. Rome Biblioteca dell'Accademia Nazionale dei Lincei, Cors. 1875, preserves a mid-14th century *Trattato di tutta l'arte dell'abacho*

---

[518] On Italian mathematical manuscripts and early printed books, see van Egmond's *Catalog*.

with, at the end, some geometry, astronomy and astrology. See problem [177] and pp. $30n$, $34n$, $86n$, $98n$, $173n$, $215n$, $217n$.

(**4e**) We find in MS. Florence Biblioteca Nazionale Centrale, Conventi soppressi G 7 1137, from the end of the 14th century, a *Libro delle ragioni d'abacho*, which expounds the arithmetical operations, then problems, and also contains algebra and geometry. See p. $136n$.

(**4f**) F. Bartoli Bentaccordi († 1425) was an Italian merchant living in Avignon at the papal court. He was no mathematician, nor did he make any contribution to mathematics, but he has left us a notebook (*Memoriale*) illustrating the particular interests of merchants: indications on the weights and coins of various regions, as well as conversion tables and elements of arithmetic; in addition, recreational problems to serve as applications of arithmetical computing. From his notebook are our problems [33], [55]. His complete notebook has since been edited in a collective work (*Il tesoro di un povero*, Roma 2016).

(**4g**) MS. Milan Biblioteca Trivulziana 90, from the early 15th century, contains as first treatise Jacopo da Firenze's *Tractatus algorismi*, which is a commercial arithmetic with monetary and interest problems and a few recreational ones at the end. See pp. $33n$, $87n$, $97n$.

(**4h**) MS. Florence Biblioteca Nazionale Centrale, Palatino 573, from the 15th century, contains a sizable *Trattato di praticha d'arismetricha*, ending with algebra, with frequent references to authors of the end of the 14th century. See pp. $5n$, $89n$, $110n$.

(**4i**) MS. New York Columbia University Library, Plimpton 167, from the mid-15th century, contains the work *Trattato di tutta l'arte dell'abacho*, followed by a short medical treatise. See p. $37n$.

(**4j**) Piero della Francesca (*c.* 1415-1492) is not only one of the most illustrious Italian painters but also the author of scientific treatises on perspective (*De prospectiva pingendi*), on polyhedra (*De corporibus regularibus*) and, in Italian, of one treatise on arithmetic, algebra and geometry. Like other similar treatises at that time, it was supposedly intended for merchants, as he himself writes at the beginning (*alcune cose de abaco necesarie a mercatanti*); but clearly some of the subject-matter goes beyond the average merchant's interests (various types of equations, problems in solid geometry involving polyhedra). For samples of (recreational) problems, see [46], [74], and p. $173nn$.

(**4k**) To Benedetto da Firenze is attributed a commercial arithmetic (*Trattato di praticha d'arismetrica*) based, we are told, on the work of

Lionardo Pisano ed altri auctori and intended for a 'teach-yourself' public (*sança usare el maestro*). It is preserved in the manuscript Siena Biblioteca Comunale degli Intronati L.IV.21, dated 1463. See pp. 30n, 113n. A *trattato d'abacho* of his is preserved in the MS. Florence Biblioteca Nazionale Centrale, Magliabechiano XI 76; see pp. 50, 142, 144n.

(**4l**) Filippo Calandri is the author of *De arimethrica opusculum*, a book dedicated to Giuliano de' Medici, published 1491/2 in Florence, republished 1518, containing the elements of commercial arithmetic. The level is rather modest, but this book comes with numerous little illustrations, many of which we have reproduced here (from the 1518 edition) for their interest as pictures of contemporary life —these woodcuts are not, however, of particular artistic value. See problems [104], [175], [176], [178], [180], and pp. 68nn, 69n, 71nn, 96, 103n, 104n, 107n, 217n, 222n, 224n.

These pictures are in no way comparable to the magnificent ones illustrating his earlier manuscript, also dedicated to Giuliano de' Medici (Florence MS. Riccardiana 2669, *c.* 1480), with numerous miniatures accompanying recreational problems. See pp. 68n ([51]), 92n, 116n, 144n, 165n, 215n, 229n, 232n, 233n, 235n, 236n, 241n, 242n. Note that there are two different foliations in use, differing by twelve leaves; that used by us, the original one, begins after the initial multiplication tables.

(**4m**) Pier Maria Calandri, brother of Filippo, wrote around the end of the 15th century a *Tractato d'abbacho* (now edited from MS. Florence Biblioteca Medicea Laurenziana, Acquisti e doni 154) teaching arithmetic with numerous applications to trade, and, at the end, some geometry. See pp. 55 (nice illustration), 146.

(**4n**) Pietro Paolo Muscarello is the author of a late 15th-century (1478) work, with the title 'Algorismus' (*nostro tractato, il quale si è decto Algorismus*, fol. 1^r), thus a commercial arithmetic, followed by elements of geometry. The problems of a recreational nature are not original, nor is the solving. Of note though are the illustrations found in its manuscript (now MS. University of Pennsylvania LJS 27 —thus from the Lawrence J. Schoenberg Collection). See pp. 49n, 53n, 64n, 67n, 69n, 70n, 80n, 92n, 96n, 102n, 171n, 175n, 236n.

(**4o**) The *Opera de arithmethica*, also called later *Libro de abacho*, of Pietro Borghi (or Piero Borgi), is a commercial arithmetic teaching successively calculating with integers and fractions, the rule of three and the usual applications to trade. Its early first print (1484) was followed by some fifteen further editions to 1567. See p. 105.

(**4p**) The Franciscan Luca Pacioli's *Summa de arithmetica, geometria,* *proportioni e proportionalità*, first published in 1494, is a book of fundamental importance. The author is not considered to have been very original, but his printed work made accessible the mathematical knowledge of the Middle Ages. For this reason, many of the problems he reports have found a place in later authors, sometimes only reproduced, sometimes modified, with Pacioli's solution changed and sometimes criticized, deservedly or not. It was also the first printed work explaining double-entry bookkeeping. Examples taken from the *Summa* are [32], [61], [62], [63], [82], [86], [100], [105], [129]; see also pp. 30*n*, 64*n*, 66*n*, 94, 95, 119, 125*n*, 139, 220*n*, 223*n*. An earlier treatise of his, written in the years 1476/1480 in Perugia, is extant in the Vatican Library (Vat. lat. 3129); see pp. 222–223. Another later work of his, *De viribus quantitatis* (MS. Bologna Biblioteca universitaria it. 250), remained in manuscript form at the time; from it are our [3], [144]; see also pp. 146–147, 154, 244.

(**4q**) Girolamo Tagliente's *Componimento di arithmetica*, also known as *Thesauro universale*, was first printed in 1515, then numerous times in the 16th century. It is the result of his studying, since an early age, various works by outstanding authors.[519] There are also a few leaves on practical geometry. These editions are illustrated with interesting woodcuts. Since they may vary from one edition to the other, we have often reproduced two or even three when they are notably different. See pp. 24, 66, 73, 74, 78, 91, 107, 129, 140, 161, 167, 170, 175, 241, and also pp. 80*n*, 141.

(**4r**) Francesco Ghaligai, of Florence, is the author of a *Summa de arithmetica*, or also *Pratica d'arithmetica*, a mercantile treatise which ends with problems of algebra (he refers to Fibonacci) published first in 1521 and twice later. See our [159], and also p. 168.

(**4s**) Giovanni Sfortunati, from Siena, published in Venice his *Nuovo lume, libro di arithmetica* (1534), which was reprinted several times. As he tells us, the title was chosen on account of the 'new light' shed by him on the wrong assertions of other authors. But he himself was once the object of severe criticism ([70]). From him is also [101].

(**4t**) Niccolò Tartaglia, with Cardan the major Italian mathematician of the mid-16th century, published a 'General treatise on numbers and measurements' (*General trattato di numeri et misure*) in three volumes (1556–1560). Of interest here are the 16th chapter of vol. I, which contains mathematical recreations and games, and the first chapter of vol. II, on

---

[519] Speaking of the science of arithmetic (introduction, ed. 1547): *di che in questa mia verde e giovenile etade ho voluto con lo aiuto di Dio vedere con ogni studio & diligentia diverse opere fabricate per eccellentissimi autori, & non con puoca mia fatica e industria ho voluto cumulare & componere la presente opera.*

progressions. See our [6], [13], [13'], [16], [53], [100'], [134], [135], [143], and pp. 29, 66n, 73–74, 109–110 ([105]), 141, 147–148, 154, 160, 214.

5. Sources in German

(5a) Johann Widman(n) studied in Leipzig in the 1480's, possibly also taught there. His early (1489) printed, later reprinted, treatise on commercial arithmetic was widely used and is now known mainly for its adoption of the + and − signs (Fig. 257: *unnd das + das ist mer*). See problem [4], and pp. 45n, 47n, 223n.

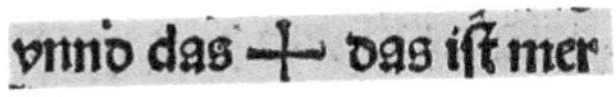

Fig. 257

(5b) Albrecht Dürer was not only a great painter and engraver but also a noteworthy mathematician. Here, though, we record only his reproduction (not: construction) of a known magic square (pp. 193, 196), for its contribution in drawing attention to the subject.

(5c) From Adam Ries(e)'s *Rechenung auff der linihen und federn* (first printed 1522), followed by numerous editions, and changes in title, is taken our [95] (ed. 1581). Here again (see 3g), we are reminded of the two ways of calculating employed by merchants: using either counters or pen (*Feder*) and ink.

(5d) Among the works written by Christoff Rudolff there is the *Coß*, the first sizable German book on algebra (1525), of a higher level than the usual text-books on arithmetic and its applications. It was revised by Stifel (above, 2v). Our [66], [69], [87], [88], [102] are all of the messenger kind. The revised *Coß* had an illustrious reader: in his autobiography, Euler tells us that his father used it to teach him the first elements of mathematics.[520]

6. Sources in Provençal

(6a) The Pamiers Arithmetic, written in Provençal (namely Languedocian) *c.* 1435 by an anonymous author in Pamiers, is a commercial arithmetic, thus one more treatise teaching arithmetic and its applications to trade. But the author was obviously an able mathematician who had

---

[520] *Bald hierauf begaben sich meine Eltern nach Riechen* (near Basel), *wo ich bey Zeiten von meinem Vater den ersten Unterricht erhielt; und weil derselbe einer von den Discipeln des weltberühmten Jacobi Bernoulli gewesen (war), so trachtete er mir sogleich die erste(n) Gründe der Mathematic beizubringen, und bediente sich zu diesem End des Christophs Rudolphs Coss mit Michaels Stiefels Anmerckungen, worinnen ich mich einige Jahr mit allem Fleiss übte* (Fellmann, *Euler*, p. 11).

other interests than such mundane ones. That seems to be the only explanation for inclusion of a problem with one negative solution, the first to be accepted, and an iterative method for root approximations leading the author to considerations on infinite sets. See [42], [161], and pp. 199–202, 211.

(6*b*) Frances Pellos is another Provençal mathematician (from Nice). His *Compendion de lo abaco*, written in the years 1456-1457 and printed in Turin in 1492, reproduced the innovative problem of his predecessor. See [42], pp. 60–61, and 216*n*.

7. Source in Spanish

The *Arte del alguarismo*, an anonymous work written in 1393, is the first known commercial arithmetic written in Castilian. The description of the arithmetical operations is followed by a collection of problems, commercial and also recreational. Among the latter, our problem [50]; see also pp. 5*n*, 31*n*, 113*n*, 162.

8. Source in Armenian

In the sixth century, Anania S̲hirakatsi was taught Greek science by the Byzantine Tychicos in Trebizond. After staying there for eight years, he returned to his native region of S̲hirak, where his books in Armenian transmitted that science. Among his numerous works there is a collection of 24 problems, one of our few sources from this late antique time. Some of them are mentioned on pp. 33, 63*n*, 111–112.

9. Sources in Arabic

(9*a*) Muḥammad ibn Mūsā al-K̲hwārizmī is the author of various general treatises ranging from mathematics to history and geography. His Arithmetic and his Algebra (written *c.* 820) greatly contributed to spreading the use of the Indian symbols and the solving of (linear and quadratic) equations, respectively, in the east and, through their Latin translation, in mediaeval Europe. He came to be known as 'the father of algebra' even though algebra existed before his time and he does not seem to have made any contribution. He nevertheless remains a leading figure in the promotion of this field. On him, pp. 118, 129, 245.

(9*b*) To the Egyptian Abū Kāmil (*c.* 890) we owe an elementary treatise on geometry, a short one on indeterminate equations of the first degree, and a sizable treatise on algebra which was partly translated into Latin (14th century) and Hebrew (15th century). Although he refers there to K̲hwārizmī as his predecessor, this treatise is of a notably higher level, with some applications to geometry (regular polygons constructible with compass and straightedge) and trade; it ends with recreational problems.[521] Abū Kāmil obviously had access to Greek sources, some of which are no longer extant (see the edition of the part of Diophant's *Arithmetica* extant in Arabic, pp. 9–10, 81–82). He is mentioned here on pp. 42n, 54n, 55, 63n, 77n, 82, 118.

(9c) Alī ibn Aḥmad al-Anṭākī (d. 987) is the author of a treatise entitled 'Commentary on the Arithmetic' (by which is meant Nicomachos's), of which only the third part has survived. The extant part must reproduce Greek material in translation: its second chapter just reproduces Mufaḍḍal ibn Thābit ibn Qurra's translation of an anonymous Greek text on magic squares (see p. 190), but without any mention of the source. Thus we may assume that our [183] — also found one century earlier in a work by al-Kindī (p. 229n)— was originally Greek.

(9d) The 10th-century astronomer Abū al-Ṣaqr al-Qabīṣī, versed in Ptolemy's *Almagest*, worked in particular on determining the distances and sizes of celestial bodies. See No. [123] and p. 119.

(9e) The 'Key to commercial transactions' (*Miftāḥ al-mu'āmalāt*) is an early-11th century work by the astronomer Muḥammad ibn Ayyūb Ṭabarī. As the title indicates (above, p. 246), it is a commercial arithmetic but, as seen on p. 24, it also includes some more theoretical questions.

(9f) 'Abd al-Raḥmān al-Khāzinī was a slave of Byzantine origin, given by his master an excellent education in science and philosophy. His most known work, the 'Balance of wisdom', was completed in 1121/22. We have used it in connection with weights problems, namely for our [14], [14'], [126]; see also pp. 25, 29n, 117–118.

(9g) A characteristic textbook for the training of civil servants is found in the MS. Paris Bibliothèque Nationale Arabic 4441, from the 16th-century. It contains information on law, weights and measures, and also elements of mathematics. The questions on the sharing-out of legacies are illustrated by various singular situations, reported here. See our [166], [168]–[170], [172], and also pp. 2n, 47n, 177n, 178n, 180n.

10. Other sources

(Mesopotamian) pp. 111, 116, 213, 215, 219, 220, 245. (Ancient Egyptian) pp. 112, 219n, 220. (Syriac) p. 127. (Russian) p. 113. (English) p. 113.

---

[521] In the introduction, Abū Kāmil begins by praising Khwārizmī's book but then goes on to claim that his own is superior and that the former may thus be disregarded (see our *Introduction to the history of algebra*, pp. 63–64, p. 67 in the French edition).

Bibliography

Abū Kāmil: كتاب فى الجبر والمقابلة (*Kitāb fi'l-jabr wa'l-muqābala*, Algebra) [photographic reproduction of the Arabic manuscript Istanbul Beyazıt Kara Mustafa Paşa 379]. Frankfurt 1986.

Abū'l-Wafā' Būzjānī: *See* Būzjānī.

A. Agostini: "Il *De viribus quantitatis* di Luca Pacioli", *Periodico di matematiche*, S. IV, 4 (1924), pp. 165–192.

Alcuin: *Propositiones ad acuendos iuvenes*, in *Patrologia, ser. latina*, ed. J. P. Migne (221 vol., Paris 1844–1864), vol. 101, coll. 667–676 [reproducing F. Forster's edition, Ratisbon 1777]; or M. Folkerts, "Die Alkuin zugeschriebenen Propositiones ad acuendos iuvenes", *Denkschriften der österreichischen Akademie der Wissenschaften, math.-naturwiss. Klasse*, 116/6 (1978).

Algorismus Ratisbonensis: *See* Vogel.

Anania S̲h̲irakatsi: *See* Kokian.

A. Anbouba: "Un mémoire d'al-Qabīsī (4e siècle H.) sur certaines sommations numériques", *Journal for the History of Arabic Science* 6 (1982), pp. 181–208.

Annales Stadenses: *See* Pertz.

G. Arrighi: *La matematica dell'Età di mezzo, scritti scelti*. Pisa 2004. *See also* F. Calandri, P. della Francesca, *Libro d'abaco*, Paolo dell'Abbaco.

El arte del alguarismo, ed. B. Caunedo del Potro & B. Córdoba de la Llave. Salamanca 2000.

Cl.-G. Bachet de Méziriac: *Problemes plaisans et delectables, qui se font par les nombres*. Lyons 1612 (other edition: 1624). *See also* Labosne.

F. Bartoli: *See* Sesiano.

B. Berlet: "Die Coß von Adam Riese", *Bericht über die Progymnasial- und Realschulanstalt zu Annaberg* 17 (1860).

Boëtius: *De institutione arithmetica libri duo, de institutione musica libri quinque*, ed. G. Friedlein. Leipzig 1867.

P. Borgi [P. Borghi]: *Nobel opera de arithmethica*. Venice 1484.

I. Bulmer-Thomas: *Selections illustrating the history of Greek mathematics* (2 vols). London 1939-1941.

P. Bungus [P. Bongo]: *Numerorum mysteria.* Bergamo 1591; reprinted 1599.

J. Buteo: *Logistica, quae & Arithmetica vulgò dicitur.* Lyons 1559.

Būzjānī: *The Arithmetic of Abu al-Wafa' al-Buzajani* (تاريخ علم الحساب العربى, 1), ed. A. S. Saidan. Amman 1971.

F. Calandri: *De arimethrica opusculum.* Florence 1491/2, 1518.

—— *Aritmetica, secondo la lezione del Codice 2669 (sec. XV) della Biblioteca Riccardiana di Firenze*, ed. G. Arrighi. Firenze 1969.

P. M. Calandri: *Tractato d'abbacho, dal Codice Acq. e doni 154 (sec. XV) della Biblioteca Medicea Laurenziana di Firenze*, ed. G. Arrighi. Pisa 1974.

G. Cardano: *Practica arithmetice, & mensurandi singularis.* Milan 1539. Reprinted in his *Opera omnia* (Lyons 1663, 10 vols), vol. IV, pp. 13–216.

E. Carruccio: "Cataldi", in *Dictionary of scientific biography*, 3 (1971), pp. 125–129.

Catalogue général des manuscrits des bibliothèques publiques de France, Départements, tome XXXIX (manuscrits de la Bibliothèque de Reims, II.1). Paris 1904.

N. Chuquet: *See* Marre.

H. Coxeter: *See* Rouse Ball.

M. Curtze: "Zur Geschichte des Josephspiels", *Bibliotheca mathematica*, N. F., 8 (1894), p. 116.

—— "Mathematisch-historische Miscellen, 6: Arithmetische Scherzaufgaben aus dem 14. Jahrhundert", *Bibliotheca mathematica*, N. F., 9 (1895), pp. 77–88.

—— "Eine Studienreise", *Centralblatt für Bibliothekswesen*, XVI, 6-7 (1899), pp. 257–306.

S. Deschauer: *Die große Arithmetik (...), eine anonyme Algorismusschrift aus der Endzeit des Byzantinischen Reiches.* Vienna 2014.

Diophant: *See* Tannery, Sesiano.

P. Duhem: *Etudes sur Léonard de Vinci* (3 vols). Paris 1906-1913.

W. van Egmond: *Practical mathematics in the Italian Renaissance: a catalog of Italian abbacus manuscripts and printed books to 1600.* Florence 1980.

A. Eisenlohr: *Ein mathematisches Handbuch der alten Aegypter* (2 vols: Commentar, Tafeln). Leipzig 1877.

L. Euler: "Solution d'une question curieuse qui ne paroit soumise à aucune analyse", *Mémoires de l'Académie Royale des Sciences et Belles-Lettres*, 15 (1759) [published 1766], pp. 310–337. Reprinted in his *Opera omnia* I/7, pp. 26–56 and in our *Euler et le parcours du cavalier*, pp. 216–243.

—— *Opera postuma mathematica et physica*, edd. P. H. Fuss & N. Fuss (2 vols). Saint Petersburg 1862.

—— *Opera omnia*. Zurich (*et al.*). 1911- .

E. Fellmann: *Leonhard Euler*. Reinbek bei Hamburg 1995.

P. (de) Fermat: *Varia opera mathematica*. Toulouse 1679.

—— *Oeuvres*, edd. P. Tannery & Ch. Henry (5 vols). Paris 1891-1922.

J. Fernel: *Cosmotheoria, libros duos complexa*. Paris 1528.

L. Fibonacci: *Scritti di Leonardo Pisano, I: Liber abbaci*, ed. B. Boncompagni. Rome 1857.

C. Flye Sainte-Marie: "Note sur un problème relatif à la marche du cavalier sur l'échiquier", *Bulletin de la Société mathématique de France* 5 (1877), pp. 144–150.

P. Forcadel: *L'Arithmeti(c)que* (3 vols). Paris 1557.

P. della Francesca: *Trattato d'abaco, dal Codice Ashburnhamiano 280* (...) *della Biblioteca Medicea Laurenziana di Firenze*, ed. G. Arrighi. Pisa 1970.

R. Franz: "Rösselsprung", *Schachzeitung* 2 (1847), pp. 341–343.

G. Galilei: *Discorsi e dimostrationi matematiche intorno à due nuove scienze*. [= *Opere, racc. da C. Manolessi* (Bologna 1655), vol. 2].

R. Gemma Frisius: *Arithmeticae practicae methodus facilis*. Paris 1545.

C. Gerhardt: *See* Planudes.

F. Ghaligai: *Pratica d'arithmetica* (...), *nuovamente rivista, & con somma diligenza ristampata*. Florence 1552.

Grand Dictionnaire universel du XIXᵉ siècle. Paris 1866-1879.

S. Günther: "Die quadratischen Irrationalitäten der Alten und deren Entwickelungsmethoden", *Abhandlungen zur Geschichte der Mathematik* 4 (1882), pp. 1–134.

H. Hankel: *Theorie der complexen Zahlensysteme*. Leipzig 1867.

T. Heath: *A history of Greek mathematics* (2 vols). Oxford 1921.

Hegesippus: *Hegesippi qui dicitur historiae libri V* [= *Corpus scriptorum ecclesiasticorum latinorum*, LXVI], ed. V. Ussani. Vienna 1932.

Heron of Alexandria: *Opera quae supersunt omnia*, edd. J. Heiberg, H. Schoene, L. Nix, W. Schmidt (5 vols). Leipzig 1899-1914 (vol. 6, Suppl., 1949).

T. von Heydebrand und der Lasa: "Notiz über ein altes, bei Janot in Paris gedrucktes Quartbändchen", *Schachzeitung* 2 (1847), pp. 317–320.

A. Hochheim: *Kâfî fîl Hisâb (Genügendes über Arithmetik) des* (...) *Alkarkhî*, I-III [3 Hefte]. Halle (Saale) 1878-1880.

J. Høyrup: *Jacopo da Firenze's 'Tractatus Algorismi' and early Italian abbacus culture*. Basel/Boston/Berlin 2007.

H. Hunger & K. Vogel: *Ein byzantinisches Rechenbuch des 15. Jahrhunderts* [= *Denkschriften der österreichischen Akademie der Wissenschaften, phil.-hist. Klasse, 78, 2*]. Vienna 1963.

K. Hunrath: *Die Berechnung irrationaler Quadratwurzeln vor der Herrschaft der Decimalbrüche*. Kiel 1884.

Iamblichos: *In Nicomachi arithmeticam introductionem liber*, ed. H. Pistelli. Leipzig 1894.

T. Ibel: *Die Wage im Altertum und Mittelalter*. Erlangen 1908.

Jacobus of Edessa: *See* Martin.

Jacopo da Firenze: *See* Høyrup.

C. F. de Jaenisch: *Traité des applications de l'analyse mathématique au jeu des échecs* (3 vols). Saint Petersburg 1862-1863.

Johannes Hispalensis: See *Liber mahameleth*.

Fl. Josephus: *Flavii Iosephi Opera omnia*, ed. E. Bekker (6 vols). Leipzig 1855-1856.

al-Karajī: *See* Hochheim, Woepcke.

N. Khanikoff [Ханыкоф]: "Analysis and extracts of كتاب ميزان الحكمة, Book of the balance of wisdom", *Journal of the American Oriental Society* 6 (1858-1860), pp. 1–128.

al-Khāzinī: كتاب ميزان الحكمة (*Kitāb mīzān al-ḥikma*). Hayderabad 1359/1940.

S. Kokian: "Des Anania von Schirak Arithmetische Aufgaben", *Zeitschrift für die deutschösterreichischen Gymnasien* 69 (1919-1920), pp. 112–117.

A. Labosne: *Problèmes plaisants & délectables qui se font par les nombres, par Claude-Gaspar Bachet, Sieur de Méziriac, troisième édition*. Paris 1874.

Liber augmenti et diminutionis: See Libri (vol. II, pp. 304–371).

Liber mahameleth (3 vols). Cham 2014.

G. Libri: *Histoire des sciences mathématiques en Italie* (4 vols). Paris 1838-1841.

Libro d'abaco, dal Codice 1754 (sec. XIV) della Biblioteca Statale di Lucca, ed. G. Arrighi. Lucca 1973.

A. van der Linde: *Quellenstudien zur Geschichte des Schachspiels*. Berlin 1881.

Livre de chiffres et de getz, nouvellement imprime. Lyons 1501.

E. Lucas: *Récréations mathématiques* (4 vols). Paris 1882-1894.

A. Maier: *Die Vorläufer Galileis im 14. Jahrhundert*. Rome 1949.

—— *Ausgehendes Mittelalter, I*. Rome 1964.

H. G. della Mantia: *Libro nel quale si tratta della maniera di giuocar'à scacchi, con alcuni sottilissimi partiti*. Turin 1597.

S. Maracchia: *Da Cardano a Galois. Momenti di storia dell'algebra*. Milan 1979.

A. Marre: "Notice sur Nicolas Chuquet et son Triparty en la science des nombres", *Bulletino di bibliografia e di storia delle scienze matematiche e fisiche* 13 (1880), pp. 555–659, 693–814.

—— "Appendice au Triparty en la science des nombres de Nicolas Chuquet parisien", *Bulletino di bibliografia e di storia delle scienze matematiche e fisiche* 14 (1881), pp. 413–460.

J. Martin: "L'Hexaméron de Jacques d'Edesse", *Journal asiatique*, 8ᵉ s., XI (1888), pp. 155–219, 401–490.

E.-E. Leon Mellema: *Arithmetique* (2 vols). Anvers 1582-1586.

M. Mersenne: *Cogitata physico mathematica, in quibus tam naturae quàm artis effectus admirandi certissimis demonstrationibus explicantur*. Paris 1644.

M. Moschopoulos: *See* Tannery, Sesiano.

Mufaḍḍal ibn Ṯhābit ibn Qurra: *See* Sesiano (*An ancient Greek treatise*).

J. Murdoch: "Infinity and continuity", in *The Cambridge history of later medieval philosophy* (Cambridge 1982), pp. 564–591.

H. Murray: *A history of chess*. Oxford 1913.

P. Muscarello: *Algorismus, trattato di aritmetica pratica e mercantile del secolo XV* (2 vols: I, reproduction of the MS. [now Pennsylvania Univ. LJS 27]; II, transcription). Verona 1972.

Nicomachos: *Nicomachi Geraseni Pythagorei introductionis arithmeticae libri II*, ed. R. Hoche. Leipzig 1866.

Ø. Ore: *Number theory and its history.* New York 1948.

N. Oresme: *Quaestiones super geometriam Euclidis*, ed. H. Busard. Leiden 1961.

J. Ozanam: *Recreations mathematiques et physiques* (2 vols). Paris 1694. [New edition, by M. Grandin, Paris 1724 (4 vols). New edition, by J.-E. Montucla, Paris 1778 (4 vols); English translation (*Recreations in mathematics and natural philosophy*), London 1803 (4 vols).]

L. Pacioli: *Summa de arithmetica, geometria, proportioni et proportionalità.* Venice 1494. [Other edition: Toscolano 1523.]

—— *De viribus quantitatis* [transcription of the MS. Bologna BU it. 250] by M. G. Peirani. Milan 1997.

Paolo dell'Abbaco: *Trattato d'aritmetica, secondo la lezione del Codice Magliabechiano XI,86 della Biblioteca Nazionale di Firenze*, ed. G. Arrighi. Pisa 1964.

R. Parker: *Demotic mathematical papyri.* Providence/London 1972.

F. Pellos: *Compendion de (lo) abaco.* Turin 1492.

J. Perott: "Sur une arithmétique espagnole du seizième siècle", *Bulletino di bibliografia e di storia delle scienze matematiche e fisiche* 15 (1882), pp. 163–169.

G. Pertz: *Monumenta Germaniae Historica, Scriptorum tomus XVI.* Hannover 1859.

Piero della Francesca: *See* Francesca.

M. Planudes: *Das Rechenbuch des Maximus Planudes*, ed. C. Gerhardt. Halle 1865.

al-Qabīṣī: *See* Anbouba, Sesiano.

Rascioni d'algorismo: *See* Vogel (*italienisches Rechenbuch*).

A. Ries(e): *Rechenbuch auff Linien unnd Ziphren, in allerley Handthierung, Geschäfften unnd Kauffmanschafft* (...). Frankfurt 1581.

E. de la Roche: *L'arismethique novellement composee.* Lyons 1520.

W. Rouse Ball & H. Coxeter: *Mathematical recreations & essays* (11th ed.). Cambridge 1947.

C. Rudolff: *Die Coß Christoffs Rudolffs. Mit schönen Exempeln der Coß durch Michael Stifel gebessert und sehr gemehrt.* Königsberg 1553/4.

A. Saidan: *See* Uqlīdisī.

W. Sartorius von Waltershausen: *Gauss zum Gedächtniss.* Stuttgart 1862.

H. Schubert: *Zwölf Geduldspiele.* Berlin 1895.

—— *Mathematische Mussestunden* (3 vols). Leipzig 1907.

J. Sesiano: *Books IV to VII of Diophantus' 'Arithmetica' in the Arabic translation attributed to Qusṭā ibn Lūqā.* New York 1982.

—— "Les problèmes mathématiques du *Memoriale* de F. Bartoli", *Physis* XXVI (1984), pp. 129–150.

—— "The appearance of negative solutions in mediaeval mathematics", *Archive for history of exact sciences* 32 (1985), pp. 105–150.

—— "A Treatise by al-Qabīṣī (Alchabitius) on arithmetical series", *Annals of the New York Academy of Sciences* 500 (1987), pp. 483–500.

—— "On an algorithm for the approximation of surds from a Provençal treatise", in *Mathematics from manuscript to print, 1300-1600* (Oxford 1988), pp. 30–55.

—— "Vergleiche zwischen unendlichen Mengen bei Nicolas Oresme", in *Mathematische Probleme im Mittelalter* (Wolfenbüttel 1996), pp. 361–378.

—— *Un traité médiéval sur les carrés magiques.* Lausanne 1996.

—— "Les carrés magiques de Manuel Moschopoulos", *Archive for history of exact sciences* 53 (1998), pp. 377–397.

—— *Une introduction à l'histoire de l'algèbre.* Lausanne 1999. [English translation: see below]

—— "Un recueil du XIIIe siècle de problèmes mathématiques", *SCIAMVS* 1 (2000), pp. 71–132.

—— "Une compilation arabe du XIIe siècle sur quelques propriétés des nombres naturels", *SCIAMVS* 4 (2003), pp. 137–189.

—— *Les carrés magiques dans les pays islamiques.* Lausanne 2004.

—— "Magic squares for daily life", in *Studies in the history of the exact sciences* [= Islamic philosophy, theology and science, LIV], Leiden 2004, pp. 715–734.

—— *An introduction to the history of algebra.* Providence 2009.

—— *Récréations mathématiques au Moyen Âge.* Lausanne 2014.

—— *Магические квадраты на средневековом Востоке*. Saint Petersburg 2014.

—— *Euler et le parcours du cavalier, avec une annexe sur le théorème des polyèdres*. Lausanne 2015.

—— *Magic squares in the tenth century. Two Arabic treatises by Anṭākī and Būzjānī*. Cham 2017.

—— *L'Arithmétique de Pamiers, traité mathématique en langue d'oc du XVe siècle*. Lausanne 2018.

—— *Magic squares, their history and construction from ancient times to AD 1600*. Cham 2019.

—— *An ancient Greek treatise on magic squares*. Stuttgart 2020.

G. Sfortunati: *Nuovo lume, libro de arithmetica*. Venice 1544/5.

F. Spinula: *De intercalandi ratione corrigenda, & de tabellis quadratorum numerorum à Pythagoreis dispositorum, διακόσμησις*. Venice 1562.

M. Stifel: *Arithmetica integra*. Nuremberg 1544. *See also* Rudolff.

Subtilitates: *See* Curtze (1895).

Ṭabarī: مفتاح المعاملات (*Miftāḥ al-mu'āmalāt*), ed. M. A. Riyāḥī. Tehran 1971.

G. Tagliente: *Componimento di arithmetica, (…) operetta (…) intitulata Tesauro universale* [various titles and editions]. Venice *et al.* (ed. used here:) 1525, 1547, 1554.

P. Tannery: "Le traité de Manuel Moschopoulos sur les carrés magiques, texte grec et traduction", *Annuaire de l'Association pour l'encouragement des études grecques en France* 20 (1886), pp. 88–118. Reprinted in Tannery's *Mémoires scientifiques*, IV, pp. 27–60.

—— *Diophanti Alexandrini Opera omnia cum Graecis commentariis* (2 vols). Leipzig 1893-1895.

—— *Mémoires scientifiques* (17 vols). Paris/Toulouse 1912-1950.

N. Tartaglia: *General trattato di numeri, et misure* (3 vols). Venice 1556-1560.

I. Thomas: *See* I. Bulmer-Thomas.

G. Toomer: *Ptolemy's Almagest*. London 1984.

J. Trenchant: *L'Aritmetique (…) departie en troys livres*. Lyons 1558. [Edition *reveuë & augmentee*: Lyons 1566 and later.]

P. Treutlein: "Das Rechnen im 16. Jahrhundert", *Abhandlungen zur Geschichte der Mathematik* 1 (1877), pp. 1–100.

J. Tropfke (*et al.*): *Geschichte der Elementarmathematik* (4th ed.). Berlin 1980.

al-Uqlīdisī: الفصول فى الحساب الهندى (*al-Fuṣūl fi'l-ḥisāb al-hindi*), ed. A. S. Saidan. Amman 1973. [Translation (by Saidan): *The Arithmetic of al-Uqlīdisī*, Dordrecht 1978].

K. Vogel: "Zur Geschichte der linearen Gleichungen mit mehreren Unbekannten", *Deutsche Mathematik* 5 (1940), pp. 217–240.

———— *Die Practica des Algorismus Ratisbonensis*. Munich 1954.

———— *Vorgriechische Mathematik*, I–II. Hannover/Paderborn 1959.

———— *Ein italienisches Rechenbuch aus dem 14. Jahrhundert*. Munich (Deutsches Museum) 1977.

———— *Ein byzantinisches Rechenbuch des frühen 14. Jahrhunderts* [= Wiener byzantinische Studien, VI]. Vienna 1968.

See also Hunger.

J. Widman: *Behende und hubsche Rechenung auff allen Kauffmanschafft-t(en)*. Leipzig 1489 (also: Pforzheim 1508, Augsburg 1526).

E. Wiedemann: "Beiträge zur Geschichte der Naturwissenschaften XIV, 4: Über das Schachspiel und dabei vorkommende Zahlenprobleme", *Sitzungsberichte der Physikalisch-Medizinischen Sozietät zu Erlangen* 40 (1908), pp. 41–58. Reprinted in his *Aufsätze*, I, pp. 440–457.

———— *Aufsätze zur arabischen Wissenschaftsgeschichte* (2 vols). Hildesheim 1970.

H. Wieleitner: "Zur Geschichte der unendlichen Reihen im christlichen Mittelalter", *Bibliotheca mathematica*, 3. F., 14 (1914), pp. 150–168.

F. Woepcke: *Extrait du Fakhrî*. Paris 1853.

A. Yushkevitsh [А. Юшкевич]: *История математики в России до 1917 года*. Moscow 1968.

Index

Addenda

Euler und die Rösselsprungaufgabe

§ 1. Gewöhnliches Schachbrett

①[1] Die Erfindung von Wegen, die durch alle Felder des gewöhnlichen 8×8 Schachbretts (Abb. 1) gehen würden, ist von Euler behandelt worden. Sein Aufsatz, 1766 erschienen, wurde am 2. März 1758 vor der Berliner Akademie gelesen. Das Problem erwähnte er schon in einem am 26. April 1757 an Christian Goldbach geschickten Brief, in welchem er auf seine 'neulich' durchgeführten Untersuchungen hinwies.[2] Nun sind große Teile dieser Untersuchungen im sechsten seiner *Notizbücher*, jetzt in der Petersburger Abteilung der russischen Akademie der Wissenschaften, erhalten geblieben (Abb. 2 ist ein Beispiel davon).[3] Aus Eulers Titel seines Aufsatzes, *Solution d'une question curieuse qui ne paroit soumise à aucune analyse*, geht hervor, daß er diese Aufgabe als 'sonderbar' ansah, welcher man von vornherein kein eigentliches Lösungsverfahren zuschreiben konnte. Daraus erklärt es sich, daß er in seinem *Notizbuch* anscheinend regellos vorgeht: aus allerlei, teils erfolgreichen teils gescheiterten Versuchen wird er allmählich Bedingungen und Einschränkungen herausfinden. Wir werden im nächsten *addendum* sehen (S. 328), wie er ähnlicherweise aus zahlreichen mühsamen Entwicklungen irrationaler Quadratwurzeln in Kettenbrüche zur Kenntnis ihrer Eigentümlichkeiten gelangte.

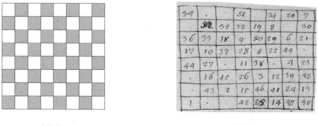

Abb. 1   Abb. 2

---

[1] Euler's studies dealing with the knight's move on the chessboard began in the years 1756-1757 and ended with his article of 1758, published in 1766. His preparatory work is mainly extant in the sixth volume of his notebooks (*Notizbücher*), preserved in St. Petersburg, and has recently been analyzed and published by us with a reproduction of Euler's article and manuscript notes. From these purely empirical attempts, some successful some others not, he finally arrived at the conditions for forming complete, sometimes also closed, paths on this and other boards.

[2] Fuss, *Correspondance*, I, S. 654–655.

[3] Untersucht, mit Nachdruck der betreffenden Seiten, sowie des Aufsatzes, in unserem *Euler et le parcours du cavalier*.

②[4] Wir wissen heute von früheren Arbeiten über diese Aufgabe; einige Beispiele haben wir im Kap. XIII oben gesehen. Die vor Eulers Zeit erreichten Lösungen waren zumeist empirisch. Zuerst hatte man festgestellt, welches die ungünstigen Felder sind. Nehmen wir nämlich in einem leeren Schachbrett ein unweit von der Mitte gelegenes Feld (•, Abb. 3). Von ihm aus kann man mittels eines Rösselsprunges acht Felder erreichen (×). Dies ist die günstigste Lage. Nähert man sich aber von den Seiten, so verringert sich die Anzahl der Möglichkeiten auf sechs, vier, und sogar zwei in den Winkeln, nämlich Zugang und Ausgang. Aus der verminderten Erreichbarkeit all dieser Felder entstand ein Grundprinzip für die Versuche: am liebsten sollte man den Verlauf in den Rändern anfangen, um diese ungünstigen Stellen zuerst zu füllen, und erst dann zur Mitte übergehen. Ein gutes Beispiel eines solchen kreisförmigen Ganges ist einer persischen Handschrift entnommen (Abb. 4).[5]

46	7	20	33	44	5	18	31
21	34	45	6	19	32	43	4
8	47	60	55	62	53	30	17
35	22	63	52	59	56	3	42
48	9	58	61	54	41	16	29
23	36	51	64	57	28	13	2
10	49	38	25	12	15	40	27
37	24	11	50	39	26	1	14

Abb. 3   Abb. 4

③[6] So einfach ist es aber in der Praxis nicht. Rückt man so versuchsweise vor, so wird man im allgemeinen zwischen 50 und 60 Feldern durchqueren, dann aber mit unerreichbaren Feldern bleiben. Und hier war

---

[4] Some previous attempts, already mentioned (Ch. XIII), were the result of a few general observations. First of all, it is readily seen that whereas the middle cells of an empty chessboard are the most accessible, such is not the case for those near the edges and, particularly, in the corners (Fig. 3). Thus a Persian MS. deals first with these less accessible cells by establishing a circular path to cover that part of the board before moving towards the middle (Fig. 4).

[5] MS. Astan Qods 12167, fol. 189; s. *Euler et le parcours du cavalier*, S. 154.

[6] This empirical procedure does not, however, guarantee a rapid solution in view of the time taken by the attempts to fill all the less accessible cells. Following a suggestion by Louis Bertrand, then in Berlin, Euler established a rule to set a path which, while keeping the cells already attained on the way, changed its end in order to be able to add further cells (Fig. 5). Repeated applications of this rule will indeed make it possible to complete the path, and even close it (with thus the last cell distant by a knight's move from the initial one).

die Einwirkung Eulers Aufsatzes entscheidend: durch wiederholte Umformungen des vorhandenen Weges wird man zu ihm schrittweise neue Felder hinzufügen können. Aus seinen handschriftlichen Notizen geht hervor, daß Euler in den Jahren 1756-1757 begann, sich mit der Aufgabe des Rösselsprunges zu befaßen.[7] Die Grundlage seiner Untersuchungen entstand aus einer Bemerkung des Genfers Louis Bertrand, 1752 in Berlin eingetroffen, gerade um dort Euler zu besuchen. Diese Bertrandsche, zu 'Eulerscher Regel' gewordene Bemerkung, ist die folgende (Abb. 5). Nehmen wir an, wir hätten einen unvollständigen Weg, der durch die Felder $1, \ldots, m$ ($m < 64$) hindurchgehe. Mitten im Weg sei das Feld x ($x \neq m - 1$), mit welchem das Endfeld m durch einen Rösselsprung verknüpft werden könne. Dann ist der Weg $1, \ldots, x, m, m - 1, \ldots, x + 1$, ein neuer Weg, der genau dieselben Felder wie der vorige durchquert, aber in einer verschiedenen Stelle endet. Kann man von diesem neuen Ende kein unbesetztes Feld erreichen, so wird man die Regel nochmals oder mehrmals anwenden, um damit einen anderen Endpunkt zu finden, ohne dabei je etwas vom früheren Ergebnis zu verlieren. Dies wird man solange wiederholen, bis man zum Schluß einen vollständigen Weg erreicht hat. Von einem vollständigen Weg kann man weiter, durch Anwendung derselben Regel, einen geschlossenen Weg erreichen.

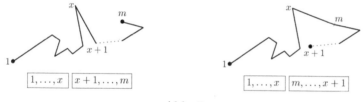

Abb. 5

④[8] Hier ist ein Beispiel aus Eulers Handschrift (Abb. 2 und, mit Buchstaben in den leeren Feldern, Abb. 6).[9] Eine Verlängerung des Weges ist weder vom Anfangsfeld 1 noch vom Endpunkt 52 durchführbar, und zwölf Felder bleiben leer. Man merke von vornherein in Abb. 6, daß, abgesehen von j, gewisse dieser Felder miteinander verknüpft werden können: a-b, c-e, f-i, k-l. Durch fünf erfolgreiche Anwendungen seiner Regel wird damit Euler imstande sein, seinen ursprünglichen Weg auf das ganze

[7] S. *Euler et le parcours du cavalier*, S. 6–7.

[8] An example from his manuscript notes, with initially twelve empty cells (Fig. 2 & 6), will help us to understand the procedure. Note that Euler does not ever depart from the rule there; he therefore does not consider that many empty cells, namely groups of cells in even number which happen to be connected to one another, could have been directly integrated into the existing path; see *remark* p. 294.

[9] *Notizbuch*, fol. 232^{r}; *Euler et le parcours du cavalier*, S. 28–32.

Schachbrett zu erstrecken — es sei jedoch bemerkt, daß wir hier nur die erfolgreichen Änderungen darlegen: in der Handschrift führen manche seiner Anwendungen der Regel zu keinem brauchbaren Ergebnis.

34	*j*	*b*	51	*f*	31	20	7
a	<u>52</u>	35	32	19	8	*g*	30
36	33	18	9	50	29	6	21
17	10	37	28	5	22	49	*h*
44	27	*e*	11	38	*i*	4	23
d	16	45	26	3	12	39	48
l	43	2	15	46	41	24	13
<u>1</u>	*c*	*k*	42	25	14	47	40

Abb. 6

39	*j*	<u>57</u>	22	*f*	42	53	12
56	21	38	41	54	11	*g*	43
37	40	55	10	23	44	13	52
20	9	36	45	14	51	24	*h*
29	46	<u>1</u>	8	35	*i*	15	50
2	19	28	47	16	7	34	25
l	30	17	4	27	32	49	6
18	3	*k*	31	48	5	26	33

Abb. 7

(1) Einschließung von *a*-*b*.

Von 52 aus kann man vier Stellen erreichen, von *a* aus drei. Den ersteren ist hier ihr unmittelbarer Nachfolger im betrachteten Weg beigefügt (wie 18 zu 17): da er dann zum Endpunkt des neuen Weges wird, kann man sogleich nachprüfen, ob er mit *a* verbunden werden kann.

$$52 \quad | \quad 17_{18} \quad 37_{38} \quad 9_{10} \quad 51$$
$$a \quad | \quad b \quad \quad 18 \quad \quad 10.$$

Eine der beiden Möglichkeiten ist also

$$1, \ldots, 17, 52, \ldots, 18, a, b.$$

(2) Einschließung von *c*-*e*.

Betrachten wir, der Anschaulichkeit halber, den umgekehrten Weg:

$$b, a, 18, \ldots, 52, 17, \ldots, 1.$$

Da

$$1 \quad | \quad 16_{15} \quad 2$$
$$c \quad | \quad d \quad \quad 45 \quad 15,$$

ermittelt man sofort die Lösung:

$$b, a, 18, \ldots, 52, 17, 16, 1, \ldots, 15, c, d, e.$$

Euler kehrt den Weg erneut um, und führt, wiederum der Anschaulichkeit halber, eine neue Numerierung ein; damit erhält er den neuen Weg $1, \ldots, 57$ der Abb. 7.

(3) Einschließung von *f*-*i*.

Betrachten wir mit ihm die Lage bei 1 (wie früher gesehen hat man um so mehr Möglichkeiten, daß man näher dem Mittelpunkt des Schachbretts steht). Nun ist, für den umgekehrten Weg $57, \ldots, 1$,

$$1 \mid 10_9 \quad 14_{13} \quad 16_{15} \quad 4_3 \quad 30_{29} \quad 2 \quad 20_{19} \quad 40_{39}$$
$$i \mid 13 \quad h \quad 25 \quad 49 \quad 27 \quad 47 \quad 45 \quad 23.$$

Aus der einzigen Wahl wird

$$57, \ldots, 14, 1, \ldots, 13, i, h, g, f.$$

(4) Einschließung von j.

Betrachten wir als Ende des Weges 57, das etwas näher bei j liegt:

$$57 \mid 54_{55} \quad 10_9 \quad 40_{41} \quad 56$$
$$j \mid 41 \quad 55 \quad 37.$$

Es gibt zwei Möglichkeiten, darunter

$$f, g, h, i, 13, \ldots, 1, 14, \ldots, 54, 57, \ldots, 55, j.$$

Führt man (wie es Euler tut) eine neue Numerierung ein, so ergibt sich Abb. 8.

43	62	59	26	1	46	57	6
60	25	42	45	58	7	2	47
41	44	61	8	27	48	5	56
24	9	40	49	18	55	28	3
33	50	17	10	39	4	19	54
16	23	32	51	20	11	38	29
l	34	21	14	31	36	53	12
22	15	k	35	52	13	30	37

Abb. 8

35	64	61	8	33	38	59	28
62	9	34	37	60	27	32	39
51	36	63	26	7	40	29	58
10	25	50	41	16	57	6	31
43	52	17	24	49	30	15	56
18	11	42	53	14	23	48	5
1	44	13	20	3	46	55	22
12	19	2	45	54	21	4	47

Abb. 9

(5) Einschließung von k-l.

Euler versucht es zuerst mit den beiden Endpunkten des neuen Weges $1, \ldots, 62$. Da aber

$$62 \mid 45_{46} \quad 61 \quad 41_{42}$$
$$1 \mid 42_{41} \quad 8_7 \quad 48_{47} \quad 2,$$

während

$$l \mid 50 \quad 32 \quad k$$
$$k \mid l \quad 23 \quad 51 \quad 31,$$

bietet sich keine Möglichkeit an; so bleibt es ihm nur übrig, einen anderen Endpunkt zu wählen. Mit $62, \ldots, 42, 1, \ldots, 41$ und zwei Umformungen gelangt Euler zum Ziel. So bekommt er nacheinander (da $41 \mid 50_{49}$ und $49 \mid 32_{31}$)

$$62, \ldots, 50, 41, \ldots, 1, 42, \ldots, 49,$$
$$62, \ldots, 50, 41, \ldots, 32, 49, \ldots, 42, 1, \ldots, 31, k, l.$$

Der vollständige Weg, nach Umkehrung und einer neuen Numerierung, ist derjenige der Abb. 9.

Bemerkung. Man könnte eigentlich die meisten dieser Felder durch einen Umweg unmittelbar anschließen: *a-b* (s. Abb. 6) zwischen 9 und 10 oder 18 und 19, *f-i* zwischen 28 und 29, *k-l* zwischen 26 und 27. Solche Vereinfachungen nimmt Euler nicht in Betracht, absichtlich oder nicht, denn er bleibt in seiner Handschrift bei der strikten Einhaltung der Regel.

Eine einfache neue Anwendung der Regel wird dann Euler erlauben, den offenen Weg in einen geschlossenen zu verwandeln: es genügt nämlich, $1, \ldots, 51, 64, \ldots, 52$ zu betrachten. Man merke nebenbei, daß der ursprüngliche Teilweg (Abb. 6) nicht mehr erkennbar ist, da gar keine Zelle ihre Zahl beibehalten hat.

§2. Andere quadratische Bretter

	$n=1$	2	3	4	5	6	7	8	9	10	...	$2k-1$	$2k$	...
$m=1$	—	—	—	—	—	—	—	—	—	—	...	—	—	...
2	•	—	—	—	—	—	—	—	—	—	...	—	—	...
3	•	•	—	×	—	—	×	×	×	⊗	...	×	⊗	...
4	•	•	•	—	×	×	×	×	×	×	...	×	×	...
5	•	•	•	•	×	⊗	×	⊗	×	⊗	...	×	⊗	...
6	•	•	•	•	•	⊗	⊗	⊗	⊗	⊗	...	⊗	⊗	...
7	•	•	•	•	•	•	×	⊗	×	⊗	...	×	⊗	...
8	•	•	•	•	•	•	•	⊗	⊗	⊗	...	⊗	⊗	...
9	•	•	•	•	•	•	•	•	×	⊗	...	×	⊗	...
10	•	•	•	•	•	•	•	•	•	⊗	...	⊗	⊗	...
...	...	...	...	...	...	...	...	...	...	...	...	...	...	...
$2k-1$	•	•	•	•	•	•	•	•	•	•	...	×	⊗	...
$2k$	•	•	•	•	•	•	•	•	•	•	...	•	⊗	...
...	...	...	...	...	...	...	...	...	...	...	...	...	...	...

Abb. 10

①[10] Während in dem gedruckten Aufsatze nur einige Beispiele anderer

---

[10] Considering in his *Notizbuch* various examples of square and rectangular figures, Euler will be able to determine the restrictions for obtaining a complete (×) or closed (⊗) path on such boards; see Fig. 10. This will of course depend on size and shape of the board; an obvious prerequisite for a closed path is that the number of cells be even since only two cells of a different colour may be connected.

rechtwinkligen Bretter betrachtet werden, sind diese Fälle in seiner Handschrift eingehender untersucht.[11] Dies wird uns zum Verständnis verschiedener Einschränkungen führen. Beim gewöhnlichen Schachbrett kann man zwar stets volle und geschlossene Wege erreichen, nicht aber bei anderen Brettern: manchmal ist ein geschlossener Weg unmöglich, gelegentlich kann man nicht einmal einen vollständigen Weg finden. Die Einschränkungen für quadratische und rechteckige Bretter ersieht man aus Abb. 10 ($\times$ steht für offene Wege, $\otimes$ für geschlossene). Da sich aber in seinem *Notizbuch* gar keine Theorie gibt, ist es aus Eulers erfolgreichen oder mißlungenen Versuchen, daß sich die Möglichkeitsbedingungen allmählich ergeben werden.

Die Diagonale in Abb. 10 zeigt uns, daß in einem Quadrat der Ordnung n ein vollständiger Weg nur dann möglich ist, falls $n \geq 5$, dazu auch geschlossen, falls n gerade ist. Letzteres ist leicht begreifbar: da sich bei jedem Sprung die Feldfarbe wechselt, sollten für einen geschlossenen Weg die Farben des Anfangs- und Endfeldes verschieden sein; dies kann aber nur bei gerader Felderzahl n^2 vorkommen, wo dann die Reihenfolge der Paare schwarz/weiß ganzzahlig ist.

②[12] Quadrate der Ordnungen $n = 3$ und $n = 4$.

13	4	11	6
10	7	14	3
1̲	12	5	8
	9	2	15̲

Abb. 11

Beim 3×3 Quadrat ist offensichtlich, daß das Mittelfeld leer bleiben muß.

Beim 4×4 Quadrat wird stets ein Feld in einem Winkel leer bleiben, wie Euler in seinem Aufsatze (S. 332) feststellt: *Car, de quelque maniere qu'on s'y prenne, il restera toujours une case angulaire vuide* (= vide)*; et on s'appercevra bientôt que toutes les transformations qu'on puisse faire ne sont pas capables de la remplir.* Tatsächlich sind einige Versuche in seinem *Notizbuch* erhalten.[13] Man kann es übrigens leicht nachprüfen:

---

[11] *Euler et le parcours du cavalier*, S. 75–126.

[12] Neither the 3×3 nor the 4×4 square may be completed: in the first case the central cell is inaccessible, in the second case it is one of the corner cells, this having to do with the limited number of connections to them.

[13] Fol. 231^{v}; *Euler et le parcours du cavalier*, S. 75–76.

jedes Winkelfeld besitzt bloß einen Zugang und einen Ausgang, die alle im mittleren Quadrat liegen. Daraus folgt, daß nur drei Ecken können besetzt werden, davon eine, oder zwei, als Anfangs- oder Endfeld, und eine andere in Durchgang, womit die gegenüberliegende leer bleiben muß (Abb. 11).

③[14] Quadrate der Ordnungen $n = 5$ und $n = 6$.

Am ausgedehntesten ist in der Handschrift die Untersuchung der Quadrate mit $n = 5$ und $n = 6$, die jetzt vollständige Wege annehmen und, was das zweite anbelangt, geschlossene. Euler versucht es auch, Mitglieder gewisser Klassen abzuzählen.

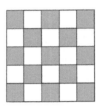

23	10	5	18	<u>25</u>
14	19	24	11	6
9	22	13	4	17
20	15	2	7	12
<u>1</u>	8	21	16	3

23	f	a	D	25
H	C	24	g	b
e	22	13	4	E
B	G	2	c	h
1	d	A	F	3

Abb. 12   Abb. 13   Abb. 14

Aus Abb. 12 ersieht man, daß die Anzahlen der schwarzen und weißen Felder im Falle $n = 5$ verschieden sind: es gibt 13 schwarze und 12 weiße. Um einen vollständigen Weg zu erhalten, muß man also wohl in einem schwarzen Feld anfangen, und (daher) mit einem schwarzen enden. Allgemein, gibt es in einer Figur eine ungerade Felderanzahl, so ist weder der Anfangspunkt beliebig wählbar noch kann, wie schon gesagt, der Weg geschlossen werden. Für einen vollständigen Weg im Falle $n = 5$ muß dazu mindestens ein Ende in einer Ecke liegen; denn jede Ecke hat einen Zugang und einen Ausgang, und diese Felder sind für zwei aufeinanderfolgenden Ecken gemeinsam. Weiter, wie Euler nachprüft, mit dem Ausgangspunkt in einer Ecke kann sich das andere Ende in jedem anderen schwarzen Feld befinden; er untersucht nämlich eingehend die verschiedenen Lagen der Endpunkte:[15]

---

[14] The 5 × 5 square, with its odd number of cells, and thus one more black cell, cannot have a closed path, nor is the point of departure arbitrary since start and end must be in a black cell. Euler then attempted in the *Notizbuch* to classify the paths according to the point of departure. He also does it for the 6 × 6 square, which is the first allowing both closed paths and a free choice of the starting point. Part of Euler's considerations, written on a separate sheet, are now lost. Indeed, there are several indications in the extant manuscript that it does not include all his early trials (below, n. 19).

[15] *Euler et le parcours du cavalier*, S. 76–88.

— liegt zuerst der Endpunkt in der Mitte, so zeigen Eulers *Variationes*, oder Umformungen, daß er in jedwede andere schwarze Zelle verlegt werden kann;

— liegt er insbesondere in derselben Randreihe wie der Ausgangspunkt, und zwar in der nächsten (schwarzen) Zelle, so erlaubt dies eine Zusammensetzung solcher Quadrate (s. unten Abb. 18, ähnlicher Fall);

— liegen Anfangs- und Endpunkt in gegenüberliegenden Ecken, so kann man ein symmetrisches Quadrat bilden, in welchem dann die Summe gegenüberliegender Felder $n^2 + 1 = 26$ beträgt. Euler findet, daß es acht verschiedene solche Anordnungen gibt, davon die der Abb. 13. Die erscheinen dann alle im gedruckten Aufsatz, werden aber in der Handschrift mit Hilfe der Abb. 14 hergestellt.

1̲	18	9	24	3	16
10	23	2	17	8	25
19	3̲6̲	11	6	15	4
22	29	20	33	26	7
35	12	31	28	5	14
30	21	34	13	32	27

Abb. 15

Während es im Aufsatz nur ein Beispiel eines Quadrates der Ordnung $n = 6$ gibt, wird diesem Falle in der Handschrift einen geräumigen Platz zugeteilt.[16] Hier kann der Weg nicht nur vollständig sein, sondern auch geschlossen und dazu von irgendeiner vorgegebenen Zelle ausgehen, wie Euler durch wiederholte Versuche feststellt. Vom Quadrate der Abb. 15 leitet er, unter Beibehaltung der Lage der Ausgangsstelle 1, vierundsechzig verschiedene Wege (*permutationes huius quadrati*); woraus unter anderem ersichtlich wird, daß der Ankunftspunkt (und daher Anfangspunkt) tatsächlich in jedweder Zelle verschiedener Farbe sein kann. Euler versucht auch, die Möglichkeiten neuer geschlossenen Wege von einem gegebenen Ausgangsfeld abzuzählen. Dies tut er zuerst von der Zelle für 20 aus, dann von derjenigen für 11. Für das erstere findet er sechs Wege, für das zweite 'sehr viele' (*plurimi casus obtinentur, ut ex charta adiecta videri licet*). Da das beigefügte Blatt verloren gegangen ist, besitzen wir heute sein Ergebnis nicht, und wissen übrigens auch nicht genau, was er damit beabsichtigte. Dies zeigt mindestens, daß Euler

---

[16] *Euler et le parcours du cavalier*, S. 88–106.

sich im Laufe seiner Betrachtungen folgende Frage gestellt hat: wieviele verschiedene Wege kann man mit Hilfe der Regel für ein vorgegebenes Ausgangsfeld bilden?

§ 3. Rechteckige Figuren

Aus Abb. 10 ersieht man, daß, für ein Rechteck mit Breite m und Länge n $(n > m)$, von $m = 5$ her Wege stets möglich sind, und zwar offene falls die Gesamtzahl der Felder ungerade ist, und geschlossene im Falle gerader Felderzahl. Sonderfälle sind $m = 3$ und $m = 4$.

①[17] Rechtecke der Breite $m = 3$ und der Länge $n \geq 4$.

Der Aufsatz gibt Beispiele für die Längen $n = 4$ und $n = 7$ ($n = 5$ und $n = 6$ sind unmöglich); im *Notizbuch* werden die Fälle $n = 4$ und $n = 9$, samt ihrer Ableitung, behandelt (keine Versuche für $n = 5$ und $n = 6$). Merke, daß ein Rechteck der Länge $n = 8$ aus demjenigen für $n = 4$ leicht ermittelbar ist.[18]

Was erstens den Fall $n = 4$ anbelangt, so gibt der Aufsatz vier Beispiele, deren (übrigens leichte) Herstellung handschriftlich vorliegt. Anfangs- und Endpunkt sind dabei stets in den Rändern: wäre der eine in der Mitte, so müßte der nächste Schritt in einer Ecke sein, und die vertikal benachbarte Ecke bliebe dann entweder leer oder wäre eine Sackgasse.

Für den Fall $n = 7$ gibt es zwei Beispiele (offener) Wege im Aufsatz, die sich eines vom anderen ableiten lassen. Das *Notizbuch* hat kein Beispiel. Daraus sollte kein Schluß gezogen werden: man erinnere sich, daß nicht alle handschriftlichen Versuche Eulers erhalten sind.[19]

Was schließlich den Fall $n = 9$ anbelangt, so gibt das *Notizbuch* ein Beispiel (Abb. 16), das dann verschiedenen Umformungen unterworfen wird.

---

[17] For a rectangle with width $m = 3$ and length n, complete paths are given for $n = 4$ and $n = 7$ in the article, and set up for $n = 4$ and $n = 9$ in the manuscript ($n = 5$ and $n = 6$ are impossible). In the article Euler tells us that with $m = 3$ a closed path is impossible. As a matter of fact, it is possible for $m \geq 10$ and $m \times n$ even. Euler was close to finding it: from his example for $n = 9$ a closed path for $n = 18$ can be constructed.

[18] *Euler et le parcours du cavalier*, S. 111.

[19] *Euler et le parcours du cavalier*, S. 7; oder hier, S. 297, 298, 301, 303, 304, 308, 309.

11	14	9	24	<u>27</u>	18	5	2	21
8	25	12	15	6	23	20	17	4
13	10	7	26	19	16	3	22	<u>1</u>

Abb. 16

Mitten in seinen Beispielen von Rechtecken der Breite $m = 3$ schiebt Euler in seinen Aufsatz eine allgemeine Bemerkung ein: *Si la largeur contient trois cases, et la longueur 5 ou 6, il est impossible de les parcourir; mais, donnant à la longueur 7 ou plusieurs cases, on pourra réussir, pourtant sans rentrer* — also: Wege sind bei Rechtecken der Breite $m = 3$ und Längen $n = 4$ und $n \geq 7$ wohl möglich, bleiben aber immer offen.

Hier liegt Eulers einzige fehlerhafte Aussage, denn für $n \geq 10$ und gerade Felderzahl können geschlossene Wege gebildet werden. Euler hätte es übrigens selber aus seinen Versuchen feststellen können. Betrachten wir nämlich sein oben angegebenes, handschriftlich erhaltenes, Rechteck für $n = 9$. Unter den Umformungen in seiner Handschrift findet man

$$1, \ldots, 20, 27, \ldots, 21,$$

dessen Ergebnis wie in Abb. 17 aussehen würde:

11	14	9	24	21	18	5	2	<u>27</u>
8	23	12	15	6	25	20	17	4
13	10	7	22	19	16	3	26	<u>1</u>

Abb. 17

Nun kann man aus diesem Rechteck und einem ähnlichen, aber mit seinen um 27 vergrößerten Zahlen, einen geschlossenen Weg bilden (Abb. 18). Hier ist $n = 18$, aber mit $n = 10$ läßt sich schon ein Beispiel angeben.[20]

11	14	9	24	21	18	5	2	27	<u>54</u>	29	32	45	48	51	36	41	38
8	23	12	15	6	25	20	17	4	31	44	47	52	33	42	39	50	35
13	10	7	22	19	16	3	26	<u>1</u>	28	53	30	43	46	49	34	37	40

Abb. 18

②[21] Rechtecke der Breite $m = 4$ und der Länge $n \geq 5$.

---

[20] *Euler et le parcours du cavalier*, S. 113.

[21] Examples of rectangles with width $m = 4$ are found both in the article ($n = 5$)

Es gibt im *Notizbuch* und im Aufsatz je ein Beispiel des Rechtecks 4×5, ohne weitere Umformungen in der Handschrift.

Für das Rechteck 4×6 wird in der Handschrift ein Teilweg gebildet (Abb. 19), dessen freigebliebene Zelle (unser a) durch die Umformung

$$5, \ldots, 1, 6, 19, \ldots, 23, 18, \ldots, 7, a$$

eingegliedert wird, worauf der jetzt vollständige (*plenus*) Weg neu numeriert wird (Abb. 20). Verschiedene Versuche müßen dann Euler überzeugt haben, daß eine Schließung des Weges nicht möglich ist: Anfangs- und Endpunkt bleiben stets in den (längeren) Rändern. Anscheinend ist hier die Unmöglichkeit aus unfruchtbaren Versuchen abgeleitet worden.

14	19	8	3	12	<u>23</u>
7	4	13	18	9	2
20	15	6	11	22	17
5	a	21	16	<u>1</u>	10

16	7	22	3	18	11
23	2	17	12	21	4
8	15	6	19	10	13
<u>1</u>	<u>24</u>	9	14	5	20

Abb. 19   Abb. 20

Für die Rechtecke 4×7 und 4×8 sind vollständige, nicht aber geschlossene, Wege möglich. Ein Beispiel des ersten, mit nebeneinanderliegenden Endpunkten, findet man im *Notizbuch*. In beiden Schriften gibt es zahlreiche Beispiele von Rechtecken 4×8, da dies auch ein Mittel zur Herstellung geschlossener Wege auf dem gewöhnlichen Schachbrett ist.[22]

③[23] Rechtecke der Breite $m = 5$ und der Länge $n \geq 6$.

Im Aufsatz wird nach der Erwähnung der Fälle $m = 3$ und $m = 4$ ausdrücklich bemerkt, daß geschlossene Wege erst von $m = 5$ an möglich sind (s. aber oben, S. 299): *Jusqu'ici les routes rentrantes en elles-mêmes ne peuvent pas avoir lieu; mais, donnant 5 cases à la largeur, et 6 à la longueur, on pourra aussi remplir cette condition, de même que dans tous les autres rectangles dont le nombre des cases est pair pourvu qu'il n'y ait pas moins de 5 cases dans un côté.*

Euler untersucht im *Notizbuch* den Fall 5×6 eingehend. Nacheinander werden verschiedene Teilwege mit, anfänglich, acht, sieben, drei leeren Feldern behandelt. Der letzte (Abb. 21) wird mit

---

and in the manuscript ($n = 5$, $n = 6$, $n = 7$; $n = 8$ is the half-chessboard, dealt with by and before Euler, see above pp. 186–188), with failed attempts in the case $n = 6$ to close the path.

[22] Siehe etwa oben, S. 188.

[23] There is just one example of a rectangle with width $m = 5$ in the article ($n = 6$), with a closed path; the preliminary steps for its construction are in the manuscript.

$$1, \ldots, 16, 27, \ldots, \ldots, 17, a, b, c$$

vervollständigt, dann mit

$$1, \ldots, 9, c, b, a, 17, \ldots, 27, 16, \ldots, 10$$

geschlossen. Das Beispiel des Aufsatzes ist dasselbe, nur wird 1 anstelle von 2 gesetzt, womit der Weg mit 29 (statt 30, 10 hier oben) beginnt und 30 (statt 1) endet. Ausgangsfeld in einer Ecke wird anscheinend vorzogen.

4	25	18	15	6	23
17	10	5	24	19	14
26	3	16	9	22	7
11	*a*	<u>1</u>	20	13	*c*
2	<u>27</u>	12	*b*	8	21

Abb. 21

Die nächsten Fälle, mit $n > 6$, werden nicht untersucht, sind ja nicht schwierig zu bilden.[24]

④[25] Rechtecke der Breite $m = 6$ und der Länge $n \geq 7$.

Kein Beispiel im Aufsatz. Im *Notizbuch* sind dagegen geschlossene Wege für $n = 7$ und $n = 8$ angegeben (Abb. 22, 23). Für den ersten Fall hat Euler zuerst einen vollständigen (*plenum*) Weg dargestellt, dann umgewandelt. Die Behandlung des zweiten ist ähnlich: der vollständige Weg wird durch Umformungen geschlossen. Beide vollständige Wege sind direkt, ohne jegliche Berichtigung, angegeben, wahrscheinlich also anderswo hergestellt worden.

28	19	40	15	30	21	38
41	12	29	20	39	14	31
18	27	16	13	8	37	22
11	<u>42</u>	25	36	5	32	7
26	17	2	9	34	23	4
<u>1</u>	10	35	24	3	6	33

Abb. 22

36	15	20	47	34	45	22	7
19	<u>48</u>	35	14	21	8	33	44
16	37	18	9	46	13	6	23
<u>1</u>	28	39	12	25	10	43	32
38	17	26	3	30	41	24	5
27	2	29	40	11	4	31	42

Abb. 23

---

[24] *Euler et le parcours du cavalier*, S. 122–123.

[25] Rectangles with width $m = 6$ do not occur in the article. In the manuscript, Euler treats the cases $n = 7$, $n = 8$, and closes his complete paths (Fig. 22, 23).

§4. Kreuzförmige Figuren

Im Aufsatz gibt es nur vier Beispiele von Kreuzen, mit geschlossenen Wegen, die Euler als 'einfachere Figuren' (*figures plus simples*) bezeichnet. Diese Aussage erklärt sich mit der Untersuchung der Handschrift, denn dort finden sich verschiedene Versuche, davon heiklere und gescheiterte.[26] Alle diese Versuche waren aber nicht umsonst, denn daraus ergaben sich allgemeine Einschränkungen zur Lösung. Eine erste ist diejenige, der wir schon in den früheren Fällen begegnet haben: ist die Anzahl weißer Felder gleich der Anzahl schwarzer Felder, so ist ein Weg *im Prinzip* ermittelbar; unterscheiden sie sich durch 1, so wird man mit der Farbe der größeren Anzahl beginnen und enden müssen; ist die Differenz größer als 1, so gibt es kein vollständiger Weg. Eine zweite Einschränkung, die wir ebenfalls schon gesehen haben, hängt von der Figur ab: je nach ihrer Gestalt können gewisse Felder, zumeist in Ecken oder (für Kreuze) am Ende von Ästen, unerreichbar werden.

①[27] **Kreuz zu 12 Feldern.**

In *Notizbuch* findet man zuerst den Weg von Abb. 24, der dann durch die Umformung $1, \ldots, 7, 12, \ldots, 8$ geschlossen wird (*perfectum*, Abb. 25). Letzterer wurde dann im Aufsatze gedruckt. Durch Umformungen in der Handschrift zeigt sich, daß sich Anfangs- und Endfeld auch Seite an Seite im mittleren Quadrat befinden können.

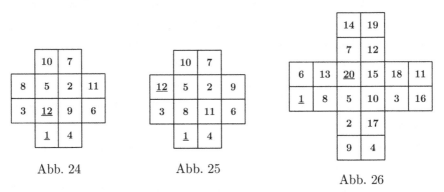

Abb. 24   Abb. 25   Abb. 26

②[28] **Kreuz zu 20 Feldern.**

---

[26] *Euler et le parcours du cavalier*, S. 127–148.

[27] Whereas the article gives four figures of crosses, all with closed paths, the manuscript has many more. These make clear how the existence of a path depends on the shape of the cross (situation at the end of the branches) and the respective numbers of white and black cells. The case of the cross with 12 cells, reported in the article, is set up in the manuscript.

[28] Fig. 26 is found in both the article and the manuscript. Since cells opposite relative

Abb. 26 findet man sowohl im Aufsatze wie im *Notizbuch*, in letzterem von einigen (unbedeutenden) Umformungen gefolgt. Man merke jedoch, daß das dargelegte Kreuz eine Besonderheit besitzt: die Differenz der Zahlen zweier zentralsymmetrisch gelegten Felder ist stets 10. Daß das *Notizbuch* keine vorherigen Versuche aufweist, nicht einmal Berichtigungen auf der vorhandenen Figur, weist erneut auf die vormalige Existenz anderer, jetzt verschollener handschriftlichen Notizen Eulers hin.

③[29] Kreuz zu 21 Feldern.

Eulers Versuche in der Handschrift blieben erfolglos. Sein erster Versuch führte zu Abb. 27, mit sechs leeren Feldern. Man sieht sofort, daß mit

$$d, c, b, a, 1, \ldots, 15$$

nur noch zwei leere Felder übrigbleiben. Eulers Umformungen führten zum gleichen Schluß, worauf seine Untersuchung dieses Falles aufhörte.

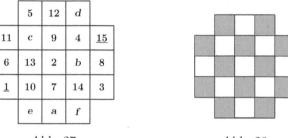

Abb. 27   Abb. 28

Es kann tatsächlich nicht anders sein, weil es zwölf schwarze und neun weiße Felder gibt (Abb. 28). Beginnt man und endet man in schwarzen Feldern, wie es sein muß, so müßen auf jeden Fall zwei schwarze Felder leer bleiben.

④[30] Kreuz zu 24 Feldern.

Ein erster Versuch im *Notizbuch* führt Euler zum Ergebnis der Abb. 29, mit drei miteinander nicht verbundenen leeren Feldern (unsere a, b, c). Mit seiner Umformung

$$a, 1, \ldots, 14, 21, \ldots, 15, b,$$

bleibt immer noch ein Feld leer.

---

to the centre display the same difference, the preparatory steps must have been noted on leaves now lost.

[29] Euler's attempts in the manuscript leave him with two empty cells. Indeed, since there are three more black cells, two must remain empty.

[30] The difference here being 2, one black cell will remain empty, as also appears from Euler's attempts in the manuscript.

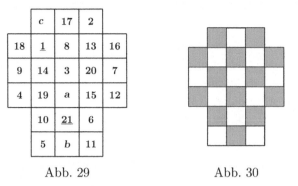

Abb. 29   Abb. 30

Die Ursache liegt erneut in den Anzahlen der Felder: es gibt dreizehn schwarze und elf weiße Felder (Abb. 30). Beginnt man also in einem schwarzen Feld, so muß jedenfalls eines frei bleiben.

⑤[31] Kreuz zu 32 Feldern.

Den geschlossenen Weg in Abb. 31 findet man sowohl im *Notizbuch* wie im Aufsatz. In der Handschrift wurde ein erster Versuch, infolge eines Schreibfehlers nach 15, teilweise auf der Abbildung selbst geändert (Abb. 32) und dann fortgeführt. Der so gewonnene offene Weg wurde dann mit der Umformung

$$1, \ldots, 19, 32, \ldots, 20$$

geschlossen und neu numeriert; daraus Abb. 31.

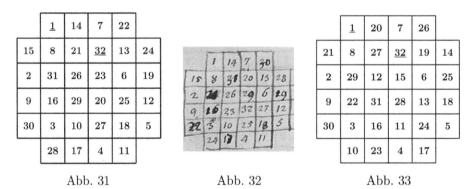

Abb. 31   Abb. 32   Abb. 33

Ein zweiter geschlossener Weg (Abb. 33) ist nur im Aufsatz vorhanden. Dabei weisen zwei zentralsymmetrisch gelegene Zellen die Differenz 16 auf. Die Herleitung muß Euler auf anderen, jetzt verlorenen, Blättern durchgeführt haben. Wie er dazu gelangte, ist uns aus anderen Beispielen

---

[31] A complete path (Fig. 32) is transformed into a closed one in the *Notizbuch*, then reproduced in the article (Fig. 31). The article gives a further example, displaying symmetry (Fig. 33). Here again the setting up must have been on separate leaves.

der Handschrift klar: beim allmählichen Füllen der Felder schreibt man jedes Mal einen Punkt in der gegenüberliegenden Zelle, um sie vorläufig auszuschließen.

Die folgenden Beispiele, obwohl sie doch einigermaßen Aufmerksamkeit verdienen würden, sind nur im *Notizbuch* vorhanden. Hatte uns doch Euler am Anfang des Aufsatzes gewarnt, er würde sich auf 'einfachere Figuren' beschränken.

⑥³² Kreuz zu 33 Feldern.

Ein erster Versuch Eulers ließ sieben Felder leer; nach mühsamen Umformungen blieb nur ein leeres Feld übrig (Abb. 34).

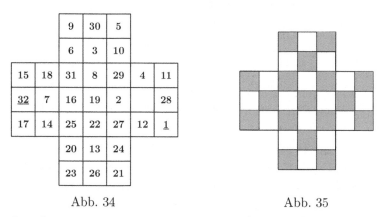

Abb. 34   Abb. 35

Anders konnte es tatsächlich nicht sein (Abb. 35). Da es siebzehn schwarze und sechzehn weiße Felder gibt, müßte man in schwarzen Feldern beginnen und enden, und alle weißen wären dann Durchgänge. Nun können die am Ende der Äste liegenden weißen Felder nicht alle Durchgänge sein: jedes hat nur zwei Zugänge, davon der eine gemeinsam mit dem benachbarten weißen Feld, auf daß für das letzte weiße Feld nur ein Zugang frei bleibt. Endet aber der Weg in einem solchen Feld, so muß wohl ein schwarzes Feld leer bleiben.

⑦³³ Kreuz zu 45 Feldern.

Hier hat Euler gleich bei seinem ersten Versuch aufgehört und seine Zeichnung gestrichen (Abb. 36). Eine Weiterführung wäre ja sinnlos, da

---

[32] Despite Euler's attempts, one cell remains empty (Fig. 34). This is not due to an excess in the difference between black (17) and white (16) cells, but to the place of the white cells at the end of the branches: all should be transit cells, but the last one (with 32) remains attainable from one side only, thus leaving one black cell empty.

[33] Impossible case since the cell difference is 3.

er keinen vollständigen Weg hätte erreichen können (drei schwarze Felder mehr als weiße, s. Abb. 37).

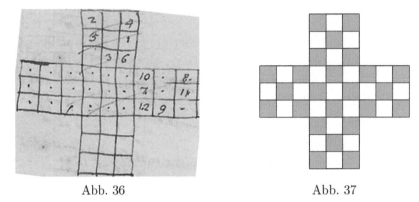

Abb. 36   Abb. 37

⑧[34] Kreuz zu 48 Feldern.

Ein erster Versuch läßt zwei Felder übrig (Abb. 38), die dann nacheinander dem vorhandenen Weg eingegliedert werden. Der erfundene Weg wird danach, erneut durch verschiedene Anwendungen der Regel, geschlossen. Es ergeben sich zwei Möglichkeiten, davon diejenige der Abb. 39.

		_1_	20	11	30		
		10	29	2	21		
39	28	19	42	31	12	3	22
18	9	40	a	_46_	43	32	13
27	38	45	b	41	34	23	4
8	17	26	37	44	5	14	33
		7	16	35	24		
		36	25	6	15		

		_1_	36	11	26		
		10	27	2	35		
19	28	37	_48_	25	12	3	34
38	9	18	21	44	47	24	13
29	20	45	40	17	22	33	4
8	39	30	43	46	5	14	23
		7	16	41	32		
		42	31	6	15		

Abb. 38   Abb. 39

⑨[35] Kreuz zu 64 Feldern.

Dieses letzte (und größte) Beispiel eines Kreuzes wird im *Notizbuch* eingehend behandelt. Nach ergebnislosen, teilweise mit Schreibfehlern behafteten Versuchen, fängt Euler erneut vom Anfang an. Diesmal wird

---

[34] From a partial way Euler obtains first a complete one, then a closed one.

[35] The largest example of a cross dealt with by Euler. See Fig. 40. First a complete path (including this time 57, . . . , 60) is constructed, then a closed one (Fig. 41).

die Behandlung auf eine besser überlegte Weise geführt: alle Kreuzäste werden nacheinander und getrennt gefüllt. Am Ende bleiben vier Felder leer, von welchen zwei können verknüpft werden (Abb. 40).[36] Nach Eingliederung von a, b-c, d erhält Euler den vollständigen Weg

$$56, \ldots, 53, a, 25, \ldots, 1, 64, \ldots, 61, 26, 27, c, b, 52, \ldots, 28, d,$$

den er dann schließt und neu numeriert. Das Ergebnis zeigt Abb. 41.

			3	6	11	8			
			12	9	2	5			
			_1_	4	7	10			
15	24	13	54	b	64	d	36	31	34
18	21	16	25	62	53	c	33	28	37
23	14	19	52	55	26	63	30	35	32
20	17	22	_61_	a	51	_56_	27	38	29
			48	45	42	39			
			43	40	47	50			
			46	49	44	41			

Abb. 40

			3	6	11	8			
			12	9	2	5			
			_1_	4	7	10			
15	24	13	28	51	_64_	59	34	55	32
18	21	16	25	62	27	52	31	58	35
23	14	19	50	29	60	63	56	33	54
20	17	22	61	26	49	30	53	36	57
			46	43	40	37			
			41	38	45	48			
			44	47	42	39			

Abb. 41

---

[36] Man merke, daß im vorläufigen Weg die vier Zahlen $57, \ldots, 60$ nicht vorkommen; Euler hatte von vorherein die Absicht, einen geschlossenen Weg zu bilden.

§5. Rautenförmige Figuren

Diese Gattung kommt nur im *Notizbuch* vor.[37]

①[38] Rautenförmiges Kreuz zu 32 Feldern.

Die vollendete Figur, die sich sicherlich anderswo herstellen ließ, gibt Abb. 42 an. Der Weg beginnt und endet am Ende von Ästen, und zwei Umformungen Eulers führten zum selben Ergebnis, womit er diesen Fall nicht weiter untersuchte.

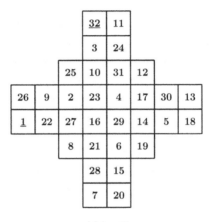

Abb. 42

Wenn man die Zellenpaare am äußersten Ende der Äste untersucht, wird man merken, daß jede Zelle genau einen Zugang und einen Ausgang besitzt, die alle am Rande des inneren 4×4 Quadrats liegen. Mittels eines Durchgangs kann man sie also alle füllen außer zwei, in aufeinanderfolgenden oder entgegengesetzten Ästen, die dann Anfangs- und Endpunkte des Weges sein müssen.

②[39] Raute zu 24 Feldern.

Nur ein Beispiel, und zwar schon in einer *Forma perfecta* (Abb. 43 & 44); daher keine weitere Untersuchung im *Notizbuch*.

---

[37] Fol. 235ʳ & 237ʳ; *Euler et le parcours du cavalier*, S. 148–150.

[38] No examples of rhomboidal figures in the article. Here only the first displays transformations, with Euler trying to reach a different path, maybe a closed one. But restricted access to the end of the branches obliges him to begin and end there.

[39] Result given directly in the manuscript. So also for the next, particularly elegant figure. The existence of separate sheets, now missing, is evident.

		17	10		
	13	8	1	18	
5	16	11	14	9	24
12	21	4	7	2	19
	6	15	20	23	
		22	3		

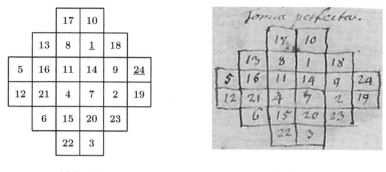

Abb. 43   Abb. 44

③ Raute zu 40 Feldern.

Wohl das schönste dieser drei Beispiele, *perfectum* wie das vorhergehende (Abb. 45; s. auch Abb. 177, p. 184 oben).

				14	31				
			30	7	2	15			
		4	13	58	17	32	1		
	40	29	6	3	8	59	16	33	
28	5	26	57	12	49	18	53	60	47
41	24	39	50	9	52	11	48	19	34
	27	42	25	56	21	54	35	46	
		23	38	51	10	45	20		
			43	22	55	36			
				37	44				

Abb. 45

Da einerseits diese drei rautenförmigen Figuren keine Berichtigungen oder Änderungen im *Notizbuch* aufweisen, und da es andererseits höchst unwahrscheinlich ist, daß das richtige Ergebnis mit einem Schlag von Euler erreicht wurde, wird man wohl wiederum auf die Existenz anderer, jetzt verschollener handschriftlichen Blätter geführt. Dank der Erhaltung derjenigen des sechsten dieser *Notizbücher* liegen aber jetzt Eulers vorbereitende Schritte zur Behandlung der Aufgabe des Rösselsprunges im wesentlichen vor.

Schrifttum

P.-H. Fuss: *Correspondance mathématique et physique de quelques célèbres géomètres du XVIIIème siècle* (2 vols). Saint Pétersbourg 1843.

L. Euler: "Solution d'une question curieuse qui ne paroit soumise à aucune analyse", *Mémoires de l'Académie Royale des Sciences et Belles-Lettres*, 15 (1759 [1766]), pp. 310–337 [reprinted in our study, pp. 216–243].

J. Sesiano: *Euler et le parcours du cavalier, avec une annexe sur le théorème des polyèdres*. Lausanne 2015.

Краткая история непрерывных дробей*
I. Конечные непрерывные дроби

§ 1. Общие понятия

①[1] Рассмотрим дробь $\frac{A}{B}$. Если A и B — целые положительные числа, и, допустим, что число A больше числа B, то, согласно алгоритму Евклида (*Начала*, VII, 2, см. ниже), имеем:

$$A = a_0 \cdot B + r_1$$
$$B = a_1 \cdot r_1 + r_2$$
$$r_1 = a_2 \cdot r_2 + r_3$$
$$\vdots$$
$$r_{n-3} = a_{n-2} \cdot r_{n-2} + r_{n-1}$$
$$r_{n-2} = a_{n-1} \cdot r_{n-1} + r_n$$
$$r_{n-1} = a_n \cdot r_n$$

где a_i, r_i натуральные числа и r_i — последовательность остатков. Так как A и B — конечные целые числа, то и алгоритм конечен. Последний не равный нулю остаток r_n является наибольшим общим делителем чисел A и B. Если же он равен единице, то A и B взаимно просты. Всё это было хорошо известно в античности (*Начала*, VII, 2 имеет следующюю цель: Δύο ἀριθμῶν δοθέντων μὴ πρώτων πρὸς ἀλλήλους τὸ μέγιστον αὐτῶν κοινὸν μέτρον εὑρεῖν, т. е.: Для двух данных чисел, не первых между собой, найти наибольшую их меру).

Если написать этот алгоритм таким образом:

$$\frac{A}{B} = a_0 + \frac{r_1}{B} = a_0 + \frac{1}{\frac{B}{r_1}} = a_0 + \frac{1}{a_1 + \frac{r_2}{r_1}} = a_0 + \frac{1}{a_1 + \frac{1}{\frac{r_1}{r_2}}} =$$

$$\cdots = a_0 + \cfrac{1}{a_1 + \cfrac{1}{a_2 + \cfrac{1}{\ddots + \frac{1}{a_n}}}},$$

* Я очень благодарен организатору Санкт-Петербургских семинаров по истории математики Г. И. Синкевич, которая поощрила меня к работе над этой статьёй и пересматривала её с терпением и усердием. Я также благодарен редактору журнала «Математика в высшем образовании» Г. М. Полотовскому за многие относящиеся замечания.

1. Considering two natural numbers A, B (say $A > B$), the Euclidean algorithm (*Elements* VII, 2) produces a sequence of integral quotients a_i and remainders r_i. Since A, B are finite, so will be the sequence of divisions, and the last remainder not zero, r_n, will be the greatest common divisor of A, B (1 if they are prime to one another). Another way to express this sequence of divisions is in the form of a continued fraction. Any rational number can be written in the form of such a finite continued fraction.

то получим «непрерывную дробь», а именно конечную непрерывную дробь в этом случае натуральных A, B. Как Ейлер пишет (*De fractionibus continuis*, с. 108; *Opera omnia*, с. 194): Omnis autem fractio finita, cuius numerator et denominator sunt numeri integri finiti, in huiusmodi fractionem continuam transformatur, quae alicubi abrumpitur (рис. 1–2 от статьи Ейлера, где соответственно $A > B$, $A < B$).

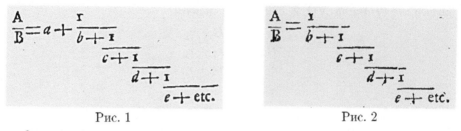

Рис. 1   Рис. 2

② [2] Отбрасывая в непрерывной дроби последовательно последнюю дробь, имеем следующий ряд дробей:

$$A_0 = a_0 = \frac{a_0}{1} \equiv \frac{P_0}{Q_0}$$

$$A_1 = a_0 + \frac{1}{a_1} = \frac{a_0 a_1 + 1}{a_1} \equiv \frac{P_1}{Q_1}$$

$$A_2 = a_0 + \cfrac{1}{a_1 + \cfrac{1}{a_2}} = \frac{(a_0 a_1 + 1)\, a_2 + a_0}{a_1 a_2 + 1} \equiv \frac{P_2}{Q_2}$$

$$\cdots$$

$$A_j = a_0 + \cfrac{1}{a_1 + \cfrac{1}{a_2 + \cfrac{1}{\ddots + \frac{1}{a_j}}}} = \cdots \equiv \frac{P_j}{Q_j}$$

$$\cdots$$

$$A_n = \cdots \equiv \frac{P_n}{Q_n} = \frac{A}{B}.$$

A_j называются *подходящими дробями*. Они все приближения к начальной дроби, а именно каждое лучше чем предыдущее. Как опять Ейлер пишет (*De fractionibus continuis*, с. 111; *Opera omnia*, с. 196): hae fractiones proxime aequales erunt expressioni $\frac{A}{B}$, eoque minus distabunt, quo remotiores fuerint a prima. А последнее значение A_n будет равно данной дроби.

③ [3] В разные исторические периоды появились пять свойств непрерывных дробей.

---

2. We may obtain a sequence of values increasingly close to $\frac{A}{B}$ by considering the so-called approximating fractions A_i obtained by breaking off at the ith integral quotient. The last of these fractions, A_n, equals the given fraction $\frac{A}{B}$.

3. Finite continued fractions display five fundamental properties, and recording the circumstances of their discovery is the object of the first part of this study.

I. The approximating fractions A_i are, with increasing accuracy, alternately smaller and larger than the exact value.

(i) Вычисляв значения подходящей дробей увидим, что

Свойство I. A_0 A_1 A_2 A_3 ...

$<$ $>$ $<$ $>$ ...

Так как можно разложить всякую рациональную дробь в непрерывную дробь, можно найти последовательность приближений поочe-
рёдно с избытком и недостатком; а, как сказано выше, последующее
приближение ближе к данной дроби чем предыдущее. В этом и заклю-
чалась главная роль конечных непрерывных дробей в истории. Как
мы позже увидим, это свойство, возможно, использовали Аристарх в
античности и Д. Швентер в начале XVII в., и, этот раз несомненно,
Х. Гюйгенс в конце XVII в.

(ii) Зная величины $a_0, a_1, a_2, \ldots$ и две последовательных пары P_i, Q_i,
можно вычислить все остальные P_k, Q_k, так как

Свойство II. $P_k = P_{k-1} \cdot a_k + P_{k-2}$

$Q_k = Q_{k-1} \cdot a_k + Q_{k-2}.$

Это правило встречается у Швентера. Иногда определяются формаль-
но $P_{-2} = 0$, $Q_{-2} = 1$, $P_{-1} = 1$, $Q_{-1} = 0$.

(iii) Из этих рекуррентных формул получаем $P_k Q_{k-1} - P_{k-1} Q_k = \pm 1$,
т. е. $\frac{P_k}{Q_k} - \frac{P_{k-1}}{Q_{k-1}} \equiv A_k - A_{k-1} = \pm \frac{1}{Q_k Q_{k-1}}$, откуда

Свойство III. Все A_k — несократимые дроби.

Возможно, было известно Швентеру, и наверняка — Гюйгенсу.

(iv) Подходящая дробь $A_k = \frac{P_k}{Q_k}$ ближе к началной дроби $\frac{A}{B}$ чем
всякая другая дробь с меньшим знаменателем, т. е., если некоторая
другая дробь лучше, то её знаменатель обязательно больше. Другими
словами,

Свойство IV. Все A_k — наилучшие приближения.

Упоминание об этом можно найти у Гюйгенса.

II. Knowing the sequence of the a_i and two consecutive approximating fractions makes
it possible to calculate all of them.

III. All the A_i are irreducible fractions.

IV. All the A_i are best approximations; that is, any approximation closer than A_i must
have a less convenient denominator (in other words : it will be larger).

V. Knowing the A_i enables us to find all best approximations. For from, say, A_j and
A_{j-2}, thus consecutive approximations on the same side, we may find the complete set
of best approximations between them.

(v) При помощи двух последовательных таких приближений с одной и той же стороны, т. е. обоих с избытком или недостатком, можно вставить другие рациональные приближения, которые сами являются наилучшими; а именно можно найти таким образом все наилучшие приближения.

Свойство V. Зная последовательность приближений A_k можно найти последовательность всех наилучших приближений.

Впервые его использовал Ж. Л. Лагранж в конце XVIII в.

§2. Аристарх Самосский (Ἀρίσταρχος, III в. до н. э.)

①[4] История использования конечных непрерывных дробей начинается, *может быть*, в греческой древности. В трактате Аристарха «О величинах и расстояниях Солнца и Луны» есть два места, где ему надо найти приближение двух больших дробей, а именно,

$$\frac{7921}{4050} \quad \text{и} \quad \frac{71755875}{61735500}.$$

О первом случае он пишет (издание греческого текста, с. 396, 1-2): Ἔχει δὲ καὶ τὰ $\overline{\zeta \lambda \kappa \alpha}$ πρὸς $\overline{\delta \nu}$ μείζονα λόγον ἤπερ τὰ $\overline{\pi \eta}$ πρὸς $\overline{\mu \epsilon}$, т. е.

$$\frac{7921}{4050} > \frac{88}{45}.$$

О втором случае он пишет (с. 406, 23-24): Ἔχει δὲ καὶ ὁ Μ$\overline{\zeta \rho o \epsilon}$ $\overline{\epsilon \omega o \epsilon}$ πρὸς Μ$\overline{\tau \rho o \gamma}$ $\overline{\epsilon \phi}$ μείζονα λόγον ἢ ὃν τὰ $\overline{\mu \gamma}$ πρὸς $\overline{\lambda \zeta}$, т. е.

$$\frac{71755875}{61735500} > \frac{43}{37}.$$

Таков текст Аристарха; т. е., в обоих случаях объяснения нет.

②[5] Рассмотрим разложение в непрерывную дробь обеих данных дробей и возьмём их первые приближения. Для первого примера имеем

$$\frac{7921}{4050} = 1 + \cfrac{1}{1 + \cfrac{1}{21 + \cfrac{1}{1 + \cfrac{1}{1 + \cfrac{1}{\ddots}}}}}, \quad \text{откуда получается}$$

---

4. The Greek Aristarchos of Samos (3rd c. BC) *may* have known how to form the approximating fractions A_i and that they are positioned alternately on either side of the exact value. He gives (but without any explanation, as if the procedure were well known) two approximate values for larger fractions, both said to be by defect.

5. Now the development into a continued fraction does indeed lead to these approximations.

$$\frac{1}{1}, \quad \frac{2}{1}, \quad \frac{43}{22}, \quad \frac{45}{23}, \quad \mathbf{\frac{88}{45}}, \quad \ldots$$

а для второго

$$\mathbf{\frac{71755875}{61735500}} = 1 + \cfrac{1}{6 + \cfrac{1}{6 + \cfrac{1}{\ddots}}}, \quad \text{откуда получается}$$

$$\frac{1}{1}, \quad \frac{7}{6}, \quad \mathbf{\frac{43}{37}}, \quad \ldots$$

Итак, если рассматривать этот вопрос относительно написания непрерывных дробей, то становится понятно, *откуда* берутся приближения и *почему* выбор Аристарха в обоих случаях является меньшим истинного значения.

③ [6] Это может означать, что древние греки уже знали форму написания алгоритма Евклида в виде непрерывной дроби. Это допустимо, так как в то время алгоритм Евклида был уже широко известен. Если это так, то можно допускать, что Аристарх знал о первом свойстве непрерывных дробей, а именно, что приближения поочерёдно то меньше, то больше истинного значения.

§3. Даниил Швентер (Daniel Schwenter, 1585-1636)

① [7] В одном месте своей *Geometria practica* (II, с. 68; или *Deliciæ*, с. 111–113; обе книги написаны на немецком языке, единственно заглавия на латыне) Швентер рассматривает отношение двух чисел. Иногда можно сокращать дробь; иногда нет, потому что числа взаимно просты. В этом случае временами окажется желательным, найти приближения меньшими числами. По словам Швентера, для этого «у вычислителей и мастеров счёта можно найти многие тонкие правила», а наилучшее, по его мнению, следующее (Wie man aber zwo groſſe Zahlen, ſo numeri primi und Arithmeticè nicht können auffgehebt werden, dem gebrauch nach kleiner machen ſoll, ſeynd (= sind) bey den Logisticis und Rechenmeiſtern viel feine Regeln zu finden. Die beſte, geheimeſte und künſtlichſte will ich hieher ſetzen. Jch ſoll die zwo Zahlen 233 und 177, alſ welche für ſich numeri (взаимно) primi, oder aber

---

6. These are the only instances in antiquity. Since the Euclidean algorithm was then well known, it is fairly plausible that how to form approximating fractions was also known.

7. In his *Geometria practica*, Schwenter (1585-1636) explains an algorithm for finding an approximation to the ratio of two (relatively prime) natural numbers, namely in his example $A = 177$ and $B = 233$. There are, he says, many methods in use among calculators, the best of which he goes on to describe (Fig. 3).

ðie Proportion $\frac{177}{233}$ in kleinern 3ahlen Mechanicè (механически = методично) auffprechen. So mache ich nun folgende Disposition oder Ordnung).

②[8] Швентер объясняет этот алгоритм на примере двух взаимно простых чисел 177 и 233 при помощи следующей таблицы (рис. 3). Что касается столбец, то их заполняем следующим образом.

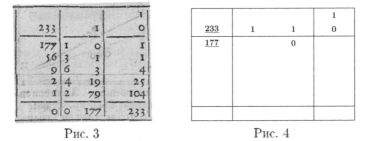

			1
<u>233</u>	1	1	0
<u>177</u>		0	

Рис. 3   Рис. 4

Сначала пишем в первом столбце наверху два данных числа (во-первых, знаменатель) и, обязательно, три раза единицу (а не два раза, как на рисунке Швентера) и два раза ноль (рис. 4). Затем используем алгоритм Евклида и пишем целые частные a_k во втором столбце, а остатки в первом (рис. 5).

			1
<u>233</u>	1	1	0
<u>177</u>	1	0	
56	3		
9	6		
2	4		
1	2		
0	0		

			1
<u>233</u>	1	1	0
<u>177</u>	1	0	1
56	3	1	1
9	6	3	4
2	4	19	25
1	2	79	104
0	0	177	233

Рис. 5   Рис. 6

Потом Швентер объясняет заполнение третьего и четвёртого столбец. Элементы третьего столбца последовательно получаются умножением числа второго столбца на число третьего столбца и добавлением к этому произведению верхнего числа третьего столбца; сумма даёт следующее число третьего столбца. Отсюда в нашем случае последовательно получается

$$1 \cdot 0 + 1 = 1, \quad 3 \cdot 1 + 0 = 3, \quad 6 \cdot 3 + 1 = 19, \text{ и т. д.}$$

---

8. One has first to draw a table with known numbers as in Fig. 4. The first two columns are filled with the numbers inferred from Euclid's algorithm (Fig. 5). As to the third and the fourth, their quantities are successively calculated using those of the second and those initially and then successively written at the top of the column considered (Fig. 6). The last two columns then give the desired approximations, namely our $A_k = \frac{P_k}{Q_k}$, with $P_k = P_{k-1} \cdot a_k + P_{k-2}$ and $Q_k = Q_{k-1} \cdot a_k + Q_{k-2}$ in the third and fourth column, respectively.

Сделаем то же самое для заполнения четвёртого столбца:

$$1 \cdot 0 + 1 = 1, \quad 1 \cdot 1 + 0 = 1, \quad 3 \cdot 1 + 1 = 4, \text{ и т. д.}$$

Числа третьего и четвёртого столбец дают последовательность приближений к нашей дроби, а именно (рис. 6)

$$\left(\tfrac{0}{1}\right), \ \tfrac{1}{0}, \ \tfrac{0}{1}, \ \tfrac{1}{1}, \ \tfrac{3}{4}, \ \tfrac{19}{25}, \ \tfrac{79}{104}, \ \left(\tfrac{177}{233}\right).$$

③ [9] Здесь нет теории, Швентер только поясняет нахождение приближений согласно с своим примером. Вычисления в первом (r_k) и втором (a_k) столбцах ясны для читателя-математика (также в то время). Величины третьего и четвёртого столбец являются нашими P_k и Q_k (§ 1), ибо, согласно указаниям Швентера, получим

$$P_k = P_{k-1} \cdot a_k + P_{k-2}, \quad Q_k = Q_{k-1} \cdot a_k + Q_{k-2}$$

итак

$$\frac{P_k}{Q_k} = \frac{P_{k-1} \cdot a_k + P_{k-2}}{Q_{k-1} \cdot a_k + Q_{k-2}}$$

Следовательно можно допускать, что во время Швентера знали о построением подходящих дробей A_i (Свойство II), начиная с $Q_{-2} = 1$, $P_{-1} = 1$, $Q_{-1} = 0$, а может быть, что эти приближения − несократимые дроби (Свойство III).

§ 4. Христиан Гюйгенс (Christiaan Huygens, 1629-1695)

① [10] Когда Гюйгенс был в Париже, он занимался построением *планетария*, т. е. механизма, который должен показывать движения шести исвестных планет солнечной системы, а, при случае, и их спутников. По правде говоря, это построение не было первым в своём роде, но здесь впервые использовалась система Коперника. И что гораздо важнее для нас, эта работа является первым применением непрерывных дробей к технике.

---

9. Although there is a single example, it appears that this is an algorithm which can be applied to any rational fraction. Since the quantities in the third and fourth columns are calculated according to the second property, starting from two given values, this property was thus known by the time of Schwenter.

10. While in Paris, Christiaan Huygens (1629-1695) devoted himself to the construction of a planetarium, that is, a mechanism representing the yearly movements of the six planets (sometimes satellites as well) in the Copernican system —this is the first use of the latter system for a planetarium. It looked like a wall clock (Fig. 7), with inside above wheels carrying the planets (Fig. 8), and below the inner mechanism (Fig. 9) comprising driving wheels attached to the axis which made a full rotation in one year. For each planet, the movement depended on the number of teeth on its two cogwheels. To find a suitable ratio for them was to be the first application of continued fractions to technics.

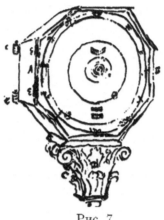

Рис. 7

Рис. 7 показывает вид планетария согласно рисунку самого Гюйгенса (*Œuvres*, XXI, с. 596). Снаружи он выглядит как стенные часы (размеры: диаметр = 2 фута; густота = 6 дюймов).

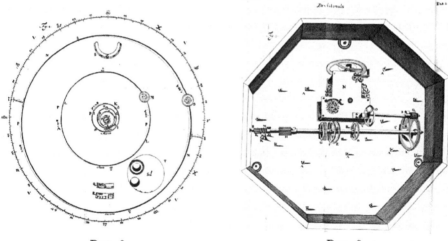

Рис. 8   Рис. 9

Если его открыть, мы увидим внутри, в первой стороне, шесть зубчатых колёс, на каждом из которых снаружи, через щель, закреплена соответствующая планета (рис. 8; *Œuvres*, XXI, с. 598). Эти высшие зубчатые колёса захватывают с ведущими зубчатыми колёсами в задней стороне; они сами прикреплены на оси, которая осуществляет полный оборот за один год, или за предварительно установленное время (рис. 9; *Œuvres*, XXI, с. 604).

②[11] Понятно, что точность движения каждой планеты зависит от

---

11. Since the longer the period of the planet, the larger the number of teeth on its

расчёта чисел зубцов её носящего и ведущего колёс. Самым трудным случаем является Сатурн, потому что его период особенно долог, так что число зубцов колеса Сатурна должно быть значительно больше числа колеса на оси. Для колеса Сатурна Гюйгенс возьмёт 206 зубцов, и 7 для ведущего колеса.

Чтобы найти эти два числа, Гюйгенс рассматривает движения Земли и Сатурна в течение полных 365 дней (т. е. чуть меньше года). Согласно измерениям той эпохи, за это время Земля перемещается на $359° 45' 40'' 31'''$, т. е. $77\,708\,431'''$, и Сатурн $12° 13' 34'' 18'''$, т. е. $2\,640\,858'''$; а эти числа взаимно просты. Следовательно, имеем

$$\frac{\text{период Сатурна}}{\text{период Земли}} = \frac{77\,708\,431}{2\,640\,858} = \frac{\text{число зубцов колеса Сатурна}}{\text{число зубцов его ведущего колеса}}.$$

③ [12] Но построение двух шестерней с такими количествами зубцов технически невозможно. Следовательно, задача состоит в том, что мы должны найти приближения этого отношения при помощи меньших чисел, и, для точности, *эти приближения должны быть наилучшими* (res tota recidit ut, datis numeris duobus magnis certam inter se rationem habentibus, alii minores inveniantur rotarum dentibus multitudine sua non incommodi, quique eandem proxime rationem ita exhibeant, ut nulli ipsis minores propius; см. *Opuscula*, с. 448; *Œuvres*, XXI, с. 627). А это и есть Свойсто IV.

Рассмотрим разложение в непрерывную дробь данной дроби (рис. 10 — в трактате Гюйгенса; изд. 1703, с. 449),

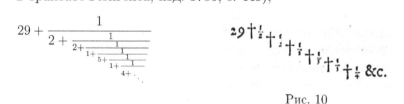

Рис. 10

wheel, the extreme case was Saturn, with a period of about 29 years. Huygens chose 206 teeth for the outer wheel and 7 for the one on the axis. To determine the number of teeth, Huygens considers the movement in 365 full days, which is, in thirds, $77\,708\,431$ for the earth and $2\,640\,858$ for Saturn. Thus

$$\frac{\text{Period of Saturn}}{\text{Period of the earth}} = \frac{\text{number of teeth on Saturn's wheel}}{\text{number of teeth on its driving wheel}} = \frac{77\,708\,431}{2\,640\,858}$$

12. Thus we are to find a pair of numbers approximating the given fraction such that the construction of the two wheels will be both technically possible (smaller numbers) and closest to this fraction (best approximation). The latter is Property IV. Incidentally, Huygens restates Property I (alternation on either side of the given value) —which is irrelevant here since we are merely to consider technical feasibility. Since 5 and 1177 teeth would hardly do, $\frac{206}{7}$ is the obvious choice.

Из последовательности величин a_k, а именно $(29\,;\,2,\,2,\,1,\,5,\,1,\,4,\,1,\,1,$ $2,\,1,\,6,\,1,\,10,\,2,\,2,\,3)$, имеются первые приближения:

$$
\begin{array}{ccccccc}
 & & 29 & 2 & 2 & 1 & 5 \\
\left(\dfrac{0}{1}\right) & \left(\dfrac{1}{0}\right) & \left(\dfrac{29}{1}\right) & \dfrac{59}{2} & \dfrac{147}{5} & \dfrac{206}{7} & \dfrac{1177}{40} \quad \ldots \; \cdot \\
< & > & < & > & < & > & <
\end{array}
$$

Гюйгенс оставляет без внимания два первых формальных члена $\frac{0}{1}$ и $\frac{1}{0}$, так и банальную дробь $\frac{29}{1}$; т. е. он начинает с первого истинного приближения $\frac{59}{2}$. Об этих приближениях он пишет, что они попеременно то больше, то меньше: это как раз наше Свойство I. Словами Гюйгенса: Sciendum vero, reductione hac nostra majorem proportionis terminum alternis majorem minoremve vero reperiri, prout a prima, tertia, quinta aut alia deinceps impari fractione reductio inchoata fuerit (*Opuscula*, с. 452; *Œuvres*, XXI, с. 633).

Но здесь меньше/больше не относится, потому что Гюйгенс просто хочет иметь что-то *технически возможное* для числа зубцов каждого колеса. А выбор Гюйгенса в случае Сатурна очевиден: так как, технически, 5 зубцов слишком мало, а 1177 зубцов слишком велико, он выбирает отношение между их дробями, т. е. $\frac{206}{7}$.

§5. Жозеф Луи Лагранж (Joseph Louis Lagrange, 1736-1813)

①[13] Эйлер написал книгу по алгебре, к которой Лагранж сделал дополнения. Первое издание дополнений Лагранжа, вместе с *Алгеброй* Эйлера, на франц̣уском языке, в двух томах, было опубликовано в Санкт-Петербурге в 1798 г. (рис. 11). Среди этих дополнениях Лагранжа есть такое (§ I.19) : Une fraction exprimée par un grand nombre de chiffres étant donnée, trouver toutes les fractions en moindres termes qui approchent si près de la vérité, qu'il soit impossible d'en approcher davantage sans en employer de plus grandes (Свойство V; кратко: Для данной

---

13. Along with the French translation of Euler's *Algebra*, Joseph Louis Lagrange (1736-1813) had various complements published (first edition : St. Petersburg 1798, Fig. 11). One problem was that of the calendar, thus the intercalation of leap years to account for the supplementary fraction of year length (0.2422 days). The addition in Caesar's Julian Calendar (47 BC) of one day every fourth year turned out to be insufficient, so Pope Gregory XIII had the Gregorian Calendar introduced (1582), with the secular years being leap years only if their number of hundreds is divisible by four. The ratio of years to leap years thus changed from $\frac{4}{1}$ to $\frac{400}{97}$. Now Lagrange's aim was to find out whether the second fraction was among the best approximations of $\frac{86400}{20929}$, which is the ratio of the seconds in one day to those in the supplementary fraction of a day according to the measurement of the year by Abbé La Caille (1713-1762).

дроби с большими числителем и знаменателем пусть требуется найти все наилучшие приближения). Следующее § I.20 является его применением: найти наилучшие соотношения между календарным и солнечным годом.

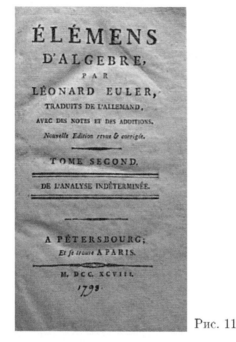

Рис. 11

По современным меркам год равен ок. 365 дн. 5 ч. 48 мин. 46 сек. = 365, 2422 дн. Так как дробь равна около четверти дня, в Юлианском календаре (реформа Юлия Цезаря, 47 г. до Р. Х.) добавляется один день в каждом четвёртом году. Но это добавление слишком большое, за четыре столетия оно доходит до около трёх дней. Исправление было сделано в Григорианском календаре таким образом (реформа папы Григория XIII, 1582 г.): сначала отмена десяти дней (5 окт. до 14 окт. 1582 г.); потом, из четырёх вековых лет 1600, 1700, 1800 и 1900 только первый год должен остаться високосным; в общем, високосным остаётся тот вековой год, количество столетий которого делится на 4. Следовательно, отношение числа лет к соответствующему числу високосных лет стало больше не $\frac{4}{1}$, но $\frac{400}{97}$ (этот раз, так как $365, 25 - \frac{3}{400} = 365, 2425 > 365, 2422$, с избытком 3 дня в 10 000 лет).

Цель Лагранжа такова: он хочет установить, является ли $\frac{400}{97}$ наилучшим отношением, или, в противном случае, то есть ли $\frac{400}{97}$ между другими наилучшими отношениями (см. с. 314, Свойство v). Но употребляющееся значение года во времена Лагранжа было измерение аббата Лакайля (N.-L. de la Caille, 1713-1762), т. е. 365 дн. 5 ч. 48 мин. 49 сек. (с избытком трёх секунд по отношению к сегодняшнему значению). Так как в сутках всего 86400 секунд и дробная часть у Лагранжа равна 20929 секунд, то отношение секунд в сутках к числу добавочных секунд, т. е. отношение числа лет к числу високосных лет, равно

$$\frac{86400}{20929}.$$

② [14] Сначала Лагранж считает приближения с помощью непрерывных дробей, и находит

		4	7	1	3	1	16	1	1	15
$\left(\dfrac{0}{1}\right)$	$\left(\dfrac{1}{0}\right)$	$\dfrac{4}{1}$	$\dfrac{29}{7}$	$\dfrac{33}{8}$	$\dfrac{128}{31}$	$\dfrac{161}{39}$	$\dfrac{2704}{655}$	$\dfrac{2865}{694}$	$\dfrac{5569}{1349}$	$\dfrac{86400}{20929}$
<	>	<	>	<	>	<	>	<	>	=

Среди них совсем нет григорианского приближения, $\frac{400}{97}$. Следовательно, Лагранж взялся за получение остальных наилучших приближений.

Швентеру, а, может быть, и Аристарху, уже был известен закон образовании подходящих дробей (Свойство II) :

$$\frac{P_k}{Q_k} = \frac{P_{k-1} \cdot a_k + P_{k-2}}{Q_{k-1} \cdot a_k + Q_{k-2}}.$$

Потом, если $a_k \neq 1$, можно вставить $a_k - 1$ новые приближения между $\frac{P_{k-2}}{Q_{k-2}}$ и $\frac{P_k}{Q_k}$. Так как обе эти дроби стоят с одной стороны от истинного значения, там же будут и новые приближения; поэтому-то Лагранж рассматривает приближения с избытком и с недостатком отдельно, и вычисляет промежуточные значения

$$\frac{P_{k-1} \cdot t + P_{k-2}}{Q_{k-1} \cdot t + Q_{k-2}} \qquad t = 1, 2, \ldots, a_k - 1.$$

③ [15] Рассмотрим во-первых все меньшие приближения непрерывной дроби (ниже, жирным шрифтом). Так как в этом случае величины a_k

---

14. Since the Gregorian value turned out not to be among the best approximations obtained from the continued fraction, Lagrange searched for the other best approximations, which led him to apply Property V. This was to be the case if, in the expression of Property II, $a_k \neq 1$.

15. He thus obtained all further best approximations on the left side (italics); but all proved to be unfeasible, being too small or too large. As to the approximations on the right side, thus by excess, there was an infinite number of them, with only a few worth considering.

имеют значения 4, 1, 1, 1, 15, Лагранж находит только четырнадцать новых приближений (первые три неинтересны):

$$\left(\tfrac{0}{1},\right) \frac{1\cdot1+0}{1\cdot0+1} = \tfrac{1}{1}, \tfrac{2}{1}, \tfrac{3}{1}, \mathbf{\tfrac{4}{1}}, \mathbf{\tfrac{33}{8}}, \mathbf{\tfrac{161}{39}}, \mathbf{\tfrac{2865}{694}}, \frac{1\cdot5569+2865}{1\cdot1349+694} = \tfrac{8434}{2043},$$

$$\tfrac{14003}{3392}, \tfrac{19572}{4741}, \tfrac{25141}{6090}, \tfrac{30710}{7439}, \tfrac{36279}{8788}, \tfrac{41848}{10137}, \tfrac{47417}{11486}, \tfrac{52986}{12835},$$

$$\tfrac{58555}{14184}, \tfrac{64124}{15533}, \tfrac{69693}{16882}, \tfrac{75262}{18231}, \tfrac{80831}{19580}, \frac{15\cdot5569+2865}{15\cdot1349+694}.$$

Так как, пишет Лагранж, последняя дробь совпадает с данной, то ясно, что последовательность невозможно продолжать (comme la dernière fraction est la même que la fraction donnée, il est clair que cette série ne peut pas être poussée plus loin).

Заметим, что числа, входящие в эту последовательность новых приближений слишком велики; следовательно промежуточные дроби не пригодны для простого отношения числа лет к числу високосных лет.

Что касается больших приближений непрерывной дроби, то так как $a_k = 7, 3, 16, 1$, имеем немедленно $6 + 2 + 15 = 23$ новых приближения:

$$\left(\tfrac{1}{0},\right) \frac{1\cdot4+1}{1\cdot1+0} = \tfrac{5}{1}, \frac{2\cdot4+1}{2\cdot1+0} = \tfrac{9}{2}, \tfrac{13}{3}, \tfrac{17}{4}, \tfrac{21}{5}, \tfrac{25}{6}, \mathbf{\tfrac{29}{7}}, \frac{1\cdot33+29}{1\cdot8+7} = \tfrac{62}{15}, \tfrac{95}{23}, \mathbf{\tfrac{128}{31}},$$

$$\frac{1\cdot161+128}{1\cdot39+31} = \tfrac{289}{70}, \tfrac{450}{109}, \tfrac{611}{148}, \tfrac{772}{187}, \tfrac{933}{226}, \tfrac{1094}{265}, \tfrac{1255}{304}, \tfrac{1416}{343}, \tfrac{1577}{382}, \tfrac{1738}{421}, \tfrac{1899}{460},$$

$$\tfrac{2060}{499}, \tfrac{2221}{538}, \tfrac{2382}{577}, \tfrac{2543}{616}, \mathbf{\tfrac{2704}{655}} \ \mathbf{\tfrac{5569}{1349}}; \ \frac{1\cdot86400+5569}{1\cdot20929+1349} = \tfrac{91969}{22278}, \tfrac{178369}{43207}, \dots.$$

Так как, пишет Лагранж, эта последовательность не заканчивается данной дробью, мы можем сколь угодно её продолжать (comme la série précédente n'est pas terminée par la fraction donnée, on peut encore la continuer aussi loin que l'on veut).

④ [16] Григорианское приближение было

$$\frac{\text{число лет}}{\text{число високосных лет}} = \frac{400}{97},$$

а пригодные результаты Лагранжа – два:

дробь $\frac{161}{39}$ ($<$), которая будет ближе к истине, чем дробь $\frac{400}{97}$ (la fraction $\frac{161}{39}$ approcherait plus de la vérité que la fraction $\frac{400}{97}$);

---

16. Whereas the Gregorian intercalation is easy to apply, Lagrange's best approximations $\frac{161}{39}$ and $\frac{450}{109}$ are obviously not. Incidentally, he wisely went no further; for, as he observes, contemporary astronomers were not in agreement as to the true length of the year.

дробь $\frac{450}{109}$ ($>$), потому что было бы гораздо точнее вставить сто девять дней в четыреста пятьдесят лет (on approcherait beaucoup plus de l'exactitude en intercalant cent neuf jours en quatre cent cinquante années).

Во всяком случае, как Лагранж сам думает, такие исследования напрасны, пока не будет известна точная длина года (au reste, comme les Astronomes sont encore partagés sur la véritable longueur de l'année, nous nous abstiendrons de prononcer sur ce sujet — впрочем, поскольку астрономы всё ещё не пришли к взаимопониманию по поводу длины года, воздержимся от выводов).

II. Бесконечные непрерывные дроби

§6. Общие понятия

① [17] Пусть α — любое положительное не целое число. Тогда, если a_0 — его целая часть, то

$$\alpha = a_0 + \frac{1}{\alpha_1}, \quad 0 < \frac{1}{\alpha_1} < 1.$$

А так как $\alpha_1 > 1$, вновь имеем

$\alpha_1 = a_1 + \frac{1}{\alpha_2}$

и, последовательно,

$\alpha_2 = a_2 + \frac{1}{\alpha_3}$,

$\dots$,

так что находим для α выражение в виде

$$\alpha = a_0 + \cfrac{1}{a_1 + \cfrac{1}{a_2 + \cdots}} = (a_0; a_1, a_2, \dots).$$

Если α *рациональное число*, тогда α_i тоже рациональные числа и алгоритм (Евклида) конечен, т. е. последовательность целых величин a_i также конечна. Но если α *иррациональное число*, тогда α_i тоже иррациональные числа, алгоритм не заканчивается и последовательность целых величин a_i бесконечна. Следовательно, в этом случае имеем бесконечную последовательность рациональных приближений,

---

17. Consider an arbitrary number α. Its non-integral part may be represented as a continued fraction with partial quotients a_i as before. If α is rational, the continued fraction will terminate; if not, the quotients $(a_0; a_1, a_2, \dots)$ will be infinite in number and the approximating fractions A_i will become closer and closer to α, once again alternately on either side, with increasing accuracy; indeed, the properties seen above do not depend upon the number of terms.

с теми же свойствами, что и конечная последовательность приближений в рациональном случае, потому что эти свойства не зависят от количества членов a_i разложения в непрерывную дробь.

Различие между этими двумя случаями описано в первой статье Эйлера о непрерывных дробях. А именно, он пишет (*De fractionibus continuis*, с. 108; *Opera omnia*, с. 194 : Omnis autem fractio finita, cuius numerator et denominator sunt numeri integri finiti, in huiusmodi fractionem continuam transformatur, quae alicubi abrumpitur (см. выше, с. 312; Ейлер продолжает:) fractio autem, cuius numerator et denominator sunt numeri infinite magni —cuiusmodi dantur pro quantitatibus irrationalibus et transcendentibus— in fractionem verē continuam et in infinitum excurrentem transibit).

② [18] Что касается значения бесконечной непрерывной дроби, то она определяет одно, и только одно, иррациональное число. И наоборот: каждая бесконечная последовательность натуральных чисел a_i определяет одно, и только одно, иррациональное число. Но нужно знать, что по данному иррациональному числу нелегко найти величины a_i, кроме случая квадратичных иррациональностей. Об этом случае подробнее.

③ [19] В исследовании этого случая принимали участие двое: Эйлер, который открыл свойства и применил их, и Лагранж, который доказал и обобщил эти свойства. Можно также упомянуть Э. Галуа, который в начале следующего века написал свою первую статью о бесконечных непрерывных дробях.

§7. Леонард Эйлер (Leonhard Euler, 1707-1783)

① [20] Спустя лет тридцать после предыдущей *De fractionibus continuis* вышла важная статья Эйлера под заголовком *De usu novi algorithmi in problemate Pelliano solvendo* (О использовании некоторого

---

18. An infinite continued fraction defines one, and only one, irrational number; likewise any infinite sequence of integers a_i considered as above defines one, and only one, irrational number. For a given α, determining the a_i is complicated, except in the case of quadratic irrationalities.

19. What follows deals with that case. Euler discovered and applied the main properties, Lagrange proved and generalized them. He also found the connection between quadratic equation and periodicity in the continued fraction, a link then specified by Galois.

20. The title of Leonhard Euler's main article already indicates its scope : solving in natural numbers the (misnamed) Pellian equation $x^2 - Ny^2 = 1$, with N a given non-square natural number. Since this involves the expansion of $\sqrt{N}$ in a continued fraction, Euler calculates it for N from 2 to 120.

нового алгоритма для решения задачи Пелля). Из названия видно, что главной целью Эйлера является решение особенной задачи; а именно, в начале статьи Эйлер пишет, какие темы он намерен последовательно объяснять (*De usu*, с. 32-33; *Opera*, с. 78): Quo ergo hoc argumentum luculentius et ordine pertractem, primum radicem quadratam ex quovis numero [подразумевается: integro] in fractionem continuam evolvere docebo, idque methodo quam minime molesta. Deinde ostendam, quomodo inde fractiones $\frac{p}{q}$ valorem irrationalem $\sqrt{l}$ [здесь пишем $\sqrt{N}$] proxime experimentes formari debeant in subsidium vocato Algorithmo novo supra explicato. Tum vero facile patebit, quomodo hinc numeros p et q definiri oporteat, ut fiat $pp = lqq + 1$ [т. е. $p^2 = lq^2 + 1$, здесь $x^2 - Ny^2 = 1$]. Denique tabulam subiungam, in qua pro omnibus numeris l centenarium non superantibus numeri bini p et q exhibentur. Так читатель сначала научится, как найти разложение в непрерывную дробь величины вида $\sqrt{N}$, где N – натуральное неквадратное число (для удобства читателя Эйлер даёт таблицу разложений величин $\sqrt{N}$ для N от 2 до 120); затем, как можно получить последовательность приближений (точно так же, как и в случае конечных непрерывных дробей, см. выше, § 1); потом (а это является его главной целью), как из этих приближений можно найти наименьшее решение в натуральных числах неопределённого уравнения $x^2 - Ny^2 = 1$ (зная, что из наименьшего решения можно получить *все* решения); наконец, Эйлер даёт таблицу наименьших решений до $N = 99$.

Замечание. Уравнение $x^2 - Ny^2 = 1$ встречается уже в *Арифметике* Диофанта (ок. 250), но между прочим (см. например издание арабской части *Арифметики*, с. 7). Для его решения Диофант положит $x^2 \equiv (my + 1)^2$, откуда $N \cdot y^2 + 1 = m^2y^2 + 2my + 1$ и $y(m) = \frac{2m}{N - m^2}$, рационально при m рациональном и положительно при подходящем выборе параметра m. Но в нашем случае речь идёт о решении в *натуральных* числах. У Архимеда (III в. до Р. Х.) есть одна задача, которая приводит к такому уравнению (это известная задача о быках, принадлежащих Солнцу); но Архимед её не решил.

Уравнение $x^2 - Ny^2 = 1$ где N натуральное неквадратное число и x, y должны быть натуральными числами называется у Эйлера *aequatio Pelliana* (John Pell, 1611-1685). По правде говоря, ему скорее заслужит носить имя Ферма (Pierre Fermat, 1601-1665), потому что, в отличие от Пелля, Ферма истинно занимался этим уравнением; возможно, Эйлер здесь перепутал Пелля с Броункером (William Brouncker, ок. 1620-1684). Первые попытки, направленные на его решение, имеются именно у Броункера и Валлиса (John Wallis,

1616-1703). Ферма и Валлис замечают, что такое уравнение всегда разрешимо в натуральных числах, и Ферма в придачу пишет, что каждое такое уравнение имеет бесконечное множество решений. Наконец, Эйлер объясняет, как можно найти решение. Для этого нужно заниматься разложением квадратной корней натуральных чисел в непрерывную дробь. Вот почему Эйлер совмещает разложение в непрерывную дробь корня N и решение уравнения в той же статье.

②[21] В первой части своей статьи Эйлер рассматривает разложение квадратичной иррациональности в непрерывную дробь. Пусть N — натуральное неквадратное число, и a_0 — целая часть корня N. Следовательно,

$$\sqrt{N} = a_0 + \frac{1}{\alpha_1}, \qquad 0 < \frac{1}{\alpha_1} < 1.$$

Так как α_1 больше единицы, имеем

$$\alpha_1 = \frac{1}{\sqrt{N} - a_0} = \frac{\sqrt{N} + a_0}{N - a_0^2} = a_1 + \frac{1}{\alpha_2} \quad \left(0 < \frac{1}{\alpha_2} < 1\right)$$

и т. д. Очевидно, что в случае квадратного корня получение последовательности членов a_k особенно легко: здесь, чтобы найти a_1, было достаточно умножать выражение для α_1 вверх и вниз на $\sqrt{N} + a_0$; надо затем, для нахождения других a_k, просто это повторять, так как вновь $0 < \frac{1}{\alpha_{k+1}} < 1$.

Первый численный пример Эйлера таков:

$$\sqrt{13} = 3 + \frac{1}{\alpha_1},$$

$$\alpha_1 = \frac{1}{\sqrt{13} - 3} = \frac{\sqrt{13} + 3}{4} = 1 + \frac{1}{\alpha_2}$$

$$\alpha_2 = \frac{4}{\sqrt{13} - 1} = \frac{\sqrt{13} + 1}{3} = 1 + \frac{1}{\alpha_3}$$

$$\dots$$

и, продолжая,

---

21. The expansion of $\sqrt{N}$ in a continued fraction is indeed not difficult, with the successive computations being elementary, as will be seen now.

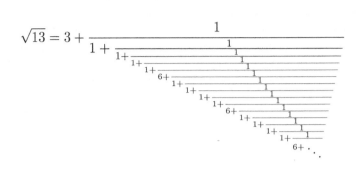

$$\sqrt{13} = 3 + \cfrac{1}{1 + \cfrac{1}{1 + \cfrac{1}{1 + \cfrac{1}{6 + \cdots}}}}$$

③ [22] Эйлер вычисляет разложения корней первых натуральных (не-квадратных) чисел, а именно величин

$$\sqrt{2},\ \sqrt{3},\ \sqrt{5},\ \sqrt{6},\ \sqrt{7},\ \sqrt{8},\ \sqrt{10},\ \sqrt{11},$$

$$\sqrt{12},\ \sqrt{13},\ \sqrt{14}, \ldots, \sqrt{118},\ \sqrt{119},\ \sqrt{120}.$$

Замечание. Люди сегодня не представляют себе, какую важную роль играла невероятная вычислительная способность Эйлера; часто он сначала вычисляет многие примеры, обнаруживает их свойства, и потом, при случае, доказывает. Этот «экспериментальный метод» также очевиден в случае задачи о ходе коня на шахматной доске, как показало неравнее издание его рукописных записок: Эйлер вывел общие правила из проб. См. выше, *Addendum* I.

Здесь из своих примеров он замечает, что у таких разложений *встречаются те же самые особенности.* Дословно (*De usu*, с. 45; *Opera*, с. 90) : In omnibus his indicum seriebus periodi deprehenduntur modo strictiores modo largiores, quae indicibus iis, qui primo duplo sunt maiores, includuntur, atque hae periodi eo clarius in oculos incidunt, si primi indices cuiusque seriei duplicantur. Deinde in qualibet periodo idem indicum ordo sive antrorsum sive retrorsum observatur ; ex quo in qualibet periodo vel unus datur index medius vel duo, prout terminorum numerus fuerit par vel impar. Следовательно Ейлер заключил из своих наблюдений два следующих свойства:

(*i*) Разложение квадратичных иррациональностей (т. е. вида $\sqrt{N}$, где N — натуральное неквадратное число) бесконечно, но периодичес-

---

22. The expression in continued fractions of such square roots leads Euler to the discovery of two properties. First, that the sequence of the integral quotients a_i is periodic, with the period beginning immediately after the integer a_0. Second, that this period is for its $n - 1$ first terms symmetrical, while its nth term equals twice the initial integer a_0. Here, as in the case of the knight's move on the chessboard (see above, *Addendum* I), various attempts led Euler to the discovery of general rules.

ки, и период начинается сразу после целой части:

$$\sqrt{N} = a_0 + \cfrac{1}{a_1 + \cfrac{1}{a_2 + \cfrac{1}{\ddots + \cfrac{1}{a_n + \cfrac{1}{a_1 + \cfrac{1}{a_2 + \cfrac{1}{\ddots}}}}}}}$$

или, более кратко,

$$\sqrt{N} = (a_0; \overline{a_1, a_2, \ldots, a_n}),$$

где a_0 — целая часть, и $\overline{a_1, a_2, \ldots, a_n}$ — период.

(ii) Период является симметричным для своих $n - 1$ первых членов, и последний член вдвое больше целой части, т. е.:

$$\sqrt{N} = a_0 + \cfrac{1}{a_1 + \cfrac{1}{a_2 + \cfrac{1}{\ddots + \cfrac{1}{a_0 + \sqrt{N}}}}},$$

или

$$\sqrt{N} = (a_0; \overline{a_1, a_2, \ldots, a_2, a_1, 2a_0}).$$

④ [23] Кроме того, замечает Эйлер, будет два случая. Во-первых, когда период состоит из n членов, n чётное (так как период симметричен в $n - 1$ членах), содержит член в середине симметричной части. Потом, когда n нечётное, такого среднего члена нет. Наконец, как частный случай, период может состоять только из одного члена; в этом случае член периода — удвоенная целая часть. Например:

— $n = 2k$: $\sqrt{54} = (7; \overline{2, 1, \mathbf{6}, 1, 2, 14})$

— $n = 2k + 1$: $\sqrt{53} = (7; \overline{3, 1, 1, 3, 14})$

— $n = 1$: $\sqrt{2} = (1; \overline{2})$; $\sqrt{37} = (6; \overline{12})$.

---

23. As to the number of terms in the period, there may be a single one or several; in the latter case there will or will not be a median term between the symmetrical parts, depending on whether the period comprises an even or an odd number of terms.

⑤ [24] А теперь читатель готов решить *problema Pellianum*. Пусть неопределённое уравнение

$$x^2 - N \cdot y^2 = 1,$$

где N — натуральное не квадратное число. По предположению, x, y — целое положительное решение. Если длина n периода — чётное число, то P_{n-1} и Q_{n-1} (т. е. члены предпоследнего приближения от первого периода) дают наименьшее решение уравнения $x^2 - N \cdot y^2 = 1$, а именно $P_{n-1} = x$ и $Q_{n-1} = y$. Например, в случае (*De usu*, с. 60; *Opera*, с. 105)

$$x^2 - 31 \cdot y^2 = 1, \quad \text{а} \quad \sqrt{31} = (5; \overline{1, 1, 3, 5, 3, 1, 1, 10}),$$

имеем

		5	1	1	3	...	1
$\left(\dfrac{0}{1}\right)$	$\left(\dfrac{1}{0}\right)$	$\dfrac{5}{1}$	$\dfrac{6}{1}$	$\dfrac{11}{2}$	$\dfrac{39}{7}$	...	$\dfrac{1520}{273}$

следовательно,

$$1520^2 - 31 \cdot 273^2 = 1.$$

Но если n — нечётное число, то, чтобы найти наименьшее решение, надо продолжать до членов P_{2n-1} и Q_{2n-1} в конце второго периода, потому что P_{n-1} и Q_{n-1} дают решение уравнения с -1 справа. Например, в случае (*De usu*, с. 61; *Opera*, с. 106)

$$x^2 - 29 \cdot y^2 = 1, \quad \text{а} \quad \sqrt{29} = (5; \overline{2, 1, 1, 2, 10}),$$

имеем сперва

		5	2	1	1	2
$\left(\dfrac{0}{1}\right)$	$\left(\dfrac{1}{0}\right)$	$\dfrac{5}{1}$	$\dfrac{11}{2}$	$\dfrac{16}{3}$	$\dfrac{27}{5}$	$\dfrac{70}{13}$

откуда

$$70^2 - 29 \cdot 13^2 = -1$$

а потом

10	2	1	1	2
$\dfrac{727}{135}$	$\dfrac{1524}{283}$	$\dfrac{2251}{418}$	$\dfrac{3775}{701}$	$\dfrac{9801}{1820}$

---

24. On this last property depends the solution of the equation $x^2 - Ny^2 = 1$: if the period has an even number of terms, finding the corresponding penultimate terms P_{n-1} and Q_{n-1} of the expansion will give the solution x, y ; if the number is odd, the two required terms will be the penultimate of the second period. In both cases, the other solutions may be obtained either by pursuing this procedure or directly from the first solution. That is why there are now tables listing the smallest solution for given N.

откуда

$$9801^2 - 29 \cdot 1820^2 = 1.$$

Как отметил Ферма (см. выше), рассматриваемое уравнение имеет бесконечное множество решений. Однако нам нужно знать только наименьшее решение x_1, y_1, так как из уравнения

$$x_k + \sqrt{N} \cdot y_k = (x_1 + \sqrt{N} \cdot y_1)^k$$

можно найти все остальные решения x_k, y_k. Поэтому сегодня имеются таблицы наименьших решений. Эйлер, как он объявил в своём введении (см. с. 326 выше), составил таблицу наименьших решений до $N = 99$ (*De usu*, с. 64–65; *Opera omnia*, с. 109–110).

§8. Жозеф Луи Лагранж (Joseph Louis Lagrange, 1736-1813)

①[25] Лагранж в своём *Sur la résolution des équations numériques* (О решении численных уравнений) использует непрерывные дроби, чтобы найти одно положительное решение полиномиального уравнения любой степени.

Пусть $\sum a_i x^i = 0$. Если r — целая часть величины x, можно положить $x = r + \frac{1}{y}$, откуда получим новое уравнение для y. Так как всегда можно найти (с помощью проб и ошибок) $s < y < s+1$, s целое число, можно вновь положить $y = s + \frac{1}{z}$, и т. д.

Первый пример Лагранжа следующий (*Sur la résolution*, §IV.25). Рассмотрим $x^3 - 2x - 5 = 0$. Так как положительный корень находится между 2 и 3, напишем $x = 2 + \frac{1}{y}$, откуда $y^3 - 10y^2 - 6y - 1 = 0$. Попробуем подставить первые целые значения до тех пор, пока выражение слева не изменит знак; так как для $y = 10$ и для $y = 11$ получим величины -61 и 54, положим $y = 10 + \frac{1}{z}$, откуда $61z^3 - 94z^2 - 20z - 1 = 0$; так как $1 < z < 2$, сделаем подстановку $z = 1 + \frac{1}{u}$, и т. д. Таким образом, найдём ряд чисел $2; 10, 1, 1, 2, 1, 3, 1, 1, 12, \ldots$, откуда можно получить выражение для x

$$x = 2 + \cfrac{1}{10 + \cfrac{1}{1 + \cfrac{1}{1 + \cfrac{1}{2 + \ddots}}}}$$

25. As Lagrange observed, continued fractions may be used to determine a positive solution of an algebraic equation of any degree $\sum a_i x^i = 0$ with an arbitrary degree of accuracy, taking each time the smallest integral value and setting out a new equation for the denominator in the fractional part : $x = r + \frac{1}{y}$, $y = s + \frac{1}{z}$, and so on.

и последовательные приближения к решению

$$\frac{2}{1}, \ \frac{21}{10}, \ \frac{23}{11}, \ \frac{44}{21}, \ \frac{111}{53}, \ \frac{155}{74}, \ \frac{576}{275}, \ \frac{731}{349}, \ \frac{1307}{624}, \ \frac{16415}{7837}, \ \ldots$$

попеременно меньшие и большие истинного значения.

②^{26} Особенно важным является результат Лагранжа о периодичности разложений в непрерывную дробь. Эйлер нашёл, что разложение $\sqrt{N}$

— периодическое

— имеет только один член перед периодом

— имеет симметричность в периоде для $n-1$ членов

— последний член периода вдвое больше целой части.

Лагранж доказал (в своих *Additions au Mémoire sur la résolution des équations numériques*), что разложение в непрерывную дробь положительного решения каждого квадратного уравнения (т. е. в виде $\frac{a+\sqrt{b}}{c}$) всегда периодично. Как он пишет (§ II.43) : On avait remarqué depuis longtemps que toute fraction continue périodique pouvait toujours se ramener à une équation du second degré, mais personne que je sache n'avait encore démontré l'inverse de cette proposition ; savoir, que toute racine d'une équation du second degré se réduit toujours nécessairement en une fraction continue périodique (Давно известно, что любая непрерывная периодическая дробь всегда может быть сведена к уравнению второй степени, но никто, насколько мне известно, ещё не доказал противоположное этому предложению; а именно, что каждый корень уравнения второй степени всегда обязательно сводится к периодической непрерывной дроби).

③^{27} На другой стороне, до периода может либо совсем не быть членов, либо быть один, либо несколько. Рассмотрим первое случае. Как мы видели с Эйлером, величина $\sqrt{N}$ выражается как

$$(a_0; \overline{a_1, a_2, \ldots, a_{n-2}, a_{n-1}, 2a_0})$$

(пренебрегаем симметричностью, так как она здесь не относится). Тогда величина $a_0 + \sqrt{N}$ выражается как

$$(2a_0; \overline{a_1, a_2, \ldots, a_{n-2}, a_{n-1}, 2a_0})$$

или, иначе,

$$(\overline{2a_0; a_1, a_2, \ldots, a_{n-2}, a_{n-1}}).$$

---

26. Lagrange further defines the case of periodicity : it characterizes roots of second-degree equations.

27. One remaining question is thus that of the number of terms preceding the period when it is not purely periodic.

Следовательно, разложение в непрерывную дробь числа $a_0 + \sqrt{N}$, где a_0 целый часть величины $\sqrt{N}$, имеет *чистый* период, т. е. совершенно без предшествующих членов. Остаётся вопрос о количеством членов до периода в общих случаях квадратичных иррациональностей.

§9. Эварист Галуа (Evariste Galois, 1811-1832)

① [28] Первая статья Галуа (1828-1829 г.) посвящена как раз вопросу чистого периода. В ней находим доказательство того, что если разложение в непрерывную дробь одного из корней некоторого уравнения второй степени имеет чистый период, то другой корень этого урав-

Démonstration d'un théorème sur les fractions continues périodiques ;

Par M. Evariste GALOIS, élève au Collége de Louis-le-Grand.

On sait que si, par la méthode de Lagrange, on développe en fraction continue une des racines d'une équation du second degré, cette fraction continue sera périodique, et qu'il en sera encore de même de l'une des racines d'une équation de degré quelconque, si cette racine est racine d'un facteur rationnel du second degré du premier membre de la proposée, auquel cas cette équation aura, tout au moins, une autre racine qui sera également périodique. Dans l'un et dans l'autre cas, la fraction continue pourra d'ailleurs être immédiatement périodique ou ne l'être pas immédiatement, mais, lorsque cette dernière circonstance aura lieu, il y aura du moins une des transformées dont une des racines sera immédiatement périodique.

Or, lorsqu'une équation a deux racines périodiques, répondant à un même facteur rationnel du second degré, et que l'une d'elles est immédiatement périodique, il existe entre ces deux racines une relation assez singulière qui paraît n'avoir pas encore été remarquée, et qui peut être exprimée par le théorème suivant :

Рис. 12

---

28. In his first published article, Evariste Galois (1811-1832) states that if one root of an equation is a purely periodic continued fraction, with $x > 1$, its other, conjugate root will equal minus 1 divided by the reverse period, with thus $-1 < x' < 0$ (Fig. 12). He proved it by taking an expansion for x with few terms and successively transferring terms from one side to the other.

нения равен минус единице, разделенной на непрерывную дробь с тем же периодом, написанным в обратном порядке. Словами Галуа (немедленно после текста рисунка 12): Si une des racines d'une équation de degré quelconque est une fraction continue immédiatement périodique, cette équation aura nécessairement une autre racine également périodique que l'on obtiendra en divisant l'unité négative par cette même fraction continue périodique, écrite dans un ordre inverse (Если один из корней некоторого уравнения какой-нибудь степени представляет собой чисто периодическую непрерывную дробь, то уравнение непременно имеет другой периодический корень, получающийся делением отрицательной единицы на непрерывную дробь, написанную в обратном порядке; см. *Сочинения*, с. 16).

То есть, если одно решение квадратного уравнения x представимо как чисто периодическая дробь

$$x = (\overline{a_0; a_1, a_2, \ldots, a_n}) \quad (\text{итак} \ \ x > 1),$$

тогда другое решение имеет форму

$$x' = -\frac{1}{(\overline{a_n; a_{n-1}, a_{n-2}, \ldots, a_0})} \quad (\text{итак} \ -1 < x' < 0).$$

И наоборот: если

$$0 < x = \frac{1}{(\overline{a_0; a_1, a_2, \ldots, a_n})} < 1, \quad \text{тогда}$$

$$x' = -(\overline{a_n; a_{n-1}, a_{n-2}, \ldots, a_0}) < -1.$$

Доказательство Галуа (и несомненно происхождение его открытия) не очень трудно; оно состоит, в простом случае периода с четырьмя членами, из последовательных переносов членов из одной стороны равенства в другую. В конце Галуа получает два различных решения того же уравнения; т. е., они сопряжённые корни того же квадратного уравнения.

Итак шаги доказательства следующие:

$$x = a_0 + \cfrac{1}{a_1 + \cfrac{1}{a_2 + \cfrac{1}{a_3 + \frac{1}{x}}}}, \qquad\qquad a_0 - x = -\cfrac{1}{a_1 + \cfrac{1}{a_2 + \cfrac{1}{a_3 + \frac{1}{x}}}}$$

$$\frac{1}{a_0 - x} = -\left(a_1 + \cfrac{1}{a_2 + \cfrac{1}{a_3 + \frac{1}{x}}}\right), \qquad\qquad a_1 + \frac{1}{a_0 - x} = -\cfrac{1}{a_2 + \cfrac{1}{a_3 + \frac{1}{x}}}$$

$$\cfrac{1}{a_1 + \cfrac{1}{a_0 - x}} = -\left(a_2 + \cfrac{1}{a_3 + \frac{1}{x}}\right), \qquad\qquad a_2 + \cfrac{1}{a_1 + \cfrac{1}{a_0 - x}} = -\cfrac{1}{a_3 + \frac{1}{x}}$$

$$\frac{1}{a_2+\cfrac{1}{a_1+\frac{1}{a_0-x}}} = -\left(a_3 + \frac{1}{x}\right), \qquad\qquad a_3 + \frac{1}{a_2+\cfrac{1}{a_1+\frac{1}{a_0-x}}} = -\frac{1}{x}$$

$$x = -\frac{1}{a_3+\cfrac{1}{a_2+\cfrac{1}{a_1+\frac{1}{a_0-x}}}} \, ,$$

c'est donc là l'autre valeur de x donnée par cette équation ; такого заключение Галуа.

Следовательно,

$$x = (\overline{a_0; a_1, a_2, a_3}) > 1, \qquad\qquad -1 < x' = -\frac{1}{(\overline{a_3; a_2, a_1, a_0})} < 0.$$

И наоборот, если $0 < x < 1$, то получится $x' < -1$:

$$x = \frac{1}{a_0+\cfrac{1}{a_1+\cfrac{1}{a_2+\frac{1}{a_3+\frac{1}{x}}}}}, \qquad\qquad \frac{1}{x} = a_0 + \frac{1}{a_1+\cfrac{1}{a_2+\frac{1}{a_3+\frac{1}{x}}}}$$

и так далее, до

$$-\frac{1}{a_3+\cfrac{1}{a_2+\cfrac{1}{a_1+\frac{1}{a_0-\frac{1}{x}}}}} = \frac{1}{x} \qquad\qquad x = -\left(a_3 + \frac{1}{a_2+\cfrac{1}{a_1+\frac{1}{a_0-\frac{1}{x}}}}\right).$$

② [29] Эта статья имеет двойной исторический интерес.

(i) Во-первых, здесь впервые появляется связь между разложениями в непрерывную дробь двух корней одного и того же квадратного уравнения. Позже в девятнадцатом веке число членов до периода будет источником дальнейших исследований. Пусть $x > 1$ (следовательно $a_0 \neq 0$) — одно решение некоторого квадратного уравнения, а

---

29. Galois' article is interesting for two reasons. Mathematically, it is the first time a link is made between the period in the representation as a continued fraction and conjugate roots x, x' of the same quadratic equation. They are purely periodic and inverted in Galois' case of $x > 1$, $-1 < x' < 0$. Later in the 19th century, the link to the number of terms before the period will be found, still considering $x > 1$: if $x' > 0$, x will have one or more terms before the period ; if $-1 < x' < 0$ (Galois' case), none ; if $x' < -1$, only one.

The other aspect has to do with Galois' life. He wrote this article while in the last year of high school, which was devoted to the study of scientific topics only and preparing for the entry into the Ecole polytechnique. The report for the year 1828-1829 is instructive : he is an outstanding student in mathematics, aware of the latest research, while taking no interest whatsoever in physics and chemistry. His performance in mathematics for the third trimester seems, however, less impressive ; events leading to the suicide of his father (2 July 1829), victim of a political intrigue, might be one reason.

x' − другое его решение. Если $x' > 0$ (так что оба решения положительны), тогда x имеет один или несколько членов до периода; если $-1 < x' < 0$, тогда x не имеет ни одного члена до периода (Галуа); если $x' < -1$, тогда x имеет один член до периода. Во всяком случае, периоды величин x и x' являются обратными. Например,

$$x = \frac{20+\sqrt{2}}{8} = (2; 1, 2, 10, \overline{1, 1, 1, 10}), \ 0 < x' = \frac{20-\sqrt{2}}{8} = (2; 3, \overline{10, 1, 1, 1});$$

$$x = \frac{1+\sqrt{3}}{2} = (1; 2, \overline{1, 2}), \qquad -1 < x' = \frac{1-\sqrt{3}}{2} = -(0; \overline{2, 1}) < 0.$$

$$x = \frac{7+\sqrt{229}}{6} = (3; \overline{1, 2, 4}), \qquad -1 > x' = \frac{7-\sqrt{229}}{6} = -(1; 2, 1, \overline{4, 2, 1}).$$

(ii) Потом, в то время Галуа ещё был в лицее, в последнем классе (где преподавались только научные дисциплины). Представляется ясным, что в возрасте восемнадцати лет Галуа уже читал работы Лагранжа, т. е. последние достижения математики. На самом деле, Галуа занимался исключительно математикой; это становится ясным из замечаний его учителей того же (1828-1829) учебного года (см. Дюпюи [Dupuy], с. 257–258; или [изд. 1903] с. 88–90; *Сочинения*, с. 322–324):

Premier trimestre (Первый квартал)

Mathématiques. Cet élève a une supériorité marquée sur tous ses condisciples
(*Математика*. Этот ученик имеет явное превосходство над своими соучениками).

Physique. Distraction ; travail : néant
(*Физика*. Рассеяность; работа: ничто).

Chimie. Distrait, travail faible
(*Химия*. Рассеянный; слабая работа).

Deuxième trimestre (Второй квартал)

Mathématiques. Cet élève ne travaille qu'aux parties supérieures des Mathématiques
(*Математика*. Этот ученик работает исключительно во высшие части математики).

Physique. Conduite passable, travail nul
(*Физика*. Сносное поведение; ничтожная работа).

Chimie. Conduite passable, travail nul
(*Химия*. Сносное поведение; ничтожная работа).

Troisième trimestre (Третий квартал)

Mathématiques. Conduite bonne, travail satisfaisant

(*Математика.* Хорошее поведение; удовлетворительная работа).

Physique. Fort distrait, travail nul

(*Физика.* Очень рассеянный; ничтожная работа).

Chimie. Fort distrait, travail nul

(*Химия.* Очень рассеянный; ничтожная работа).

Замечание его учителя математики в третьем квартале, кажется, менее похвальное. Это время явилось трудным для Галуа, потому что отец стал жертвой политических интриг, из-за которых спустя месяц он покончил с собою. Это было начало несчастий Галуа.

ЛИТЕРАТУРА

Аристарх : T. Heath, *Aristarchus of Samos, new Greek text with translation and notes.* Oxford 1913. НА РУССКОМ ЯЗЫКЕ: «Аристарх, О величинах и расстояниях Солнца и Луны» (пер. с комментариями И. Н. Веселовского в статье «Аристарх Самосский»), *Историко-астрономические исследования,* Вып. VII, 1961, с. 11–70.

Диофант : J. Sesiano, *Books IV to VII of Diophantus' 'Arithmetica' in the Arabic translation attributed to Qusṭā ibn Lūqā.* New York 1982.

Галуа Э. (E. Galois) « Démonstration d'un théorème sur les fractions continues périodiques », *Annales de mathématiques pures et appliquées,* 19 (1828-1829), с. 294–301. Или : *Œuvres mathématiques d'Evariste Galois,* Paris 1897, с. 365–377. НА РУССКОМ ЯЗЫКЕ: Э. Галуа, Сочинения, пер. Н. Н. Меймана с примечаниями Н. Г. Чеботарёва (Москва-Ленинград, 1936), с. 16–25.

Гюйгенс Х. (Chr. Huygens) : « Descriptio automati planetarii », *Christiani Hugenii (...) Opuscula postuma,* Lugduni Batavorum [Лейден] 1703, с. 429–460. Или : *Œuvres complètes de Christiaan Huygens,* т. 21 (La Haye [Гаага] 1944), с. 579–652.

Дюпюи П. (P. Dupuy) : « La vie d'Evariste Galois », *Annales scientifiques de l'Ecole Normale,* 3ᵉ s., XIII (1896), с. 197–266. Или : *Cahiers de la quinzaine,* 5ᵉ s., 2ᵉ cahier, Paris 1903. НА РУССКОМ ЯЗЫКЕ: Дюпюи П., Жизнь Галуа, см.: Галуа Э., Сочинениа, пер. Н. Н. Меймана с примечаниями Н. Г. Чеботарёва (Москва-Ленинград, 1936), с. 255–316.

Лагранж Ж. Л. (J. L. Lagrange) : « Sur la résolution des équations numériques », *Mémoires de l'Académie royale des Sciences et Belles-Lettres de Berlin,* XXIII (1769/1767), с. 311–352. Или : *Œuvres de Lagrange,* т. 2 (Paris 1868), с. 539–578.

———— : « Additions au Mémoire sur la résolution des équations numériques », *Mémoires de l'Académie royale des Sciences et Belles-Lettres de Berlin*, XXIV (1770/1768), c. 111–180. Или : *Œuvres de Lagrange*, т. 2 (Paris 1868), c. 581–652.

———— : *Élémens d'algebre, par Léonard Euler, traduits de l'allemand, avec des notes et des additions*, т. I-II, Pétersbourg 1798. Или (дополнения) : « Additions aux Éléments d'algèbre d'Euler, analyse indéterminée », *Œuvres de Lagrange, publiées par les soins de M. J.-A. Serret*, т. 7 (Paris 1877), c. 5–179.

Швентер Д. (D. Schwenter) : *Geometriæ practicæ novæ et auctæ tractatus*, т. I-III. Nürnberg 1623-1627.

———— : *Deliciæ physico-mathematicæ, oder mathemat. und philosophische Erquickstunden*. Nürnberg 1636.

Эйлер Л. (L. Euler) : « De fractionibus continuis dissertatio », *Commentarii Acad. imp. scient. Petrop.*, IX (1744/1737), c. 98–137. Или : *Opera omnia*, s. I, vol. XIV (Leipzig/Berlin 1925), c. 187–215.

———— : « De usu novi algorithmi in problemate Pelliano solvendo », *Novi commentarii Acad. imp. scient. Petrop.*, XI (1767/1765), c. 28–66. Или : *Opera omnia*, s. I, vol. III (Leipzig/Berlin 1917), c. 73–111.

Achevé d'imprimer en février 2023 par Corlet Imprimeur — 14110 Condé-en-Normandie
Dépôt légal : février 2023 — N° d'imprimeur : 23010659 — *Imprimé en France*